Artificial Intelligence for Cyber Security and Industry 4.0

Artificial Intelligence for Cyber Security and Industry 4.0 offers a comprehensive exploration of the intersection of artificial intelligence (AI) and cyber security, providing readers with a thorough understanding of both the advantages and risks posed by AI technologies in modern industries. Covering a wide array of topics, from data anonymization and intrusion detection to AI's role in cloud security, border surveillance, and healthcare, this book addresses current challenges and proposes innovative solutions. It also highlights ethical concerns related to AI's use in weapon autonomy and border migration. This book is ideal for researchers, industry professionals, policy makers, and students looking to deepen their knowledge of AI's impact on cyber security and its applications in the evolving landscape of Industry 4.0. Through practical insights and forward-thinking discussions, readers will gain a well-rounded perspective on how AI can be leveraged for security while being mindful of emerging risks.

Key Features:

- Explores the dual role of AI in strengthening and threatening cyber security in the context of Industry 4.0.
- Provides an in-depth analysis of AI-driven cyber security techniques, including machine learning-based intrusion detection and data anonymization.
- Investigates the malicious use of AI, addressing both expanded existing threats and the emergence of novel vulnerabilities.
- Discusses advanced software design for privacy preservation in big data environments.
- Covers the use of AI in specific security domains, such as border surveillance, healthcare, and the Internet of Things.
- Highlights AI applications in cloud security, data integrity, and privacy protection.
- Introduces Quantum Machine Learning algorithms and their relevance to cyber security.

- Explores the ethical concerns surrounding AI technologies, particularly in the context of weapon autonomy and border migration.
- Includes real-world scenarios and methodologies, bridging the gap between academic research and industry practice.
- Offers forward-looking insights into the role of AI in future cyber security challenges and solutions.

Artificial Intelligence for Cyber Security and Industry 4.0

Edited by
Dinesh Sharma, Geetam Singh Tomar, and Anand Jha

CRC Press
Taylor & Francis Group
Boca Raton London New York

CRC Press is an imprint of the
Taylor & Francis Group, an **informa** business

Designed cover image: Shutterstock image 2302759565
First edition published 2025

by CRC Press
2385 NW Executive Center Drive, Suite 320, Boca Raton FL 33431

and by CRC Press
4 Park Square, Milton Park, Abingdon, Oxon, OX14 4RN

CRC Press is an imprint of Taylor & Francis Group, LLC

ISBN: 9781032621463 (hbk)
ISBN: 9781032657257 (pbk)
ISBN: 9781032657264 (ebk)

DOI: 10.1201/9781032657264

Typeset in Sabon
by KnowledgeWorks Global Ltd.

Contents

About the Editors

Dr. Dinesh Sharma Dr. Dinesh Sharma obtained his Ph.D. in wireless sensor networks with a focus on artificial intelligence (AI)/machine learning (ML) and the Internet of Things (IoT) from Uttarakhand Technical University, Uttarakhand, India, in 2021. Presently, he serves as an Associate Professor in the Department of Data Science Engineering at Manipal University Jaipur, Rajasthan, India. With a remarkable track record, he has published seven patents, two of which center on the intersection of AI and agriculture.

Dr. Sharma has been recognized for his exceptional contributions to the academic field. Amity University Madhya Pradesh awarded him the title of Best Faculty in both 2012 and 2019. Moreover, he was consistently acknowledged as the Best Innovator by the university from 2019 to 2022. These accolades reflect his outstanding prowess in innovation and research.

Dr. Sharma is a Master Trainer in High-Performance Computing. Having published over 15 research articles in esteemed journals such as SCI, Scopus, Web of Sciences, and IEEE, Dr. he has established himself as a prolific author. His expertise has led to numerous invitations as a resource person for various international conferences, seminars, Faculty Development Programs, webinars, and short-term courses. Additionally, he has actively contributed to the academic community by serving as a Technical Program Committee member and reviewer for esteemed journals and international conferences. Beyond his scholarly pursuits, Dr. Sharma has made significant strides in software development. He has successfully designed and developed numerous Android and web applications. Under his guidance, several students have secured funding for their projects and received prestigious awards at both the national and international levels. His research interests encompass wireless sensor networks, the IoT, AI/ML, and web/android application development. His multifaceted expertise in these domains further enriches his contributions to the academic and technological landscapes.

Prof. Geetam Singh Tomar Prof. Dr. Geetam Singh Tomar holds the position of Director at Rajkiya Engineering College in Sonbhadra, Uttar Pradesh, India. He obtained his undergraduate, postgraduate, and Ph.D. degrees in electronics engineering from prestigious universities in India. In addition to his role as Director, he also serves as the Principal of Malwa Institute of Technology & Management in Gwalior, India, and as an R&D Advisor at MPCT Group of Colleges, Gwalior. Prof. Dr. Tomar is actively engaged in research and consultancy in various fields, including wireless sensor networks, advanced communication networks, artificial intelligence/machine learning, agriculture, and the Internet of Things.

He regularly participates in international conferences held in India and other countries and has successfully organized over 24 IEEE International Conferences (CICN 2009–2023, CSNT 2010–2023). His expertise and knowledge have led him to his appointment as a Visiting Professor at renowned institutions such as Hannam University in Korea, Thapar University in Patiala, India, and several other prestigious institutes. He has also contributed his services to IIITM and other Institutes of National Importance. In recognition of his academic excellence, Prof. Dr. Tomar received the International Pluto Award in 2009 from Cambridge, UK.

Furthermore, he was honored as one of the top 100 academicians in the world in 2009. He was also consistently listed in global rankings for the years 2008, 2009, and 2010.

Prof. Dr. Tomar actively contributes to the academic community as the Chief Editor of five international journals. He has published more than 150 research articles in esteemed international journals and conferences. Additionally, he has authored ten books, with four of them published by Springer Verlag, Germany. His primary focus lies in promoting and expanding research activities in regions of India where research is currently limited. Prof. Dr. Tomar's dedication to advancing knowledge and his contributions to the field have also led him to serve as the Chairman of the IEEE Madhya Pradesh Section.

Mr. Anand Jha Anand Jha is an Assistant Professor in the Department of Information Technology at Rustamji Institute of Technology, located within the Border Security Force Academy in Tekanpur, Gwalior, Madhya Pradesh, India. As an academic professional, he is involved in teaching, research, and mentoring students in the field of Information Technology. His specific areas of expertise and research interests include artificial intelligence and machine learning, data science and analytics, etc.

List of Contributors

Raoof Altaher
Electrical and Computer Engineering
Department, Altinbas University,
Istanbul, Turkey

Kirti Raj Bhatele
Department of Computer Science
and Engineering, Rustamji
Institute of Technology, BSF,
Tekanpur, Gwalior, Madhya
Pradesh, India

Anand Singh Bisen
Vikrant University, Gwalior,
Madhya Pradesh, India

Neha Chauhan
Department of CSE, Babu Banarasi
Das Institute of Technology and
Management, Lucknow, Uttar
Pradesh, India

Deepika Dubey
Department of CSE, Shriram
College of Engineering &
Management, Morena, Madhya
Pradesh, India

Shitanshu Jain
Department of Data Science and
Engineering, Manipal University
Jaipur, Jaipur, Rajasthan, India

Atharva Jaiswal
Manipal University Jaipur, Jaipur,
Rajasthan, India

Chirag Joshi
Department of Data Science
Engineering, School of
Information Technology,
Manipal University Jaipur,
Jaipur, Rajasthan, India

Premanand K. Kadbe
Department of Electronics
and Telecommunication
Engineering, G H Raisoni
College of Engineering
and Management, Pune,
Maharashtra, India

Prashant Sai Kalepu
Amity University Gwalior, Madhya
Pradesh, India

Hakan Koyuncu
Computer Engineering Department,
Altinbas University, Istanbul,
Turkey

Bagesh Kumar
Department of Information
Technology, Manipal University
Jaipur, Jaipur, Rajasthan, India

Ajay Kumar
Department of Computer Science
  and Engineering, Manipal
  University Jaipur, Jaipur,
  Rajasthan, India

Sagar Onkarrao Manjare
Mahatma Gandhi University,
  Byrnihat, Meghalaya, India

Varun Kumar Mishra
GLA University, Mathura, Uttar
  Pradesh, India

Balasaheb H. Patil
Savitribai Phule Pune University,
  Pune, Maharashtra, India

Rupali B. Patil
Department of Electronics and
  Telecommunication Engineering,
  G H Raisoni College of
  Engineering and Management,
  Pune, Maharashtra, India

Sanjay Patsariya
Department of Information
  Technology, Rustamji Institute
  of Technology, BSF, Tekanpur,
  Gwalior, Madhya Pradesh, India

Rohit S. Piske
Department of Electronics and
  Telecommunication Engineering,
  Vidya Pratishthan's Kamalnayan
  Bajaj Institute of Engineering
  and Technology, Pune,
  Maharashtra, India

Abhishek Raghuvanshi
Department of Computer
  Engineering, Mahakal Institute
  of Technology, Ujjain, Madhya
  Pradesh, India.

Nirmal Rajkumar
Manipal University Jaipur, Jaipur,
  Rajasthan, India

Ranjeet K. Ranjan
Department of Computer Science
  and Engineering, PDPM
  Indian Institute of Information
  Technology, Design and
  Manufacturing, Jabalpur,
  Madhya Pradesh, India

Vijaypal Singh Rathor
Department of Information
  Technology, ABV-Institute
  of Information Technology
  and Management, Gwalior,
  Madhya Pradesh, India

Shweta Sharma
Department of Information
  Technology, Manipal
  University Jaipur, Jaipur,
  Rajasthan, India

Hamed Taherdoost
University Canada West,
  Vancouver, BC, Canada
Westcliff University, Irvine,
  CA, United States

Devanshu Tiwari
Department of Computer
  Science and Business
  Systems, Madhav Institute
  of Technology and Science -
  Deemed University, Gwalior,
  Madhya Pradesh, India

Girraj Kumar Verma
Department of Applied
  Mathematics, Amity
  University, Gwalior,
  Madhya Pradesh, India

Nahida Majeed Wani
Department of Applied
    Mathematics, Amity
    University, Gwalior, Madhya
    Pradesh, India

Anjali Yadav
Department of Computer Science
    and Engineering, Manipal
    University Jaipur, Jaipur,
    Rajasthan, India

Chapter 1

AI for cyber security and cyber security for AI

Complementary role

Hamed Taherdoost

1.1 INTRODUCTION

The link between artificial intelligence (AI) and cybersecurity has become more and more complicated in the modern digital environment, and it now serves as a crucial ally in the continuous fight against cyber threats (Bécue et al., 2021). In other words, the role of AI in cybersecurity is expanding. The increasing reliance of our society on digital technology for various requirements has also heightened our exposure to hazards and vulnerabilities. Both defenders and attackers can use AI as a powerful cybersecurity tool due to its exceptional data processing, pattern recognition, and decision-making powers (Hosseini & Parvania, 2021; Lopes Antunes & Llopis Sanchez, 2023; Taherdoost, 2023b).

Understanding the unique importance of each discipline is crucial before exploring the synergistic link between AI and cybersecurity. AI has made tremendous progress in recent years by creating intelligent systems that mimic human cognitive capacities (Taherdoost, 2023a; Vigo et al., 2022). A wide range of AI approaches, such as machine learning (Zhou et al., 2017), deep learning (Mohammadi et al., 2018), and natural language processing (Olivetti et al., 2020), among others, have led to significant increases in capacity to handle massive datasets and extract insights in recent years.

Similar to this, the cybersecurity threat landscape has always been changing. Malicious actors are constantly devising new ways to attack using the same technology, despite the positive and negative applications of AI. Individuals and businesses need to have access to advanced security solutions that can react instantly to emerging threats in order to combat these dangers.

There is no direct relationship between cybersecurity and AI. However, utilizing AI technology has expedited incident response, increased threat detection, and boosted security procedures (Maezo & Rey, 2023; Taherdoost, 2022; Zhang et al., 2022). On the other hand, the integrity and security of AI systems have recently raised serious concerns due to the wide variety of attacks that AI models and algorithms are subject to. It is crucial to make sure that AI technologies are reliable and robust for their deployment to be successful (Hasija & Esper, 2022).

DOI: 10.1201/9781032657264-1

This chapter aims to delve deeper into this intricate relationship between AI and cybersecurity, its noteworthy impact on digital security, and the challenges it presents. These challenges include those related to integration, resource constraints, privacy issues, and legislative limits. This will emphasize the crucial role that AI and cybersecurity play in preserving the integrity of the digital ecosystem and provide insight into future trends and directions, such as the effect of quantum computing on cybersecurity and advanced AI tactics for defending AI systems.

1.1.1 Objectives of the chapter

The objectives of the chapter are:

- To elucidate the pivotal role of AI in bolstering cybersecurity measures.
- To highlight the reciprocal relationship between AI and cybersecurity, emphasizing their complementary roles in modern digital environments.
- To explore the importance of cybersecurity in safeguarding AI systems against vulnerabilities and ensuring model robustness.
- To discuss the challenges and considerations inherent in the integration of AI and cybersecurity, including resource gaps, privacy concerns, and regulatory hurdles.
- To analyze future trends and directions in AI and cybersecurity, including advancements in quantum computing, adversarial attack mitigation, and quantum-safe AI.

1.1.2 Organization of the chapter

The chapter begins with an introductory section that sets the stage for understanding the intertwined relationship between AI and cybersecurity. It highlights the critical role each plays in contemporary digital landscapes and foreshadows the exploration of their symbiotic connection. Following the introduction, the chapter delves into the multifaceted role of AI in cybersecurity. It explores how AI powers threat detection, incident response, and security analytics, utilizing techniques such as anomaly detection, behavioral analysis, and predictive analysis to fortify defense mechanisms against evolving cyber threats. Subsequently, the narrative shifts to elucidate the significance of cybersecurity for AI systems. This section probes vulnerabilities inherent in AI architectures, the imperative of ensuring model robustness, and the complex regulatory and ethical considerations that underpin AI development and deployment. As the discussion progresses, the chapter emphasizes the complementary nature of the relationship between AI and cybersecurity. It delineates how AI strengthens cybersecurity measures by enhancing threat detection and response capabilities while concurrently underscoring how robust cybersecurity frameworks safeguard AI systems against exploitation and compromise. Amidst these analyses, the chapter confronts the challenges and considerations intrinsic to the integration of AI and cybersecurity. It navigates through the hurdles of integration and implementation, resource and skill shortages, privacy and ethical dilemmas, and the labyrinth of legal and regulatory frameworks governing digital ecosystems. Looking ahead, the chapter prognosticates future trends and directions in AI and cybersecurity integration. It contemplates the convergence of AI and quantum computing, the imperative of mitigating adversarial attacks, and the development of quantum-safe AI solutions to fortify digital infrastructures against emerging threats. In its concluding remarks, the chapter synthesizes key insights gleaned from the discourse, reiterating the symbiotic relationship between AI and cybersecurity and underscoring the imperative of collaborative initiatives and ongoing research endeavors to fortify digital resilience.

Embedded within the chapter's narrative are references to scholarly works and research findings, providing readers with avenues for further exploration and a deeper understanding of the intricate interplay between AI and cybersecurity in contemporary digital ecosystems.

1.2 THE ROLE OF AI IN CYBERSECURITY

An equally adaptable and strong defensive system is required due to the growth of cyber threats, which may range from smart and covert incursions to large-scale, disruptive assaults. This chapter explores the mutually beneficial link between AI and cybersecurity, emphasizing the crucial role AI plays in bolstering companies' and people's defenses against a constantly changing threat environment.

To remain one step ahead of attackers, the modern cybersecurity ecosystem requires a proactive, data-driven, and intelligent strategy. With its capacity to quickly analyze enormous information, identify trends, and adjust to new threats, AI has emerged as a key component in this project. AI is no longer just a fad in technology; it is now a crucial ally in protecting digital assets, confidential data, and privacy (Timan & Mann, 2021).

1.2.1 AI-powered threat detection

The cutting edge of contemporary cybersecurity is AI-powered threat detection, which gives businesses the tools to recognize and counter a variety of cyber threats, from malware and phishing attempts to sophisticated intrusions (Dasgupta et al., 2023; Kumar & Kumar, 2023).

1.2.1.1 Anomaly detection

To identify known dangers, traditional security procedures may depend on pre-established guidelines or indicators. When it comes to addressing fresh and emerging dangers, this technique is ineffective. Machine learning in particular

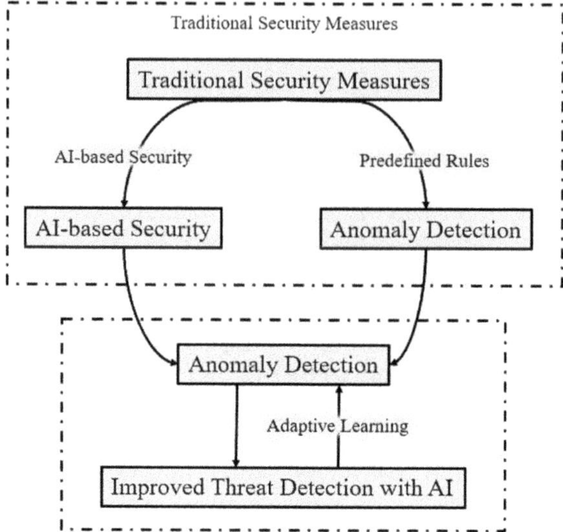

Figure 1.1 Anomaly detection in security measures.

and AI in general provide a solution for anomaly detection (Rousopoulou et al., 2022).

Anomaly detection refers to the process of establishing a baseline for what is considered usual behavior within a system or network using AI models. AI systems may provide warnings when they see unexpected patterns or activities by continually scanning for departures from this baseline and looking for deviations (Patcha & Park, 2007). This approach is very good at recognizing previously undetected threats since it does not depend on established signatures.

AI excels in anomaly detection because of its capacity to adjust to and learn from fresh data. Over time, the system will get better at distinguishing between benign and harmful abnormalities. This will reduce false positives and improve the threat detection process overall. The shift from traditional security measures to AI-based security with anomaly detection, which improves threat detection via adaptive learning, is shown in Figure 1.1.

1.2.1.2 Behavioral analysis

Behavioral analysis is an additional key component that needs to not be overlooked in AI-powered threat identification (Illiashenko et al., 2023). It entails keeping an eye out for patterns and irregularities in user, software, and device behavior that can point to a system security vulnerability. AI systems can establish the typical behavior of each object in a network using historical data and machine learning algorithms. Any behaviors that differ from the norm, such as a person accessing resources they do not regularly use or a device displaying odd network activity, may be recognized by the AI system and flagged as suspicious (Ertoz et al., 2004; Verma et al., 2019).

Identification of insider threats, which happen when people with authorized access to systems and data misuse their rights or act in unanticipated ways, is one of the most effective uses of behavioral analysis. Organizations may identify abnormalities, such as persistent security violations and illegal activity, by continually observing and analyzing behavior.

1.2.1.3 Predictive analysis

AI has developed to the point where it can be utilized for threat identification and predictive analysis (Telo, 2017). It entails making use of historical data and machine learning algorithms to foresee possible security threats or vulnerabilities before attackers can take advantage of them. Businesses may improve their security posture by taking preventative action with the help of this proactive technique.

AI makes it possible to analyze historical attack trends and patterns, find security flaws, and predict the possibility of future assaults (Haider et al., 2020). A predictive analysis may also consider outside variables, such as emerging threats and dangers that are specific to a certain business.

Organizations may more efficiently prioritize security measures and spend resources to address expected vulnerabilities and threats by using predictive analysis. This data-driven approach enables more efficient and preemptive solutions to cybersecurity issues. Predictive analytic applications in cybersecurity are briefly summarized in Table 1.1, along with use cases, related AI techniques, and tools.

Table 1.1 Applications of predictive analysis in cybersecurity

Application	Use cases	AI techniques	Tools and technologies
Threat forecasting	Predicting malware outbreaks, identifying vulnerabilities	Machine learning, data mining	Security information and event management (SIEM), Threat Intelligence Platforms
UEBA (User Entity and Behavior Analytics)	Detecting unauthorized access, abnormal user behavior	Machine learning, behavioral analytics	UEBA Tools
Zero-Day Vulnerability Detection	Identifying vulnerabilities, patch prioritization	Machine learning, anomaly detection	Vulnerability Assessment Tools
Advanced persistent threat (APT) Detection	Identifying targeted attacks, mitigating APTs	Machine learning, threat intelligence	Network Security Tools, APT Detection Systems
Security Automation	Automating tasks, real-time threat response	Robotic process automation, AI orchestration	SOAR (Security Orchestration, Automation, and Response) Platforms

1.2.2 AI-Enhanced incident response

Cybersecurity's cornerstone is effective incident response, and AI is vital in improving this important facet of security operations. Identification, containment, elimination, and recovery from security events and breaches are all parts of incident response (Ahmad et al., 2020). The use of AI technology helps to improve this process' efficiency, accuracy, and proactivity.

1.2.2.1 Automated remediation

One of the most innovative methods of reacting to cybersecurity problems nowadays is automated remediation enabled by AI capabilities (Kadel et al., 2022). There are occasions when using traditional, manual methods to resolve situations results in delays and the possibility of human mistakes. On the other hand, automated remediation provides a framework for a proactive and efficient response that may carry out predetermined measures in real time when security concerns are discovered. This responsiveness is crucial in the context of cybersecurity, where threats evolve quickly and firms need to be able to quickly manage crises and lessen their effects.

One of the key benefits of this strategy is automatic remediation's capacity to maintain consistency in incident response. This is achieved by following specified security protocols and carrying out data- and algorithm-based activities, which lessens the inherent unpredictability of human decision-making. Furthermore, it lessens the possibility of human mistakes, which is beneficial in demanding and high-stakes circumstances. Automated systems are furthermore available 24 hours a day, 7 days a week, enabling organizations to react to occurrences at any time of day or night, especially those that take place beyond regular business hours.

Automated repair often entails isolating infected systems, limiting harmful network activity, blocking questionable user accounts, permitting data restoration, and turning on alerts as necessary. The future of automated remediation is anticipated to grow more sophisticated as machine learning and AI technologies continue to progress (Johnphill et al., 2023). This necessitates the creation of algorithms with higher levels of intelligence, better integration with threat information sources, and more flexibility to threats that are always changing.

1.2.2.2 Threat intelligence

AI is a key component in the gathering, processing, and distribution of this vital information, which is essential for effective incident response in the cybersecurity field. Threat actors' tactics, techniques, and procedures (TTPs), vulnerabilities that may be exploited, indications of compromise (IoCs), and other information concerning possible security risks are all included in threat intelligence (Kim et al., 2021; Mavroeidis & Bromander, 2017). Threat information generated by AI is essential for preventative cybersecurity actions.

Organizations may use AI to spot new dangers and weaknesses that might not be well-known yet. AI systems regularly scan a variety of data sources, such as dark web forums, malware archives, and hacker talk, using machine learning and natural language processing to identify new hazards. AI also adds context by assessing a multitude of threat data from many sources and linking it to the unique environment of an organization. This enables incident responders to comprehend the gravity and repercussions of an occurrence in the context of their systems and data. By automating analysis and quickly spotting parallels with well-known attack patterns or the existence of IoCs, AI also accelerates the threat assessment process, allowing quicker and more effective responses to security issues.

As threat actors continuously adapt their strategies and targets, real-time updates made possible by AI-driven threat intelligence keep enterprises well-prepared to fight against the most recent attacks. By ranking threats based on their relevance and possible impact, AI improves incident response. This allows security teams to better deploy resources and customize their remediation efforts to particular threat characteristics. By doing forensics analysis and trying to identify a particular threat actor or group as the perpetrator of the attack, AI may also help in the post-incident phase. While AI-driven threat intelligence has many benefits, several issues need to be addressed, including data privacy and ethics, the possibility of false positives and negatives, and the need for adaptation in the face of threat actors' evasion tactics.

1.2.3 AI-driven security analytics

The foundation of an organization's capacity to comprehend, evaluate, and react to the changing panorama of cybersecurity risks is security analytics. Traditional techniques of security analysis often fall short in the modern digital world, where data quantities and complexity are constantly increasing. By using the power of machine learning and data processing to provide deeper insights and more effective threat mitigation, AI-driven security analytics represent a fundamental paradigm change.

Data mining is the first step of AI-driven security analytics, where enormous amounts of structured and unstructured data are combed through to find significant patterns and trends. AI systems can quickly examine gigabytes of data, unlike human analysts. They are excellent at finding trends in past data and spotting abnormalities that might be signs of security problems. These abnormalities might include odd file transfers, unwanted access attempts, and strange network traffic spikes. Early threat identification is made possible by AI's ability to recognize these patterns and abnormalities, which minimizes the potential harm brought on by breaches.

In AI-driven security analytics, machine learning takes center stage and offers numerous significant benefits for the cybersecurity industry. As they analyze additional data and gain knowledge from fresh dangers, these algorithms adapt and change. They provide predictive analysis, enabling businesses to foresee potential attack surfaces or weaknesses. Machine learning models are excellent at detecting dynamic risks by adjusting to the changing threat environment and finding novel, previously undiscovered hazards. By identifying adjustments in user and system behavior that could point to a security breach, they also participate in behavioral analysis. Additionally, machine learning systems may use natural language processing methods to efficiently understand textual data, assisting in the discovery of possible security risks hidden inside unstructured data (Kumari et al., 2021; Sarker et al., 2021).

1.3 THE IMPORTANCE OF CYBERSECURITY FOR AI

From healthcare and banking to autonomous cars and smart cities, AI provides previously unheard-of capabilities and efficiencies (Chan et al., 2022; Harry, 2023; Khan & Nazir, 2023; Varian, 2018). But as AI systems proliferate and grow more sophisticated, cybersecurity is crucial to ensure their dependability, privacy, and integrity.

1.3.1 Vulnerabilities and threats to AI systems

Despite substantial advancements in capability, AI systems are still vulnerable to dangers that need care. The dependability and security of AI technology might be threatened by these flaws. Attacks known as "data poisoning," in which attackers alter the training data needed to build AI models, are a serious problem (Sun et al., 2021). Attackers may skew the learning of AI models and provide inaccurate or biased predictions by adding false or misleading data during the training process. For instance, in the context of categorizing spam emails, a data poisoning attack might cause the AI model to incorrectly identify dangerous emails as safe, endangering security.

Another danger is adversarial assaults. These assaults entail complex input data changes intended to trick AI algorithms, which is especially worrying for applications that rely on data such as image recognition, natural language processing, and other data-driven applications. Risks might arise in applications like autonomous cars when slight, undetectable changes to a picture cause the AI model to categorize it incorrectly. Furthermore, since AI systems routinely analyze substantial quantities of private data, privacy issues emerge. Unauthorized access or data leaks may result in serious legal, moral, and reputational repercussions for privacy violations. Strong security measures need to be in place to prevent possible privacy violations and breaches since protecting people's data privacy is of the utmost importance.

A broad approach is required to address these threats and weaknesses. This strategy needs to include technological solutions, strict security procedures, and adherence to moral and regulatory requirements. To detect and reduce these risks and ensure that AI systems continue to be reliable and trustworthy in a variety of applications, proactive actions are essential.

1.3.2 Ensuring AI model robustness

Model robustness in the context of AI refers to a system's capacity to operate consistently under a variety of circumstances, withstand hostile manipulation, and provide reliable and accurate predictions. A key component of cybersecurity and a vital step in reducing threats and vulnerabilities is ensuring the resilience of AI models (Saeed et al., 2023).

AI model resilience is ensured by a multidimensional procedure that centers on exhaustive model testing and validation. This crucial stage attempts to pinpoint weaknesses, evaluate performance, and improve the adaptability of AI systems to various difficulties. Testing AI models for robustness entails exposing them to a variety of conditions, such as noisy data, environmental unpredictability, and even malicious inputs. Since real-world data seldom completely matches training data, the goal is to assess how well the model performs in less-than-ideal circumstances. A significant part of evaluating the model's resilience to manipulative assaults is adversarial testing. Adversarial attacks manipulate input data subtly to trick the AI model, hence it is crucial to assess the model's vulnerability to such attacks in security-critical applications.

Quality control processes are crucial in assuring the dependability of AI models in addition to these testing approaches. The stringent quality assurance procedures also include security audits, continuous integration, unit testing, and code reviews. To address recently found vulnerabilities and maintain the model's resilience over time, regular upgrades and patching are also necessary. Figure 1.2 shows the integrated procedures for guaranteeing the robustness of AI models, incorporating quality control, robustness testing, and maintenance to fix flaws and assess performance.

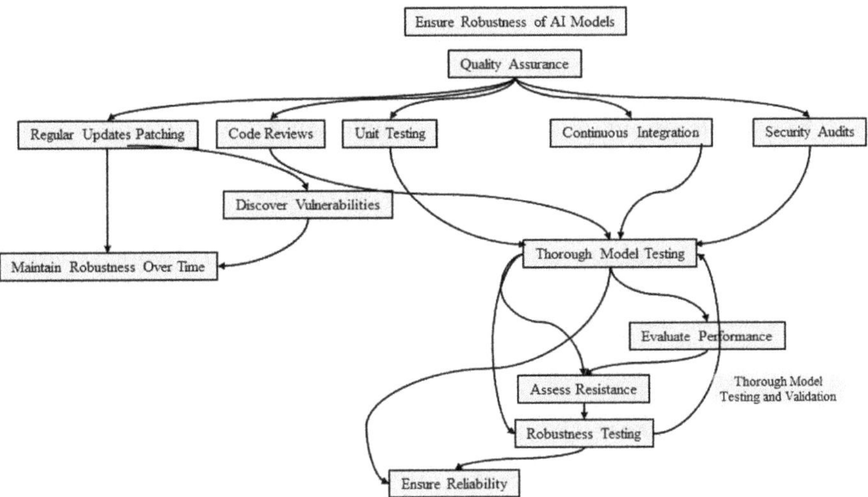

Figure 1.2 Ensuring robustness of AI models process.

1.3.3 Regulatory and ethical considerations

Governments and businesses are realizing the necessity for comprehensive frameworks and standards to control the creation and implementation of AI technology as AI applications continue to spread across sectors. The General Data Protection Regulation (GDPR), which establishes stringent regulations on the management of personal data in AI systems, is one such framework. The "right to explanation" guaranteed by GDPR provides openness and accountability in AI algorithms, especially where such choices have a significant influence on people's lives, such as when determining credit scores or loan approvals (Brkan, 2019; Liu et al., 2019). The GDPR's key provisions on data minimization, consent, the "right to be forgotten," and data portability have an impact on how AI models access, utilize, and safeguard personal data (Forcier et al., 2019).

Ethical criteria that guide the proper development and use of AI technology complement these regulatory issues. Fairness, transparency, accountability, and privacy are ethical principles in AI. Preventing prejudice and discrimination in algorithmic decision-making is necessary to ensure justice in AI. Since users and stakeholders should be able to understand how AI systems make choices, transparency is crucial to fostering confidence. Accountability implies that organizations and developers are accountable for fixing mistakes, biases, and unintended outcomes. Data protection and privacy-preserving methods are required due to privacy concerns, particularly concerning sensitive data. Last but not least, beneficence and non-maleficence demand that AI systems work to assist people and society as a whole while reducing damage. The fusion of AI and cybersecurity highlights the necessity to strike a careful balance between legal compliance and moral concerns.

1.4 THE COMPLEMENTARY RELATIONSHIP

Cybersecurity and AI integration is a truly symbiotic connection that goes beyond using AI for threat detection or protecting AI models. AI's skills and cybersecurity needs are fundamentally complementary, which makes them an effective team in the fight against online threats.

AI and cybersecurity have a connection that goes beyond simple integration to include mutual assistance. Real-time monitoring, quick incident response, and predictive capabilities offered by AI help cybersecurity, while cybersecurity secures AI by defending it from threats, maintaining regulatory compliance, and preserving trust and ethical principles. A strong defense against the changing panorama of cyber threats is built on this synergy.

1.4.1 How AI strengthens cybersecurity

AI has a set of special talents that greatly improve different facets of cybersecurity. Real-time threat monitoring, where AI's ability to quickly digest massive amounts of data comes to the forefront, is one of its most notable features. Whether it is a sudden increase in network traffic, a change in user behavior, or the introduction of new attack vectors, it excels at spotting abnormalities, odd patterns, and possible security breaches. By allowing security professionals to react quickly and effectively to new threats, real-time monitoring capabilities help companies remain one step ahead in the cybersecurity cat-and-mouse game.

AI also helps with cybersecurity via quick incident response. AI-driven systems can take quick steps like isolating damaged systems, stopping malicious activities, or implementing access restrictions without the need for human participation by automating incident response. Speed is essential for controlling threats and reducing damage. Human response teams, on the other hand, often need hours or even days to go through logs, find security problems, and come up with solutions. Due to the huge reduction in reaction times brought about by AI's automation of these operations, it is an indispensable tool in the battle against cyber attackers. AI strengthens cybersecurity with better prediction ability. Traditional cybersecurity techniques could depend on recognized attack signatures and patterns, which might not be enough to ward against fresh or changing attack tactics. Cybersecurity is improved by AI, especially machine learning, which offers predictive capabilities (Taddeo et al., 2019). AI can spot new dangers and foresee probable future assaults by examining past data. This proactive strategy reduces exposure to unanticipated cyber-attacks by empowering firms to plan and modify their defenses.

1.4.2 How cybersecurity safeguards AI

Despite being a potent weapon, AI is not exempt from flaws and assaults, and it often makes a good target for nefarious individuals. To guarantee the dependability, integrity, and moral application of AI, cybersecurity procedures are essential.

1.4.2.1 Protection from attacks

Despite being a potent weapon, AI is not exempt from flaws and assaults, and it often makes a good target for nefarious individuals. Cybersecurity procedures are essential for assuring the dependability, honesty, and moral use of AI (Rjoub et al., 2023). Like all digital assets, AI systems are vulnerable to several kinds of attacks. Defending AI against these assaults is a crucial component of cybersecurity. For example, adversarial assaults include changing input data to trick AI systems. Attackers deceptively change the input data to make the AI system guess the wrong things. Strict input validation should be used to identify and reject data that seems suspect to counteract these dangers. The AI model may be strengthened and made more resistant to these assaults by using data augmentation, which entails including hostile cases in the training data. It is also crucial to continuously assess a model's resilience using methods like stress testing and model stress testing.

The integrity and confidentiality of the system may be jeopardized by unauthorized access to AI models, datasets, or the infrastructure that supports AI. Strict access restrictions should be implemented to guarantee that only authorized individuals may alter the AI system, access private information, or communicate with the AI model to avoid this. Access may be further restricted to authorized users by using strong authentication techniques, such as multi-factor authentication. Another essential measure of protection is the deployment of security monitoring systems that can identify and notify users of illegal access attempts or suspicious activity in real-time.

Large datasets are often used by AI for training, and data leaks might jeopardize the confidentiality and integrity of these datasets. Data encryption, which renders data unreadable without the proper decryption keys, should be used to safeguard data at rest and in transit to reduce this risk (James & Rabbi, 2023). Data masking lessens the effects of a prospective data breach by anonymizing or pseudonymizing sensitive data in the training dataset. AI datasets need to be protected by secure data storage procedures that adhere to best practices in data storage security, including access limits and monitoring. Table 1.2 lists the many forms of AI assaults, including adversarial attacks, unauthorized access, and data breaches, as well as the related attack paths, outcomes, and protective safeguards.

1.4.2.2 Regulatory compliance

Securing AI systems requires ensuring regulatory compliance above everything else. These systems need to abide by data protection and privacy laws such as GDPR, Health Insurance Portability and Accountability Act (HIPAA), and sector-specific norms, especially when processing sensitive or personal data (Dove, 2018; Politou et al., 2022). When it comes to ensuring that AI applications are compliant with regulations, cybersecurity procedures are crucial. Data encryption, which protects data while it is at rest or in transit and makes sure it cannot be read by unauthorized parties, is one of the essential components.

Table 1.2 AI cybersecurity protection measures

Type of attack	Definition	Attack vectors	Consequences	Safeguard measures
Adversarial attacks	Manipulating input data to deceive AI	Evasion, poisoning	Incorrect predictions, biases	Input validation, data augmentation, model robustness testing
Unauthorized access	Unauthorized access to AI resources	Unauthorized login, insider threats	Data theft, system compromise	Access controls, strong authentication, security monitoring
Data breaches	Compromising data privacy and integrity	Hacking, insider threats, third-party breaches	Data exposure, loss of trust	Data encryption, data masking, secure data storage

Secure data handling procedures, which include data reduction, transparent data retention guidelines, and strict data access restrictions, are also essential for regulatory compliance. To promote accountability and transparency, AI developers and organizations need to provide transparent audit trails that show who uses the AI system and what actions are taken. In addition, the idea of privacy by design is accepted, including data security and privacy concerns in the fundamental design of AI systems. To develop safe and legally compliant AI systems that respect user privacy and rights, collaboration between cybersecurity and AI teams is crucial (Andraško et al., 2021).

Compliance also includes handling data subject rights, such as the ability to view, correct, or erase personal information, and consent management for processing user data. Meeting these conditions is both a moral obligation and a legal duty.

1.4.2.3 *Trust and ethical considerations*

To guarantee responsible and ethical AI activities, cybersecurity and AI practitioners need to work together to address trust and ethical considerations in AI. The role of cybersecurity in encouraging the ethical usage of AI is complex. It entails the use of fairness-aware algorithms to reduce biases and allow AI systems to make fair and impartial judgments. Cybersecurity measures may be used to impose ethical standards and principles, ensuring that AI models and systems function ethically (Taddeo et al., 2021).

Cybersecurity aids in transparency by enabling the construction of thorough audit trails that document the actions and judgments of AI systems. Professionals in cybersecurity may also encourage the development of explainable AI (XAI) approaches, which make AI models easier to understand and more transparent for users and auditors (Charmet et al., 2022; Rawal et al., 2021). Finally, it is crucial to promote security awareness and education. This includes teaching both AI users and developers how AI works, its capabilities, possible dangers, and the ethical and security issues that should be taken into account.

1.5 CHALLENGES AND CONSIDERATIONS

Although the incorporation of AI into cybersecurity and the protection of AI systems via cybersecurity measures have many benefits, they are nevertheless not without their share of difficulties and issues that call for careful study. For organizations to create efficient and responsible AI-driven cybersecurity strategies that ensure both the realization of AI's potential in cybersecurity and the ethical use of AI systems in an increasingly interconnected digital landscape, they need to acknowledge and address these complex issues.

1.5.1 Integration and implementation challenges

There are several difficulties involved with integrating and using AI in cyber-security. The quality and variety of the data is one important concern. The accuracy and representativeness of the training data are crucial components of AI models used in cybersecurity. Maintaining accurate and trustworthy data is a constant problem, particularly in the face of huge and diversified datasets. Due to compatibility concerns resulting from various legacy technologies, integrating AI tools into current security systems may be challenging. Building a strong cybersecurity framework requires making sure AI technologies function in harmony with other security solutions.

Organizations face a difficulty as they expand: scalability. AI systems need to effectively adapt to the expanding data volume and new dangers. AI solutions often need to be customized and tuned to fit the specific threat environment of an organization, but these procedures may be time- and resource-consuming. Additionally, the use of AI in cybersecurity necessitates particular knowledge and abilities. Recruitment and training challenges may result from a lack of specialists in cybersecurity and AI. Continuous education and training are required for the administration, interpretation, and fine-tuning of AI models.

Using AI in cybersecurity entails expenses for staff, hardware, and software that might be prohibitive for smaller businesses. Regulatory compliance is a crucial factor since AI systems employed in cybersecurity need to abide by regional and industry-specific standards, and non-compliance has serious legal and financial repercussions. Enterprises need to constantly maintain AI systems that are successful in the face of quickly changing attack routes and techniques. This is because the flexibility of AI-based security systems to emerging threats is a constant issue.

1.5.2 Resource and skill gaps

It takes sufficient resources and qualified personnel to integrate AI into cyber-security and protect AI systems via cybersecurity measures. However, several issues arise because of resource and talent shortages. AI-driven cybersecurity may have significant computing needs. Complex algorithms and deep learning models need high-performance technology, which may be out of the price range of smaller enterprises with tighter budgets, possibly exposing them to cutting-edge attacks. A Security Operations Center (SOC) needs to be set up since AI-based security solutions need continual monitoring and maintenance. However, sustaining such teams may be time-consuming and difficult for firms, especially smaller ones.

The dearth of specialists with knowledge of both AI and cybersecurity is one urgent issue. These two industries need to be more closely integrated, thus experts in cybersecurity and AI technology are needed. Such talent is increasingly difficult to find, hire, and keep in firms of all sizes. Additionally, for some people, the price of purchasing efficient AI technologies for cybersecurity may be too high. Smaller firms could find it difficult to invest in these solutions, resulting in differences in cybersecurity protection across various industries and business sectors.

Organizations may use a variety of techniques to fill these resource and talent shortages. First off, investing in training and skill development programs may assist in upskilling current staff members so they can comprehend the complexities of both AI and cybersecurity. Smaller firms may be able to combine resources and expertise via collaborative initiatives like industry alliances or consortia. Resource shortages may also be filled by adopting cloud-based AI cybersecurity solutions or outsourcing cybersecurity to managed service providers. Additionally, public-private collaborations may help with training, resource sharing, and information distribution, resulting in a more secure and resilient digital ecosystem and leveling the cybersecurity playing field.

1.5.3 Privacy and ethical concerns

A variety of privacy and ethical considerations that need careful consideration are brought to the fore by the integration of AI into cybersecurity and the protection of AI systems. It is crucial to strike a balance between the promise of AI and the protection of individual rights and ethical standards.

As AI is used in threat detection, worries about data privacy and security are heightened. The handling of huge data sets, which often include sensitive and personal information, calls for a careful balance between the need for cybersecurity and the preservation of individual privacy. With AI involved, the potential of data breaches increases, bringing legal and financial repercussions should personal or security data be compromised.

Both the defensive and offensive uses of AI in cybersecurity raise ethical issues. AI has the potential to be used in offensive cyber operations, which has serious ethical ramifications. These include the creation of cyberweapons and difficulties with attribution in light of AI's ability to muddle the origin of assaults. It becomes crucial to follow ethical rules and principles, such as proportionality, discrimination, and damage minimization. Complying with international rules, such as GDPR, and data protection legislation in the context of transnational cyber threats is further complicated by knowledge of export limitations for certain AI technology. Organizations continue to struggle with finding a balance between the need to improve cybersecurity via AI and the protection of privacy and moral norms.

1.5.4 Legal and regulatory hurdles

Organizations need to masterfully handle the myriad legal and regulatory obstacles associated with the integration of AI into cybersecurity to assure compliance and sustain effective security procedures. Data protection laws are the most significant of these difficulties, with the GDPR in Europe acting as a notable example. The handling and processing of personal data by enterprises need to adhere to these standards' stringent criteria, which places heavy compliance burdens on AI-powered cybersecurity solutions. Another difficulty is navigating the complicated web of foreign rules and jurisdictional difficulties. Cyber threats often cross international boundaries, making it challenging to assign blame and execute laws. Although the actual implementation and enforcement of the Budapest Convention on Cybercrime remain challenging, it aims to enhance international collaboration in prosecuting cybercriminals.

Export constraints increase complexity since certain AI technologies employed in cybersecurity may be classified as having both military and civilian applications, subjecting them to export regulations. When implementing AI-driven cybersecurity technologies across borders, organizations need to be aware of these rules and maintain compliance. Governments implement legislation in the name of national security, which also affects national security laws and laws relating to surveillance. It may be difficult to navigate the legal and ethical issues of working with government organizations while maintaining local laws and international human rights norms. Additionally, as AI tools and algorithms often use private IP, intellectual property rights need to be taken into account. IP infringement or disagreements might result in legal or financial repercussions, mandating careful IP protection and consideration for others' rights. Furthermore, because attackers often hide behind many layers of anonymity, identifying legal responsibility for cyberattacks is a challenging task. In the case of a cyber-incident, accurate attribution is important for recourse in court and financial recompense, but it is still a difficult process.

Finally, businesses typically explore larger ethical issues including responsible AI usage, transparency, justice, and reducing damage since ethical concerns often go beyond legal frameworks. To effectively navigate these legal and regulatory hurdles, a multidisciplinary strategy that includes legal knowledge, compliance methods, and adherence to international agreements is necessary. To guarantee that businesses operate legally while efficiently fending off cyber threats and safeguarding AI systems, it is essential to stay up to date on the changing legal environment in the context of AI and cybersecurity.

1.6 FUTURE TRENDS AND DIRECTIONS

1.6.1 AI and quantum computing in cybersecurity

Cybersecurity faces tremendous potential as well as difficulties with the coming age of quantum computing. Data security might be seriously jeopardized by quantum computers' capabilities to decrypt data and execute complicated computations at rates significantly faster than those of conventional computers. The convergence of AI and quantum computing, however, also throws up new possibilities for improving cybersecurity.

The creation of quantum-resistant encryption algorithms is one of the main goals of AI in quantum computing security. These algorithms have been developed to survive the quantum computer's processing capacity. AI may be essential in developing and refining such encryption techniques. Quantum-resistant encryption data may be analyzed by machine learning algorithms for patterns, enhancing its efficiency and security. Data protection is a priority with this proactive approach to security, even in the post-quantum computing era.

To improve encryption methods in the post-quantum age, academics are investigating AI-driven strategies. AI models may be used to find weaknesses in already-used encryption techniques and improve their robustness. Utilizing deep learning algorithms to find flaws in cryptographic systems and provide suggestions for enhancing security are examples of this. AI and quantum computing are working together to create encryption techniques that can withstand the enormous computational power of quantum computers.

Using the concepts of quantum physics, the secure communication technique known as quantum key distribution (QKD) exchanges cryptographic keys. By enhancing key management, monitoring quantum channel security, and fusing QKD with traditional encryption techniques, AI can speed up the implementation of QKD systems. To ensure the durability of QKD systems, machine learning models may assist in the detection of abnormalities and intrusions. Additionally, AI may use the quantum computers' enhanced processing power for real-time threat detection, making it more difficult for hostile activity to go undetected. The importance of AI in bolstering cybersecurity against the quantum threat will increase as quantum computing develops, delivering a more secure digital environment. With AI systems assuring the appropriate and ethical use of quantum technology in the field of cybersecurity, ethical issues will also gain prominence.

1.6.2 Advanced AI techniques for AI security

As AI systems grow more and more prevalent in a variety of industries, including banking, healthcare, and autonomous cars, their security is of the utmost significance. The creation and use of cutting-edge AI methods specifically designed for AI security are required for protecting AI systems.

1.6.2.1 Adversarial attack mitigation

The danger from adversarial assaults, in which bad actors influence AI models by adding minute changes to data, is still very real. Advanced AI approaches with real-time detection and reaction capabilities are required for mitigating such threats. Utilizing generative adversarial networks (GANs) as a protection mechanism is one strategy. When GANs are used to create adversarial examples that are included in training data, AI models gain the ability to detect and reject aggressive inputs, making them more robust. Further enhancing its security, reinforcement learning methods are used to train AI systems how to adapt to new hostile strategies.

1.6.2.2 XAI

Because it may be used by attackers, the opacity of many AI models is a security concern in and of itself. Understanding how and why AI systems make judgments is crucial in crucial cybersecurity scenarios. This challenge's frontline is XAI. By allowing human operators to understand AI judgments, XAI systems seek to promote transparency. This not only assists in identifying security flaws in AI models but also promotes control and confidence in AI systems. Attention mechanisms and decision trees are two cutting-edge XAI techniques that are constantly developing to provide complete insights into AI decision-making processes.

1.6.2.3 Federated learning for privacy-preserving AI

In the context of AI security, protecting privacy is of utmost importance, especially when dealing with sensitive data. Federated learning is a sophisticated AI approach that enables the training of AI models across several distributed devices or data sources without using shared raw data. This strategy protects data privacy while allowing AI models to develop and learn together. Federated learning may be used in AI security to perform threat detection and other security-related activities, protecting sensitive data and promoting cross-organizational cooperation without sacrificing data privacy.

1.6.2.4 AI-Enhanced deception technologies

By using dummy systems and data, deception technologies may keep attackers away from valuable assets. These deception technologies are presently being improved by advanced AI approaches by making the decoys more lifelike and dynamic. AI-driven deception systems may adjust to attacker behavior in real time, learning from prior encounters and steadily increasing the efficacy of their defensive strategy. This reduces the dangers to real assets by drawing opponents into a never-ending web of deceit.

1.6.2.5 Quantum-Safe AI

The need for quantum-safe AI security solutions has increased with the danger that quantum computing poses to current encryption techniques. AI has a key role to play in developing encryption methods that can withstand quantum assaults. This involves using AI to create post-quantum encryption techniques that can protect data against quantum-powered decoding attempts. The advancement of this field's study and development is essential for guaranteeing the long-term security of AI systems and data.

1.6.3 Collaborative initiatives and research areas

Cooperative efforts and continuing research in numerous fields will determine the future of AI in cybersecurity and the interaction between cybersecurity and AI security. To handle new dangers, improve security procedures, and make the digital world safer, these coordinated efforts are being made.

The adoption of an "AI-First Security" strategy, in which AI serves as both a major defensive mechanism and a tool, is one important development in cybersecurity. From threat detection to remediation, AI is incorporated into every layer of cybersecurity as part of AI-first security. It makes it possible for businesses to respond to threats more quickly and proactively, thereby thwarting sophisticated assaults.

To combat cyber threats, enterprises, governments, and security providers need to collaborate and share information more effectively. AI-driven threat detection, which offers real-time analysis of threats and vulnerabilities, is crucial in this respect. Sharing threat information provides a collective defense against changing threats as firms grow more linked. To enable quick responses to new risks, collaborative platforms, and standards for exchanging this information are being established. Internet of Things (IoT)-specific security solutions driven by AI are being developed via collaborative research. In IoT networks and endpoints, machine learning algorithms may be used to spot unusual activity, assisting in the prevention of illegal access, data breaches, and other security risks. Securing these devices will be crucial to preserving data integrity and privacy as more companies use IoT technology.

IoT device proliferation creates new security difficulties. AI algorithms are used by autonomous security systems to identify, examine, and react to security problems in a highly automated way. Without human interference, these systems make instantaneous judgments and respond to threats. To make autonomous security solutions effective and efficient, collaborative activities in this field are focused on optimizing these systems to reduce false positives and false negatives. The legal and ethical ramifications of AI making security and incident response choices will also be covered by research on autonomous security systems. In conclusion, joint projects and research in cybersecurity and AI are paving the path for a more secure and robust digital environment. Figure 1.3 illustrates the interconnected focus areas of collaborative initiatives in AI and cybersecurity, including AI-driven security solutions, enhancing security practices, security assurance, responsible AI, privacy considerations, and data security.

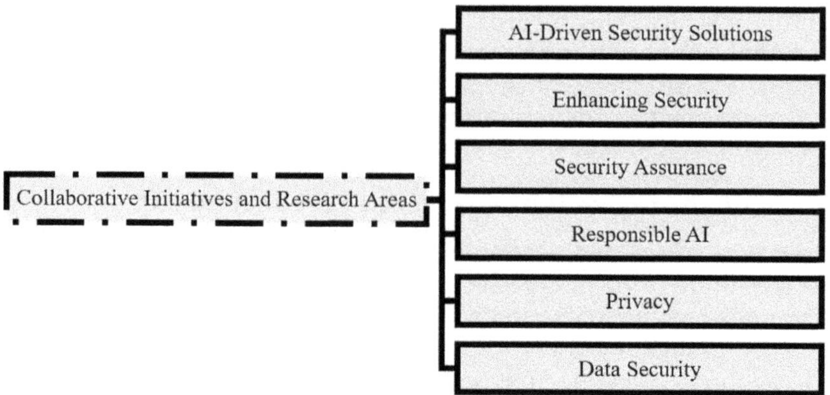

Figure 1.3 Key collaborative initiatives and research areas in AI and cybersecurity.

1.7 CONCLUSION

We've explored the complex interrelationships between AI and cybersecurity in this chapter, revealing the significant influence they have on the current state of our digital world. The effectiveness of AI in threat detection, incident response, and security analytics, which all work together to reinforce our digital defenses, makes it clear how important it is to cybersecurity. The usefulness of AI in securing our digital infrastructure and ensuring that our enterprises remain ahead of ever-evolving threats is shown by real-world instances. In addition, we've emphasized how crucial cybersecurity is for AI systems given their susceptibility to adversarial assaults and data tampering. To ensure the reliability and trustworthiness of AI, rigorous testing, safe data management, and steadfast respect for ethical and legal norms are necessary.

The interwoven, symbiotic connection between AI and cybersecurity is the main finding of this investigation. Real-time monitoring, quick responses, and predictive skills provided by AI make cybersecurity more effective, while cybersecurity, in turn, protects AI from outside attacks, assures regulatory compliance, and helps it traverse tricky ethical dilemmas. Cooperative efforts and continuous research endeavors will be crucial in determining the future as we face obstacles including integration barriers, resource shortages, and ethical issues. The future chapters in the history of AI and cybersecurity offer innovation and advancement while respecting security and ethical norms, with quantum computing on the horizon and sophisticated AI approaches constantly emerging.

REFERENCES

Ahmad, A., Desouza, K. C., Maynard, S. B., Naseer, H., & Baskerville, R. L. (2020). How integration of cyber security management and incident response enables organizational learning. *Journal of the Association for Information Science and Technology, 71*(8), 939–953.

Andraško, J., Mesarčík, M., & Hamuľák, O. (2021). The regulatory intersections between artificial intelligence, data protection and cyber security: Challenges and opportunities for the EU legal framework. *AI & Society*, 1–14.

Bécue, A., Praça, I., & Gama, J. (2021). Artificial intelligence, cyber-threats and Industry 4.0: Challenges and opportunities. *Artificial Intelligence Review*, 54(5), 3849–3886.

Brkan, M. (2019). Do algorithms rule the world? Algorithmic decision-making and data protection in the framework of the GDPR and beyond. *International Journal of Law and Information Technology*, 27(2), 91–121.

Chan, L., Hogaboam, L., & Cao, R. (2022). Artificial intelligence in finance. In *Applied Artificial Intelligence in Business: Concepts and Cases* (pp. 101–118). Springer.

Charmet, F., Tanuwidjaja, H. C., Ayoubi, S., Gimenez, P.-F., Han, Y., Jmila, H., & Zhang, Z. (2022). Explainable artificial intelligence for cybersecurity: A literature survey. *Annals of Telecommunications*, 77(11-12), 789–812.

Dasgupta, S., Yelikar, B. V., Naredla, S., Ibrahim, R. K., & Alazzam, M. B. (2023). AI-powered cybersecurity: Identifying threats in digital banking. In Proceedings of the 2023 3rd International Conference on Advance Computing and Innovative Technologies in Engineering (ICACITE).

Dove, E. S. (2018). The EU General Data Protection Regulation: Implications for international scientific research in the digital era. *Journal of Law, Medicine & Ethics*, 46(4), 1013–1030.

Ertoz, L., Eilertson, E., Lazarevic, A., Tan, P.-N., Kumar, V., Srivastava, J., & Dokas, P. (2004). Minds-Minnesota intrusion detection system. *Next generation data mining*, 199–218.

Forcier, M. B., Gallois, H., Mullan, S., & Joly, Y. (2019). Integrating artificial intelligence into health care through data access: Can the GDPR act as a beacon for policymakers? *Journal of Law and the Biosciences*, 6(1), 317–335.

Haider, N., Baig, M. Z., & Imran, M. (2020). Artificial intelligence and machine learning in 5G network security: Opportunities, advantages, and future research trends. *arXiv preprint arXiv:2007.04490*.

Harry, A. (2023). The future of medicine: Harnessing the power of AI for revolutionizing healthcare. *International Journal of Multidisciplinary Sciences and Arts*, 2(1), 36–47.

Hasija, A., & Esper, T. L. (2022). In artificial intelligence (AI) we trust: A qualitative investigation of AI technology acceptance. *Journal of Business Logistics*, 43(3), 388–412.

Hosseini, M. M., & Parvania, M. (2021). Artificial intelligence for resilience enhancement of power distribution systems. *The Electricity Journal*, 34(1), 106880.

Illiashenko, O., Kharchenko, V., Babeshko, I., Fesenko, H., & Di Giandomenico, F. (2023). Security-informed safety analysis of autonomous transport systems considering AI-powered cyberattacks and protection. *Entropy*, 25(8), 1123.

James, E., & Rabbi, F. (2023). Fortifying the IoT landscape: Strategies to counter security risks in connected systems. *Tensorgate Journal of Sustainable Technology and Infrastructure for Developing Countries*, 6(1), 32–46.

Johnphill, O., Sadiq, A. S., Al-Obeidat, F., Al-Khateeb, H., Taheir, M. A., Kaiwartya, O., & Ali, M. (2023). Self-healing in cyber–physical systems using machine learning: A critical analysis of theories and tools. *Future Internet*, 15(7), 244.

Kadel, R., Shrestha, H., Shrestha, A., Sharma, P., Shrestha, N., Bashyal, J., & Shrestha, S. (2022). *Emergence of AI in Cyber Security*. IRJMETS.

Khan, H. U., & Nazir, S. (2023). Assessing the role of AI-based smart sensors in smart cities using AHP and MOORA. *Sensors*, 23(1), 494.

Kim, K., Shin, Y., Lee, J., & Lee, K. (2021). Automatically attributing mobile threat actors by vectorized ATT&CK matrix and paired indicator. *Sensors*, 21(19), 6522.

Kumari, S., Vani, V., Malik, S., Tyagi, A. K., & Reddy, S. (2021). Analysis of text mining tools in disease prediction. Hybrid Intelligent Systems: 20th International Conference on Hybrid Intelligent Systems (HIS 2020), December 14-16, 2020,2023

Kumar, D., & Kumar, K. P. (2023). Artificial intelligence based cyber security threats identification in financial institutions using machine learning approach. In 2023 2nd International Conference for Innovation in Technology (INOCON),

Liu, H.-W., Lin, C.-F., & Chen, Y.-J. (2019). Beyond State v Loomis: Artificial intelligence, government algorithmization and accountability. *International Journal of Law and Information Technology*, 27(2), 122–141.

Lopes Antunes, D., & Llopis Sanchez, S. (2023). The age of fighting machines: the use of cyber deception for adversarial artificial intelligence in cyber defence. In Proceedings of the 18th International Conference on Availability, Reliability and Security,

Maezo, R. G., & Rey, A. E. (2023). Boosted CSIRT with AI powered open source framework. In 2023 JNIC Cybersecurity Conference (JNIC),

Mavroeidis, V., & Bromander, S. (2017). Cyber threat intelligence model: An evaluation of taxonomies, sharing standards, and ontologies within cyber threat intelligence. In 2017 European Intelligence and Security Informatics Conference (EISIC),

Mohammadi, M., Al-Fuqaha, A., Sorour, S., & Guizani, M. (2018). Deep learning for IoT big data and streaming analytics: A survey. *IEEE Communications Surveys & Tutorials*, 20(4), 2923–2960.

Olivetti, E. A., Cole, J. M., Kim, E., Kononova, O., Ceder, G., Han, T. Y.-J., & Hiszpanski, A. M. (2020). Data-driven materials research enabled by natural language processing and information extraction. *Applied Physics Reviews*, 7(4).

Patcha, A., & Park, J.-M. (2007). An overview of anomaly detection techniques: Existing solutions and latest technological trends. *Computer Networks*, 51(12), 3448–3470.

Politou, E., Alepis, E., Virvou, M., & Patsakis, C. (2022). *Privacy and Data Protection Challenges in the Distributed Era* (Vol. 26). Springer.

Rawal, A., McCoy, J., Rawat, D. B., Sadler, B. M., & Amant, R. S. (2021). Recent advances in trustworthy explainable artificial intelligence: Status, challenges, and perspectives. *IEEE Transactions on Artificial Intelligence*, 3(6), 852–866.

Rjoub, G., Bentahar, J., Wahab, O. A., Mizouni, R., Song, A., Cohen, R., & Mourad, A. (2023). A Survey on Explainable Artificial Intelligence for Network Cybersecurity. *arXiv preprint arXiv:2303.12942*.

Rousopoulou, V., Vafeiadis, T., Nizamis, A., Iakovidis, I., Samaras, L., Kirtsoglou, A., & Tzovaras, D. (2022). Cognitive analytics platform with AI solutions for anomaly detection. *Computers in Industry*, 134, 103555.

Saeed, S., Altamimi, S. A., Alkayyal, N. A., Alshehri, E., & Alabbad, D. A. (2023). Digital transformation and cybersecurity challenges for businesses resilience: Issues and recommendations. *Sensors*, 23(15), 6666.

Sarker, I. H., Furhad, M. H., & Nowrozy, R. (2021). Ai-driven cybersecurity: An overview, security intelligence modeling and research directions. *SN Computer Science*, 2, 1–18.

Sun, G., Cong, Y., Dong, J., Wang, Q., Lyu, L., & Liu, J. (2021). Data poisoning attacks on federated machine learning. *IEEE Internet of Things Journal*, 9(13), 11365–11375.

Taddeo, M., McCutcheon, T., & Floridi, L. (2019). Trusting artificial intelligence in cybersecurity is a double-edged sword. *Nature Machine Intelligence*, 1(12), 557–560.

Taddeo, M., McNeish, D., Blanchard, A., & Edgar, E. (2021). Ethical principles for artificial intelligence in national defence. *Philosophy & Technology*, 34, 1707–1729.

Taherdoost, H. (2022). Cybersecurity vs. information security. *Procedia Computer Science, 215,* 483–487. https://doi.org/10.1016/j.procs.2022.12.050

Taherdoost, H. (2023a). Deep learning and neural networks: Decision-making implications. *Symmetry, 15*(9), 1723.

Taherdoost, H. (2023b). Towards artificial intelligence in sustainable environmental development. *Artificial Intelligence Evolution,* 49–54.

Telo, J. (2017). AI for enhanced healthcare security: An investigation of anomaly detection, predictive analytics, access control, threat intelligence, and incident response. *Journal of Advanced Analytics in Healthcare Management, 1*(1), 21–37.

Timan, T., & Mann, Z. (2021). Data protection in the era of artificial intelligence: Trends, existing solutions and recommendations for privacy-preserving technologies. In *The Elements of Big Data Value: Foundations of the Research and Innovation Ecosystem* (pp. 153–175). Springer International Publishing.

Varian, H. (2018). Artificial intelligence, economics, and industrial organization. In *The Economics of Artificial Intelligence: An Agenda* (pp. 399–419). University of Chicago Press.

Verma, K. K., Singh, B. M., & Dixit, A. (2019). A review of supervised and unsupervised machine learning techniques for suspicious behavior recognition in intelligent surveillance system. *International Journal of Information Technology,* 1–14.

Vigo, R., Zeigler, D. E., & Wimsatt, J. (2022). Uncharted aspects of human intelligence in knowledge-based "intelligent" systems. *Philosophies, 7*(3), 46.

Zhang, Z., Ning, H., Shi, F., Farha, F., Xu, Y., Xu, J., & Choo, K.-K. R. (2022). Artificial intelligence in cyber security: Research advances, challenges, and opportunities. *Artificial Intelligence Review,* 1–25.

Zhou, L., Pan, S., Wang, J., & Vasilakos, A. V. (2017). Machine learning on big data: Opportunities and challenges. *Neurocomputing, 237,* 350–361.

Chapter 2

Malicious usage of artificial intelligence

Expansion of existing threats and novel threats

Premanand K. Kadbe, Balasaheb H. Patil, Rohit S. Piske, and Rupali B. Patil

2.1 INTRODUCTION

In an era where the lines between reality and technology blur, deepfakes have emerged as a groundbreaking yet profoundly controversial advancement. Deepfakes are a manifestation of the remarkable capabilities of artificial intelligence (AI), particularly deep learning algorithms to manipulate and generate highly convincing audio and video content. While they offer immense potential in various fields, such as entertainment and computer graphics, they also raise alarming ethical, social, and security concerns when used maliciously or irresponsibly. The deepfake technology allows the creation of highly realistic, computer-generated audio and video content, making it appear as though someone is saying or doing something they never did. While it has legitimate uses in entertainment and special effects, it poses significant risks when employed for malicious purposes [1]. Public figures can use deepfakes to create convincing fake news videos or speeches, spread false information, and manipulate public opinion. To extort money or information from victims, criminals may use deepfakes to create compromising videos or audio recordings.

2.1.1 The mechanics of deepfakes

Deepfakes are born from a combination of deep learning algorithms, neural networks, and large datasets. The fundamental process of creating a deep fake entails data collection, model training, and deep fake generation. First, we gather a substantial dataset of images and audio clips featuring the target individual or subject. Next, we employ deep neural networks, typically generative adversarial networks (GANs) or variational autoencoders (VAEs), to scrutinize and reproduce patterns within the data. Ed model generates the new content by merging elements from the target dataset to create a realistic video or audio that may feature the subject doing or saying things they never actually did.

The term "deepfake" has gained notoriety for its ability to create highly convincing yet entirely synthetic audio and video content. At the core of this technology lie the intricate mechanics of deep learning algorithms. In this

DOI: 10.1201/9781032657264-2

section, we will delve into the mechanics of deepfakes, exploring how these AI systems work and the steps involved in creating realistic deepfake content.

2.1.1.1 Deep learning and neural networks

Deep learning forms the foundation of deepfake technology. It relies on the neural networks of a replica of the human brain, which are made up of electronic components. These networks belong to layers of interconnected nodes, each processing and transforming data in increasingly complex ways. Deep learning algorithms are particularly adept at recognizing patterns and learning from vast datasets. Deepfake creation often employs convolutional neural networks (CNNs) for image-related tasks [2]. CNNs excel at extracting features from images, such as facial expressions, and can generate realistic face swaps. The recurrent neural networks are suitable for tasks involving sequences, making them valuable for generating realistic speech and audio in deepfakes.

2.1.1.2 Data collection and preprocessing

The process of creating a deepfake begins with collecting and preprocessing a substantial dataset. For facial deepfakes, this typically involves collecting images and videos of the target individual from various angles, lighting conditions, and facial expressions. The audio deepfakes require audio recordings of the target speaking, capturing their unique voice patterns, intonations, and speech mannerisms. We use data augmentation techniques such as image rotation, scaling, and noise addition to diversify the dataset and enhance the model's generalization ability. Occasionally, we add annotations such as facial landmarks or phonetic transcriptions to the data to aid in the training process [3].

2.1.1.3 Training deep learning model

The heart of deepfake creation lies in training a deep learning model, typically a GAN or a VAE. One can summarize the training process using a generator network, a discriminator network, adversarial training, and an iterative process. The generator network learns to create synthetic content, such as images or audio, by attempting to replicate the patterns and characteristics observed in the target dataset. The discriminator network finds out how to distinguish between real and synthetic content. It functions as a critic, providing feedback to the generator in order to improve its ability to create realistic content. The generator and discriminator networks engage in a competitive process where the generator strives to produce content that is indistinguishable from existent data, whereas the discriminator attempts to identify fakes, and after that, the training involves multiple iterations (epochs) during which the networks continuously refine their abilities. As training progresses, the quality of the generated content improves [4].

2.1.1.4 Generating the deepfake

After training the model sufficiently, it can produce deepfake content such as image deepfakes, audio deepfakes, and combinations of audio and video deepfakes. The generator network synthesizes images by superimposing the facial features of the target individual onto existing video frames. This process involves aligning the facial landmarks, adjusting lighting, and ensuring consistent motion. For audio, the generator network produces speech that mimics the intonation and speech patterns of the target voice. This often involves text-to-speech synthesis with the learned voice characteristics. The final step combines the audio and video content to produce a coherent deepfake video, matching the target individual's lip movements with the synthesized audio [5].

2.1.1.5 The positive applications

Deepfakes offer promising applications in areas such as entertainment, education, and accessibility. Movies and video games have used them to create lifelike characters and scenes, thereby reducing production costs and expanding creative possibilities [6]. The deepfake technology enables the creation of more engaging and interactive educational content. Generating realistic sign language interpretations or synthetic voices can assist individuals with disabilities.

2.1.1.6 The dark side of deepfakes

The deepfakes have become synonymous with deception, and their malicious uses are a growing concern. We can weaponize deepfakes to create fake news, misinformation, and fabricated content, which can rapidly spread through social media, influencing public opinion and trust. For extortion, fraud, or political manipulation, malicious actors can use deepfakes to impersonate individuals, public figures, or business leaders. The deepfake technology can violate an individual's privacy by superimposing their likeness into compromising or explicit content, often for revenge or harassment [7].

2.1.1.7 Ethical and legal dilemmas

The rise of deepfakes has prompted discussions about the ethical and legal frameworks needed to govern their use which includes consent and privacy, defamation and libel, and regulations. The questions about obtaining consent for using someone's likeness in a deepfake and the right to control one's digital image have come to the forefront. The deepfakes can potentially lead to complex legal disputes regarding defamation and libel as fabricated content can harm an individual's reputation. Some countries and states have started to enact legislation aimed at restricting or regulating the creation and distribution of deepfakes although the enforcement remains a challenge [8].

2.1.2 Objectives

1. Discuss privacy, trust, and the overall well-being of individuals and communities.
2. Propose Countermeasures and Defense Strategies: We provide and explore effective countermeasures and defense strategies against AI-driven threats, including both technical solutions and policy considerations to mitigate risks and enhance cyber security resilience.
3. Examine Regulatory Frameworks: To address the challenges posed by malicious AI usage, we evaluate the existing regulatory frameworks and propose potential enhancements. We also explore the role of policymakers in creating a secure environment for the responsible development and deployment of AI technologies.
4. Discuss Future Trends and Scenarios: We anticipate future trends in the malicious use of AI, speculate on potential scenarios that may arise, and provide recommendations for adapting cyber security practices to stay ahead of evolving threats.

This chapter is organized as follows: Sections 2.2 and 2.5 provide a detailed overview of AI-powered cyberattacks, while Section 2.3 discusses autonomous weapon systems (AWS), and Section 2.4 delves into AI-enabled social engineering. The data poisoning and manipulation are discussed in Section 2.6, followed by a discussion on adversarial attacks in Section 2.7, data tampering in Section 2.8, supply chain attacks in Section 2.9, and finally intellectual property (IP) theft in Section 2.10.

2.2 AI-POWERED CYBERATTACKS

AI can enhance cyberattacks in multiple ways, making them more sophisticated, faster, and difficult to detect. AI-driven malware can adapt to changing network conditions, identify vulnerabilities, and exploit them in real time, making traditional cyber security measures less effective.

2.2.1 Adaptive malware

The adaptive malware represents a significant evolution in cyber threats, leveraging AI and machine learning (ML) to heighten its susceptibility and evade traditional cyber security defenses. These malicious programs adapt and respond to the changing conditions, making them formidable adversaries in the cyber security landscape. The adaptive malware differs from traditional malware in its ability to learn, evolve, and execute attacks in a more dynamic and intelligent manner. The key characteristics include ML integration, real-time adaptation, evasion techniques, and dynamic attack strategies. The adaptive malware incorporates ML algorithms, which enable it to analyze its environment, learn from interactions, and adjust its behavior accordingly. These malware variants can adapt to changes in their operating environment in real time, making them more challenging to detect and mitigate. The adaptive malware employs evasion techniques to avoid detection, such as polymorphism, which means changing its code structure, or obfuscation, which means hiding its true intent, making it harder for signature-based antivirus solutions to identify. These malware strains can adjust their attack vectors and tactics based on their observations of network traffic, system configurations, and security defenses [9].

The presence of adaptive malware has significant implications for cyber security, such as reduced detection rates and increased attack sophistication and resilience. Traditional security solutions are less likely to detect adaptive malware due to its ability to constantly alter its appearance and behavior to evade signature-based detection. The adaptive nature of this malware allows it to carry out more polished and targeted attacks, potentially causing more damage or exfiltrating sensitive data. Adaptive malware is more resilient and can survive longer within a compromised system, making it more challenging to remove and eradicate completely.

Deterring adaptive malware requires proactive and multilayered approaches like behavior-based detection, ML defenses, anomaly detection, regular patching and updates, security awareness, and an incident response plan. Rather than relying solely on signature-based methods, we should implement behavior-based detection systems that concentrate on identifying unusual or suspicious behaviors. This use of ML algorithms helps develop adaptive defenses that can detect and respond to evolving threats, employing anomaly detection mechanisms to identify abnormal network or system behavior that may indicate the presence of adaptive malware. There is a need to keep systems and software up to date to patch known vulnerabilities, reducing potential entry points for adaptive malware. Then the training personnel recognize signs of malware infection and practice safe computing habits to minimize the risk of malware infiltration. Finally, develop a robust incidental response plan that outlines routines for detecting, containing, and eradicating adaptive malware in case of attack.

2.2.2 AI-powered phishing attacks

AI can craft individualized phishing emails by analyzing a victim's online behavior, making them more credible and increasing the possibility of success. Phishing attacks are a prevalent form of cyberterror in which attackers use delusory techniques to trick people into divulging erogenous information or taking malicious action. The integration of AI has given rise to AI-powered phishing attacks, making them more persuasive, personalized, and difficult to detect. AI-powered phishing attacks leverage AI and ML to automate and optimize various aspects of the phishing process. The key characteristics include personalization, spear phishing, content generation, timing, and delivery optimization. AI analyzes vast amounts of data about potential victims, including their online activity and predilection, to craft highly individualized phishing messages that appear more convincing because they mimic the interests and context of the recipient. AI-powered phishing often takes the form of spear phishing attacks targeting specific individuals or organizations with tailored messages that exploit their unique characteristics or vulnerabilities. AI-generated content, such as text and voice, can make phishing messages more sophisticated and realistic, increasing the chances of success. AI can determine the most opportune times to send phishing messages and the best channels to reach potential victims, increasing the likelihood of engagement.

AI-powered phishing attacks pose significant implications for cyber security, like higher security rates, increased volume, evasion of traditional defenses, etc. The personalization and sophistication of these attacks increase the likelihood of victims falling for them, leading to compromised accounts, data breaches, or financial losses. AI enables attackers to launch a large number of phishing attempts simultaneously, broadening the attack surface and putting more individuals and organizations at risk. AI-powered phishing attacks frequently circumvent traditional email and security filters, posing a challenge for conventional detection methods [10].

The defense against AI-powered phishing attacks needs an aggregation of technological solutions and user awareness, like advanced email filters, user training, multifactor authentication (MFA), AI-powered detection and reporting, and incident response. This employs advanced email filtering solutions that incorporate ML and natural language processing (NLP) to detect phishing emails with a higher degree of accuracy. Subsequently, the implementation of MFA is necessary to grant access to sensitive systems or data, providing an additional layer of security against compromised credentials. AI-powered threat detection tools are used to analyze email content, sender activity, and context to determine anomalies indicative of phishing attacks. This establishes clear reporting mechanisms for employees to report suspicious emails promptly, and an incident response plan is in place to address and mitigate phishing incidents.

2.2.3 Automated attacks

AI can automate the process of vulnerability scanning, brute force attacks, and data exfiltration, allowing malicious actors to scale their operations efficiently. Automated attacks are a category of cyber threats that leverage AI and automation technologies to carry out malicious activities at scale and speed. These attacks aim to exploit vulnerabilities, compromise systems, and accomplish their objectives with minimal human intervention. Automated attacks are characterized by their ability to operate autonomously and adapt to changing conditions. The key elements to consider include scalability, speed, targeted exploitation, and data exfiltration. The automated attacks can target a large number of systems simultaneously, making them suitable for launching widespread distributed attacks. These attacks operate at high speeds, allowing attackers to exploit vulnerabilities rapidly and evade detection or response. Attackers can use automation to scan and identify specific vulnerabilities in systems or networks, and then exploit them for unauthorized access or data theft. Automated attacks often include data exfiltration capabilities, allowing attackers to steal sensitive information and transfer it to remote servers [11].

Automated attacks have significant implications for cyber security, such as increased attack volume, reduced time to exploit, resource intensiveness, and data breaches. Attack techniques are automated, allowing adversaries to launch a large number of attacks, increasing the likelihood of successful compromises. The automated attacks can identify and exploit vulnerabilities within minutes or even seconds, leaving little time for defenders to respond. The automated attacks can strain network and system resources, potentially causing service disruptions or slowdowns. The speed and scale of these attacks make them effective at breaching security controls and stealing sensitive data.

The defense against automated attacks requires multifaceted approaches like patch management, intrusion detection systems, rate-limiting and access controls, behavioral analysis, incident response, and ML-based defense. Systems and software are kept updated with the latest security patches to decrease vulnerabilities that attackers might exploit. Next, implement Integrated Detection System (IDS) solutions that can detect unusual patterns of network traffic and behavior indicative of automated attacks. Next, implement rate limiting and access controls to restrict the number of login attempts, reducing the risk of brute force attacks. Then enforce strong user authentication mechanisms, including MFA, to mitigate the risk of unauthorized access. Behavioral analysis is used to identify deviations from normal user behavior, helping to detect automated attacks. An incident response plan is developed that outlines procedures for identifying, mitigating, and recovering from automated attacks and finally implements ML-based security solutions that can adapt to emerging threats and detect automated attack patterns.

2.3 AUTONOMOUS WEAPON SYSTEMS

The development and deployment of AWS represent a transformative shift in military technology and warfare. These systems equipped with AI and advanced automation have the potential to redefine the nature of armed conflict. This section explores the concept of AWS, their capabilities, ethical considerations, and the implications for international security. AWS are military technologies capable of independently selecting and engaging targets without direct human intervention. They encompass a spectrum of systems, from drones to ground-based robots armed with AI algorithms that enable them to make decisions based on predefined criteria and sensor inputs.

AWS have a variety of capabilities, including target identification and engagement, adaptability, and lethality. The AWS can autonomously identify and engage targets using sensors, ML, and pattern recognition, potentially with greater precision and speed than human operators. These systems can adapt to dynamic and evolving situations, adjusting tactics and targets based on real-time data, and changing battlefield conditions. The AWS can carry a range of lethal payloads, from small arms to precision-guided munitions, increasing their potential impact on the battlefield [12].

The ethical and legal considerations include lack of human judgment, accountability, and risk of autonomous escalation. One of the primary ethical concerns surrounding AWS is the absence of human judgment in target selection and engagement, potentially leading to unintended harm to civilians and civilian infrastructure. Determining the responsibility for actions taken by AWS is challenging, making it difficult to hold individuals or entities accountable for violations of international humanitarian law. The autonomous systems could lead to rapid decision-making and escalation without human oversight, increasing the risk of conflicts spiraling out of control.

The international legal framework includes laws of armed conflict and conventions on certain conventional weapons (CCWs). The AWS must adhere to the principles of distinction, which means differentiating between combatants and non-combatants; proportionality, which means using force proportionate to the military objective; and precautions, which means taking measures to minimize harm to civilians. The CCW, which is a United Nations framework, discusses AWS regulation, including discussions on meaningful human control over these systems.

The implications for international security include the arms race, deterrence, stability, and human rights concerns. The development and deployment of AWS could lead to an arms race as nations seek to maintain or gain a competitive edge in military technology. The AWS could alter the dynamics of deterrence and strategic stability, requiring a reevaluation of existing military doctrines and strategies. The use of AWS in policing and law enforcement could raise concerns about human rights violations, privacy, and surveillance.

Finally, the roles of ethics and governance include ethical development and international cooperation. The AWS developers should prioritize ethical considerations, promote transparency, and engage in public discourse on the ethical implications of these systems. Multilateral efforts and international cooperation are required to establish norms and guidelines for the responsible use of AWS, with a focus on minimizing civilian harm and maintaining human control.

2.3.1 Lethal AI, ethics, and challenges

The development of AWS powered by AI raises grave ethical and security concerns. AI systems can equip autonomous drones and military robots to decide when and who to engage in combat, potentially resulting in unintended casualties and escalation.

The advent of lethal AI presents a profound ethical dilemma at the intersection of advanced technology, warfare, and human morality. This section delves into the concept of lethal AI and examines its capabilities, ethical considerations, and the pressing challenges it poses for society and international security. Lethal AI refers to the integration of AI into AWS designed to cause physical harm or death to human beings or destruction of infrastructure. These systems can operate without direct human control or intervention in the decision-making process [13].

The major capabilities of lethal AI include precision targeting, rapid decision-making, and data integration. The lethal AI systems are capable of highly precise targeting, minimizing collateral damage, and reducing civilian casualties. AI algorithms enable quick and adaptive decision-making, potentially outpacing human reaction times in fast-paced combat scenarios. These systems can process vast amounts of data from sensors and sources, enhancing situational awareness and aiding in target selection.

The ethical and moral considerations for this include a loss of human-like control, civilian protection, and escalation risk. The primary ethical concern is the latent loss of human-like control over life and death decisions, raising questions about responsibility and accountability for actions taken by AI. The protection of civilians during armed conflict becomes more challenging because AI systems may have difficulty distinguishing combatants from non-combatants. The rapid decision-making capabilities of AI systems could inadvertently lead to the escalation of conflicts without human oversight or restraint.

The international legal framework includes laws of armed conflict where lethal AI systems must adhere to the established principles, including distinction, proportionality, and precautions to minimize harm to civilians during armed conflicts. The discussions within the CCW framework aim to address concerns related to lethal AI and establish guidelines for responsible use.

The implications for international security include strategic and tactical changes, as the lethal AI has the potential to alter military strategies and tactics, necessitating a reassessment of doctrines and approaches to warfare. The development of deadly AI could trigger an arms race among nations, heightening global security concerns. AI's influence on deterrence and strategic stability requires careful examination and consideration of unintended consequences.

Ethical development and governance include ethical frameworks in which developers and users of lethal AI systems must prioritize ethical considerations, promote transparency, and engage in ongoing ethical assessments. Multilateral cooperation and international agreements are crucial for establishing norms, standards, and guidelines for the responsible development and use of lethal AI.

2.3.2 Lack of accountability in AI and autonomous systems

The development of AI and autonomous systems has raised concerns about accountability in various domains, including technology, law, ethics, and governance. This section explores the concept of accountability in the context of AI and autonomous systems, examines the challenges it presents, and discusses potential solutions to address the lack of accountability. Accountability refers to the obligation and responsibility of individuals, organizations, and systems for their actions and decisions [14].

In the context of AI and autonomous systems, accountability encompasses human accountability, technical accountability, and legal and regulatory accountability. Humans play a crucial role in the creation, implementation, and functioning of AI systems. Technical accountability involves the ability of AI systems to explain their actions and decisions, particularly in complex or critical scenarios, and adherence to existing laws and regulations governing AI use. The legal and regulatory accountability involves the establishment of legal frameworks to address new challenges.

The various challenges of accountability in AI include opaque decision-making, a lack of human oversight, and ambiguity in legal responsibility. Many AI algorithms function as black boxes, making it challenging to understand the rationale behind their judgment, especially in high-stakes uses like healthcare and criminal justice. Autonomous systems, particularly in warfare and critical infrastructure, can operate with minimal human intervention, raising questions about human control and accountability. Determining legal responsibility for AI-related harms is complex, particularly when multiple parties are involved, like developers, operators, and users.

The ethical considerations for this include bias, discrimination, and transparency. The lack of accountability can lead to biased and discriminatory outcomes in AI systems, perpetuating societal inequalities. For ethical AI, transparency is essential. The accountability mechanisms should ensure that AI systems are transparent about their decision-making processes.

The legal frameworks and regulations include product liability laws and regulatory oversights. The evolving product liability laws may need to include provisions for AI systems' accountability in case of harm or malfunction. The governments and regulatory bodies must establish and enforce accountability standards for AI in various sectors. To address this, various technological solutions include explainable AI and auditing and monitoring tools. Advancements in explainable AI can empower AI systems to offer comprehensible explanations for their decisions, necessitating the development of auditing and monitoring tools that can track AI system behavior and performance, thereby facilitating accountability assessments. Human oversight and decision-making authority involves humans, retaining ultimate authority and control over AI and autonomous systems. Additionally, ethics boards or committees oversee the deployment of AI in critical areas, ensuring its ethical and accountable use.

2.3.3 Proliferation of AI and autonomous systems

The widespread proliferation of AI and autonomous systems across various sectors has ushered in a novel era of technological advancement and socio-economic change. This section explores the concept of proliferation in the context of AI and autonomous systems, examines the driving forces behind it, and discusses the resulting implications for society, ethics, and governance.

The proliferation refers to the rapid and widespread adoption and integration of AI and autonomous systems into diverse applications, industries, and domains. It includes technological proliferation, global spread, and democratization. For example, AI technologies are spreading across industries like healthcare, finance, transportation, and defense. AI and autonomous systems are also growing on a global scale, crossing geographical and political boundaries. AI tools, platforms, and knowledge are widely available and easy to access, letting many people use these technologies [15].

The driving forces behind proliferation are technological advancements, market demand, commercial interests, and global competition. The continuous breakthroughs in AI research and development have accelerated the proliferation of AI capabilities and applications. The growing market demand for AI-driven solutions in sectors such as healthcare, finance, and manufacturing fuels the proliferation of AI systems. Tech companies and startups are keen to capitalize on AI's potential, leading to the rapid proliferation of AI-based products and services. Nations and organizations engage in AI research and development to maintain or gain a competitive edge in technology, economy, and security.

The major implications of proliferation are societal impact, ethical concerns, security risks, and governance challenges. The proliferation has the potential to reshape society, affecting employment, education, healthcare, and human interaction, among other facets of daily life. The ethical dilemmas related to AI, including bias, privacy, and discrimination, become more pronounced as AI proliferates. The widespread use of AI introduces new cyber security challenges, including AI-powered cyberattacks and vulnerabilities in autonomous systems. Policymakers and regulatory bodies must grapple with the task of creating and enforcing laws and standards for AI use. The ethical considerations for this include equity and accessibility, transparency, and accountability. The proliferation should ensure equal access to AI benefits while avoiding exacerbating existing inequalities. Ethical AI proliferation requires transparency in algorithms and accountability mechanisms for AI systems.

Global cooperation and governance include international collaborations and regulatory frameworks. The collaboration among nations and organizations is vital to address the global dimensions of AI proliferation. Robust and adaptable regulatory frameworks must be established to govern AI use and mitigate its risks. As a result, encouraging responsible innovation is essential to harnessing AI's potential while addressing its challenges. The proliferation necessitates adaptive strategies to keep pace with AI advancements and their evolving implications.

Following are the case studies highlighting the proliferation and usage of specific tools in various sectors involving AI and autonomous systems.

2.3.3.1 Financial sector algorithmic trading

In the financial sector, hedge funds and trading firms increasingly rely on AI-driven algorithmic trading systems to execute trades at high speeds and make complex decisions in real time. The quantitative trading algorithms leveraging ML models analyze market data, identify patterns, and execute trades autonomously. This technology enables faster and more efficient trading, but it also raises concerns about market manipulation and the potential for algorithmic trading to exacerbate financial market volatility.

2.3.3.2 Healthcare medical imaging diagnosis

Healthcare applies AI to medical imaging for more accurate and timely diagnosis. AI algorithms analyze medical images such as X-rays and MRIs to detect abnormalities and assist healthcare professionals in making diagnostic decisions. Deep learning models trained on large datasets help identify patterns and anomalies in medical images, improving diagnostic accuracy. This technology enhances the speed and accuracy of medical diagnoses, but challenges include ensuring robustness, interpreting ability, and addressing concerns related to patient privacy.

2.3.3.3 Transportation autonomous vehicles

The transportation sector has witnessed the development and testing of autonomous vehicles that use AI to navigate and make real-time decisions. ML algorithms and sensor fusion technologies enable autonomous vehicles to interpret their surroundings, make driving decisions, and adapt to changing conditions. The autonomous vehicles have the potential to improve road safety, reduce accidents, and enhance transportation efficiency. However, the challenges include regulatory hurdles, ethical considerations, and the need for robust safety mechanisms.

2.3.3.4 Manufacturing robotics and automation

Manufacturing is increasingly deploying AI-powered robotics and automation systems to enhance efficiency, precision, and adaptability in production processes. Industrial robots equipped with AI algorithms optimize production workflows, handle complex tasks, and adapt to variations in manufacturing conditions. The use of AI in manufacturing leads to increased productivity, reduced costs, and improved quality control. However, the concerns include potential job displacement and the need for workforce upskilling.

2.3.3.5 Cyber security threat detection

The cyber security sector employs AI to detect and respond to evolving cyber threats in real time. The ML algorithms analyze network traffic, user behavior, and system logs to identify patterns indicative of potential security breaches or anomalies. AI-driven threat detection enhances the speed and accuracy of identifying cyber threats, helping organizations respond proactively. The challenges include adversarial attacks attempting to deceive AI-based defenses.

These case studies illustrate the diverse applications of AI and autonomous systems across sectors, showcasing both the benefits and challenges associated with their proliferation.

2.4 AI-ENABLED SOCIAL ENGINEERING

AI-enabled social engineering represents a sophisticated evolution of traditional social engineering tactics in which malicious actors manipulate individuals or organizations into divulging sensitive information or taking harmful actions. This section explores the concept of AI-enabled social engineering, its methods, implications, and strategies for mitigation. AI-enabled social engineering refers to the use of AI and automation to enhance and optimize the tactics employed in social engineering attacks. These attacks exploit human psychology, trust, and cognitive biases to deceive individuals or entities [16].

Various methods of AI-enabled social engineering include personalization, phishing and spear phishing, deepfakes and voice synthesis, chatbots, and conversational agents. Figure 2.1 displays such a variety of methods. AI algorithms analyze extensive data to craft highly personalized messages or interactions that resonate with the target, making the attacker's approach seem credible. AI can automate the creation and distribution of phishing emails or messages, increasing the volume and sophistication of these attacks. AI-generated deepfake videos and voice synthesis can impersonate trusted individuals, enhancing the credibility of fraudulent communication. AI-powered chatbots simulate human conversation, allowing attackers to engage with targets at scale, gather information, or propagate malicious links.

The implications of AI-enabled social engineering include increased effectiveness, deeper intrusion, and trust erosion. AI enhances the effectiveness of social engineering attacks, leading to higher success rates in manipulating targets. The malicious actors can gain deeper access to systems and data, potentially leading to data breaches, financial losses, or unauthorized access. Successful AI-enabled social engineering attacks can erode trust in digital communications, impacting online interactions and cyber security awareness.

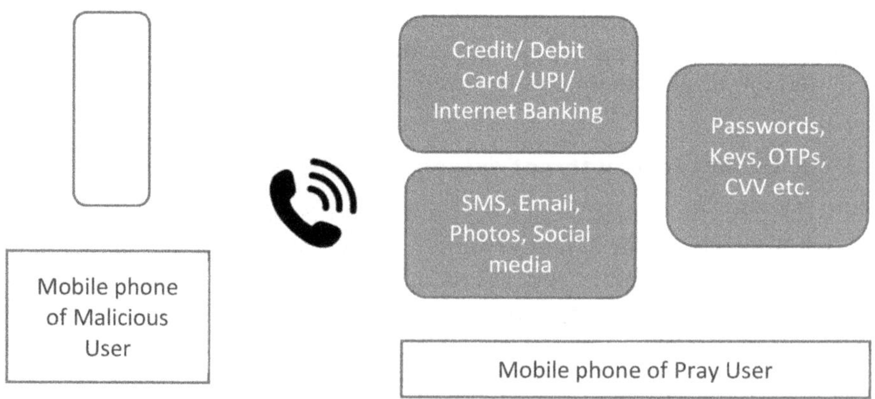

Figure 2.1 AI enabled social engineering.

2.4.1 AI-enabled impersonation attacks

AI-enabled impersonation attacks represent a significant cyber security threat where malicious actors leverage AI and deep learning techniques to impersonate individuals, organizations, or entities for fraudulent purposes. This section explores the concept of AI-enabled impersonation, its methods, potential consequences, and strategies for defense. AI-enabled impersonation refers to the use of AI and ML to mimic the appearance, behavior, or communication style of a specific individual or entity with the intent to deceive or defraud. These attacks can take various forms, from deepfake videos to AI-generated text messages.

The various methods of AI-enabled impersonation include deepfakes, voice-cloning chatbots, conversational AI, and email spoofing. AI-generated deepfake videos use neural networks to manipulate and superimpose one person's likeness onto another, creating convincing but fabricated videos. AI-powered voice synthesis can mimic an individual's voice, enabling attackers to impersonate someone over phone calls or voice messages. AI chatbots can simulate human conversation, making it hard for users to differentiate between real and fake interactions. AI can automate the creation of convincing email messages that happen to come from trustworthy origins, which are frequently used in phishing attacks.

Various implications of AI-enabled impersonation include fraud and deception, privacy invasion, misinformation, and disinformation. The impersonation attacks can lead to financial fraud, identity theft, and reputational damage. Deepfake videos and voice cloning can invade personal privacy by creating fabricated content that appears to depict real individuals. People can use the impersonation to spread false information, potentially influencing public opinion or confusing.

The various mitigation strategies aid in digital verification, media forensics, education and awareness, MFA, and content moderation. We must encourage the use of digital verification tools and techniques, like digital signatures or blockchain-based authentication, to confirm the authenticity of content or communication. The media forensics tools that can detect AI-generated or manipulated content must be developed, helping to identify impersonation attempts. The cyber security awareness and education of users must be promoted about the existence of AI-enabled impersonation threats and how to identify them, and also implement MFA for sensitive transactions or communications to add an extra layer of security. Social media platforms and content-sharing websites can implement AI-driven content moderation to detect and remove fake or harmful content. AI-enabled impersonation attacks can infringe upon individuals' privacy rights by using their likeness or voice without consent. Impersonation attacks have the potential to contribute to the spread of misinformation, influencing public discourse and trust in online content. AI detection tools, regulations, and standards are the technological countermeasures. For this, there is a need to develop AI-based detection tools that can identify AI-generated or manipulated content in real time. This explores regulatory frameworks and standards that govern the responsible use of AI, particularly in content generation and impersonation prevention.

2.4.2 AI-powered spear phishing attacks

AI-enabled impersonation attacks represent a significant cyber security threat where malicious actors leverage AI and deep learning techniques to impersonate individuals, organizations, or entities for fraudulent purposes. This section explores the concept of AI-enabled impersonation, its methods, potential consequences, and strategies for defense. AI-enabled impersonation refers to the use of AI and ML to mimic the appearance, behavior, or communication style of a specific individual or entity with the intent to deceive or defraud. These attacks can take various forms, from deepfake videos to AI-generated text messages.

The various methods of AI-enabled impersonation include deepfakes, voice-cloning chatbots, conversational AI, and email spoofing. AI-generated deepfake videos use neural networks to manipulate and superimpose one person's likeness onto another, creating convincing but fabricated videos. AI-powered voice synthesis can mimic an individual's voice, enabling attackers to impersonate someone over phone calls or voice messages. AI chatbots can simulate human conversation, making it difficult for users to differentiate between real and fake interactions. AI can automate the creation of convincing email messages from trustworthy origins, a common tactic in phishing attacks.

AI can analyze huge data to craft highly personalized spear phishing attacks targeting individuals with tailored messages. AI-powered spear phishing attacks represent a potent and increasingly sophisticated form of cyberthreat where malicious actors leverage AI and ML techniques to craft highly targeted and convincing phishing campaigns. This section delves into the concept of AI-powered spear phishing, its methods, implications, and strategies for defense. AI-powered spear phishing is a specialized form of phishing attack that employs AI and ML to tailor phishing messages and tactics to specific individuals or organizations. Highly targeted, these attacks aim to deceive recipients into divulging sensitive information or taking malicious actions [17].

The various methods of AI-powered spell phishing include personalization, contextual awareness, timing, and delivery optimization. AI algorithms analyze vast datasets to create highly personalized phishing emails that mimic the writing style, interests, or behaviors of the target, making the attack more convincing. AI systems can gather information about the target's recent activities, social connections, or ongoing projects to craft phishing messages that align with the target's current situation. AI can determine the most opportune times to send phishing emails and the best channels for delivery, increasing the likelihood of engagement.

The various implications of AI-powered spear phishing include targeted compromises, evasion of detection, and loss of trust. AI-powered spear phishing attacks are more likely to succeed in compromising specific individuals or organizations, potentially leading to data breaches, financial losses, or reputational damage. The personalization and sophistication of these attacks make them challenging to detect with traditional email filtering and security solutions. The successful spear phishing attacks erode trust in digital communications, making it more difficult for individuals and organizations to discern legitimate from malicious messages.

Different mitigation strategies for this include user training and awareness, email filtering, MFA, anomaly detection, and reporting mechanisms. Regular cyber security awareness training must be conducted to educate individuals about the risks of spear phishing and how to recognize suspicious emails. Employ advanced email filtering solutions that integrate ML and NLP to identify spear phishing attempts. Subsequently, implement MFA to enhance security, even if attackers manage to obtain login credentials. Next, implement anomaly detection mechanisms to identify unusual email patterns or behaviors indicative of spear phishing. And finally, establish clear reporting mechanisms for employees to report suspicious emails promptly, facilitating a swift response to potential spear phishing incidents. The different ethical considerations, including the use of AI for spear phishing attacks, raise concerns about privacy violations, especially when attackers gather and misuse personal information. The spear phishing attacks exploit human psychology and trust, leading to ethical concerns about deception and manipulation. Hence, the technological countermeasures include AI-powered detection, regulation, and compliance. To achieve this, there is a need to develop AI-powered threat detection tools that can analyze email content, sender behavior, and context to identify anomalies indicative of spear phishing, as well as explore regulatory frameworks and compliance standards that govern the responsible use of AI in spear phishing prevention and detection.

2.4.3 AI-powered manipulation of public opinion

Automated bots can amplify divisive narratives on social media, manipulate public sentiment, and influence elections. The manipulation of public opinion has evolved with the advent of AI and ML. This section explores the concept of AI-powered manipulation of public opinion, its methods, implications, and strategies for countering the spread of disinformation and manipulation in the digital age. AI-powered manipulation of public opinion refers to the use of AI and ML techniques to create, disseminate, or amplify false or misleading information with the intent to shape public perceptions, beliefs, or behaviors. These manipulative campaigns often exploit the vulnerabilities of social media and online platforms [18].

The different methods of AI-powered manipulation include automated bots, deepfakes, content recommendation algorithms, and sentiment analysis. AI-powered bots can generate and disseminate vast amounts of content, including fake news articles, tweets, and comments, to influence public discourse. AI-generated deepfake videos and audio can be used to create convincing but fabricated content featuring political figures, celebrities, or public figures who influence public opinion. AI algorithms used by social media platforms can amplify and promote divisive or misleading content, shaping users' perceptions and beliefs. AI can analyze social media posts and comments to gauge public sentiment and tailor manipulation strategies accordingly.

The different implications of AI-powered manipulation include erosion of trust, political influence, and social divisions. Manipulative AI campaigns erode trust in online content and information sources, undermining the credibility of journalism and experts. AI-powered manipulation can influence elections, political narratives, and public policy decisions, posing threats to democratic processes. The manipulation campaigns can exacerbate social divisions, polarizing communities and fostering hostility.

The various mitigation strategies include media literacy, transparency, fact checking, regulations, and legislation. To help individuals distinguish credible sources from misinformation, we must promote media literacy and critical thinking skills among the public. Media literacy includes transparency in the operation of algorithms on social media platforms, as well as recommendations for clear labeling of manipulated content. Support and fund fact-checking organizations that need to verify and debunk false information circulating online and explore regulatory frameworks and legislation that address the spread of disinformation and manipulation while balancing freedom of speech and expression.

AI-powered manipulation of public opinion refers to the use of AI techniques, algorithms, and tools to influence and shape the beliefs, attitudes, and perceptions of the general public. This manipulation often occurs through various digital channels and social media platforms.

2.4.3.1 Social media bots and amplification

The malicious actors use AI-driven bots to create and amplify content on social media platforms. These bots can generate fake accounts, post messages, share content, and engage with real users to create the illusion of widespread support or dissent. ML algorithms enable these bots to mimic human behavior, learning from patterns in engagement and content to make their activities more convincing and difficult to detect. By artificially boosting the visibility of certain opinions or narratives, these manipulative campaigns can sway public perception, create a false sense of consensus, and contribute to the spread of misinformation.

2.4.3.2 Deepfake technology

The deepfake technology utilizes AI to create highly realistic and often deceptive audio, video, or text content. This can involve manipulating the words or actions of public figures to spread false information or influence public opinion. Deep learning algorithms analyze and synthesize large datasets of facial expressions, voices, and language patterns to generate convincing fake content. Impersonating political figures, celebrities, or influential individuals with deepfakes can cause public confusion, mistrust, and potential reputational harm. The spread of fakes challenges the authenticity of digital content.

2.4.3.3 Algorithmic personalization

Online platforms use AI algorithms to personalize content recommendations for users based on their preferences and past behavior. However, these algorithms can unintentionally create filter bubbles, reinforcing existing beliefs and limiting exposure to diverse perspectives. Recommender systems leverage ML to predict user preferences, creating echo chambers where individuals are primarily exposed to content that aligns with their existing views. By tailoring content to individual preferences, AI algorithms contribute to the polarization of public opinion, reducing exposure to diverse viewpoints and fostering a fragmented information landscape.

2.4.3.4 Automated disinformation campaigns

AI-driven tools are employed to create and disseminate disinformation at scale. This involves generating false narratives, fake news articles, and misleading content to manipulate public sentiment. NLP algorithms can be used to generate coherent and contextually relevant disinformation, making it more convincing to unsuspecting readers. Automated disinformation campaigns can exploit social and political divisions, sow confusion, and influence public opinion on critical issues, potentially affecting elections, public policy, and social cohesion.

2.4.3.5 Sentiment analysis for manipulation

Malicious actors use AI-powered sentiment analysis tools to gauge public sentiment on specific topics. By identifying key emotional triggers, they can tailor their messaging to evoke desired reactions. Sentiment analysis algorithms process vast amounts of textual data from social media, news articles, and online forums to understand public sentiment trends. By strategically crafting messages that resonate with prevailing emotions, manipulators can amplify divisive narratives, exploit fears, and influence public opinion in a targeted manner.

Addressing the challenges associated with AI-powered manipulation of public opinion requires a multifaceted approach, including improved detection methods, transparency in algorithmic processes, media literacy education, and responsible AI development practices.

The technological countermeasures are to develop AI-powered detection tools that can identify manipulated content, deepfakes, and bot-driven campaigns and also advocate for greater transparency in the algorithms used by social media platforms, allowing researchers and the public to scrutinize their impact on content dissemination.

2.5 AI-POWERED CYBERATTACKS

As the world becomes more digital, the landscape of cyber threats is rapidly changing. AI has emerged as a double-edged sword, offering both defenders and attackers the powerful tools to advance their agendas. In this section, we will explore the domain of AI-powered cyberattacks, scrutinizing how AI amplifies cyber threats and the consequences for cyber security. AI has enabled the creation of adaptive malware that can evolve and respond to changing circumstances, making traditional cyber security measures less effective. AI-powered malware can employ evasion techniques like polymorphism or obfuscation to change its code, continually evading signature-based detection systems. AI can identify and exploit zero-day vulnerabilities more quickly and effectively than human hackers, making it challenging for organizations to defend against these unknown threats. The malware equipped with AI can mimic normal user behavior, making it difficult for intrusion detection systems to distinguish between legitimate and malicious actions [19].

2.5.1 Adaptive malware

The adaptive malware represents a new breed of cyber threats that utilize AI and ML techniques to evolve and adapt in real time. This section explores the concept of adaptive malware, its capabilities, implications, and strategies for defense in the ever-evolving landscape of cyber security. Adaptive malware refers to malicious software that employs AI and ML algorithms to continuously adapt and evolve in response to changing cyber security defenses and tactics. These threats have the ability to learn, modify their behavior, and evade detection in real time. The different capabilities of adaptive malware include stealth and evasion, self-propagation, data exfiltration, and distributed control. Adaptive malware can change its code, behavior, or appearance to evade detection by traditional antivirus and intrusion detection systems. These threats can autonomously spread across networks, exploiting vulnerabilities and compromising systems. The adaptive malware can exfiltrate sensitive data while avoiding detection, potentially leading to data breaches. Some adaptive malware leverages decentralized or peer-to-peer control mechanisms, making it difficult to disrupt [20].

The different implications of adaptive malware include persistent threats, cyber espionage, and financial losses. The adaptive malware can maintain a persistent presence on compromised systems, remaining undetected for extended periods. The state-sponsored actors and cybercriminals can use adaptive malware for espionage, stealing valuable information from individuals, organizations, or governments. Adaptive malware can lead to financial losses, including theft of funds, fraud, and damage to an organization's reputation.

The various mitigation strategies for this include behavioral analysis, threat intelligence, zero trust architecture, patch management, and AI-powered defense. Advanced behavioral analysis techniques are used to detect deviations from normal system behavior, which may indicate the presence of adaptive malware. Threat intelligence feeds and information sharing are utilized to stay updated on emerging threats and attack techniques. A zero-trust security model is adopted where trust is not assumed and verification is required for all network and system activities and needs regular updates and patches of software and systems to minimize vulnerabilities that adaptive malware might exploit. For this, AI-powered security solutions must be utilized that can adapt and respond to evolving threats in real time. Adaptive malware can compromise the privacy of individuals and organizations by exfiltrating sensitive data. The use of adaptive malware in cyberattacks leads to legal consequences for the attackers, including criminal charges.

2.5.2 Phishing attacks

With the help of AI, phishing attacks have become more sophisticated, enabling cybercriminals to craft highly convincing and personalized phishing campaigns. Phishing attacks, which are a form of cyber threat that involves deceptive tactics to trick individuals into revealing delicate information or picking harmful actions, remain a persistent and evolving cyber security challenge. This section delves into the concept of phishing attacks, their techniques, implications, and strategies for defense in an increasingly deceptive digital landscape. Phishing attacks are malicious attempts to deceive individuals or organizations into disclosing sensitive information such as login credentials, financial data, or personal details. These attacks often take the form of deceptive emails, messages, or websites that impersonate trusted entities. AI analyzes huge amounts of data from social media and other sources to craft convincing phishing messages that appear to come from trusted sources, increasing the likelihood of success. AI can tailor phishing emails to specific individuals, leveraging information about their preferences, habits, and relationships, making the emails appear more legitimate. AI can create realistic-sounding text and voice messages, making phishing emails and calls even more convincing [21].

The various methods and techniques of phishing attacks include email phishing, spear phishing, vishing (voice phishing), and smishing (SMS phishing). The attackers send deceptive emails impersonating trusted entities, often containing links or attachments that lead to malicious websites or malware downloads. Attackers tailor their messages to specific individuals in a highly targeted form of phishing, frequently taking advantage of personal or professional connections. Attackers conduct phishing attacks over the phone, impersonating legitimate organizations or authorities to extract information or manipulate victims. The phishing attempts via SMS or text messages often contain links to malicious websites or requests for sensitive information. Phishing attacks can result in various consequences such as data breaches, financial losses, and reputation damage. Successful phishing attacks can lead to data breaches exposing sensitive information, financial data, and personal records. The phishing attacks can result in financial losses due to fraudulent transactions or identity theft. If phishing campaigns use their trusted identities, organizations and individuals may face reputational damage.

Obviously, this requires mitigation strategies such as user training and awareness, email filtering and URL scanning, MFA, reporting mechanisms, and email authentication protocols. Regular cyber security awareness training must be conducted to educate individuals about the jeopardy of phishing and how to recognize phishing attempts. Advanced email filtering solutions and URL scanning tools must be employed to detect and block phishing emails and malicious links. The MFA implementation is needed to aid an excess layer of security even if attackers succeed in acquiring login credentials. Establish clear reporting mechanisms for employees or users to promptly report suspicious emails or phishing attempts. The email authentication protocols like Sender Policy Framework (SPF), DomainKeys Identified Mail (DKIM), and Domain-based Message Authentication, Reporting & Conformance (DMARC) must be implemented to prevent email spoofing. Continuously, authentication mechanisms must be enhanced to make it harder for attackers to impersonate trusted entities. AI-driven phishing detection tools must be developed that can examine email content, sender activity, and context to determine phishing attempts.

2.5.3 Automated attacks

AI automation tools have revolutionized cyberattacks by sanctioning attackers to scale their operations and launch attacks with greater speed and efficiency. AI-powered bots can execute brute force attacks more rapidly, attempting multiple password combinations to gain unauthorized access. AI can automate the scanning network process and vulnerability identification, reducing the time required to find potential entry points. AI can analyze and filter stolen data, enabling attackers to exfiltrate valuable information more efficiently without triggering alarms. Automated attacks, a category of cyber threats in which malicious actors use autonomous tools and scripts to exploit vulnerabilities and compromise systems, have become increasingly prevalent in the digital landscape. This section explores the concept of automated attacks, their methods, implications, and strategies for defense in an era of rapid automation and cybercrime. Automated attacks are cyber threats in which attackers use autonomous tools, scripts, or bots to carry out malicious activities without direct human intervention. These attacks often target vulnerabilities in systems, networks, or applications [22].

The various methods and techniques of automated attacks include brute force attacks, credential stuffing, botnets, scanning, and enumeration. The automated scripts attempt to guess passwords or encryption keys by systematically exploiting weak or default credentials. The attackers use stolen or leaked username and password combinations to gain unaccredited access to multiple accounts. The autonomous tools scan networks, identifying open ports, vulnerable services, and potential targets for exploitation. Attackers can launch distributed automated attacks, such as distributed denial of service attacks, from the networks of compromised devices under their control.

The implications of automated attacks include data breaches, financial losses, and system compromise. The automated attacks can lead to data breaches, exposing delicate content and personal data. Attacks such as credential stuffing can lead to financial losses due to fraudulent transactions or account takeovers. The compromise of vulnerable systems and devices can potentially lead to further attacks or unauthorized access.

The mitigation strategies include patch management, intrusion detection systems, rate limiting and account lockout, bot detection, and MFA. To mitigate known vulnerabilities that automated attacks may exploit, regularly update and patch software, systems, and applications. Additionally, implement IDS solutions that can detect and respond to unusual network activity indicative of automated attacks. To prevent brute force attacks, implement rate limiting and account lockout policies, and bot detection and mitigation solutions to identify and block malicious bots or automated scripts. This necessitates MFA to provide an extra layer of security, making it difficult for attackers to gain unauthorized access. Automated attacks can compromise individuals' privacy by stealing sensitive information. The perpetrators of automated attacks may face legal consequences, including criminal charges. The technological countermeasures include AI-powered threat detection and behavioral analysis to identify patterns of automated attacks, adapt to evolving threats, and identify unusual patterns of activity indicative of automated attacks.

2.6 DATA POISONING AND MANIPULATION

AI can poison or manipulate datasets, resulting in erroneous decisions and jeopardizing the integrity of AI-driven systems. Data poisoning and manipulation represent insidious threats in the digital age where malicious actors deliberately tamper with data to undermine trust, mislead decision-making, or compromise the integrity of information. This section explores the concept of data poisoning and manipulation, their methods, implications, and strategies for defense in an era where data is the lifeblood of technology and decision-making. Data poisoning and manipulation refer to the deliberate act of corrupting or manipulating data with malicious intent. These attacks can target various forms of data, including datasets, ML models, or information systems [23].

The different methods and techniques of data poisoning and manipulation include injection attacks, model poisoning, misleading information, and supply chain attacks. The malicious actors inject false or malicious data into databases or data streams, potentially compromising the integrity of the entire data set. The attackers manipulate training data used in ML models to compromise the performance or decision-making of these models. Disseminating manipulated or false information, like deepfake videos or fabricated news articles, serves to deceive and mislead. The data supply chain undergoes manipulation of software or hardware components to create vulnerabilities or contaminate the data.

Various implications of data poisoning and manipulation aim at misguided decision-making, erosion of trust, and security risks. The manipulated data can lead to misguided decisions in various domains, from finance and healthcare to autonomous systems. The data poisoning attacks erode trust in data sources and decision-making processes, impacting user confidence and data integrity. The compromised data can lead to security risks, including data breaches and system vulnerabilities.

The mitigation strategies include data validation and verification, behavioral analysis, model monitoring, and information authentication. The data validation and verification mechanisms are implemented to identify anomalies or signs of data manipulation, and behavioral analysis and anomaly detection are employed to identify unusual patterns in data or system behavior. The ML models are continuously monitored for signs of model poisoning or degradation. The cryptographic techniques are used to ensure the authenticity and integrity of data sources. Data manipulation can compromise individuals' privacy by altering or misusing their personal data. The data manipulation attacks often involve the spread of misinformation, raising ethical concerns about deception and manipulation. The technological countermeasures include the development of AI-powered detection tools that can identify anomalies and patterns of data manipulation or poisoning, as well as the implementation of blockchain technology to create transparent and tamper-evident records of data transactions.

2.7 ADVERSARIAL ATTACKS

The attackers using AI generate malicious inputs that deceive ML models, causing them to make incorrect predictions or classifications. The adversarial attacks represent a potent and evolving facet of AI-powered cyber threats. These attacks exploit the vulnerabilities in ML models and neural networks, demonstrating the intricate dance between offense and defense in the world of AI security. Adversarial attacks are a class of techniques wherein malicious actors intentionally handle input data to delude AI systems. The objective is to induce errors or misclassification in the output of the target AI model. The key characteristics of adversarial attacks include adversarial examples of generation, transferability, and stealth. The attackers generate adversarial examples by making slight, often imperceptible, modifications to input data such as images or text to mislead AI models [24]. The attackers carefully design these alterations to exploit the model's weaknesses. The adversarial examples crafted to deceive one AI model are often effective against others that share similar architectures or training data. This transfer ability property amplifies the potential impact of adversarial attacks. Humans design adversarial perturbations to be inconspicuous, making them difficult to detect visually or audibly. This ensures that the attacks can bypass human review or oversight.

There are several types of adversarial attacks targeting different AI domains, which include image, text, and audio-based attacks. In computer vision, the attackers manipulate images with imperceptible alterations to deceive image recognition systems. For instance, an autonomous vehicle perception system could subtly modify a "stop" sign to appear as a "yield" sign. In NLP, the adversaries may inject carefully crafted sentences or words into text data to trigger incorrect sentiment analysis, misinformation detection, or other AI-powered language processing tasks. These target voice recognition systems by subtly altering audio recordings to cause misinterpretations or unauthorized voice command execution. To create adversarial audio, attackers can use techniques such as waveform synthesis or voice synthesis.

The various implications of adversarial attacks include security vulnerabilities, privacy concerns, and trust erosion. The adversarial attacks can exploit vulnerabilities in critical AI systems, potentially leading to unauthorized access, data breaches, or manipulation of AI-driven decisions. Attackers can reveal sensitive information by exploiting AI models that rely on sensitive data, such as medical records or financial transactions. As adversarial attacks become more prevalent, trust in AI systems may erode, impacting their acceptance and adoption in various applications.

The defending against adversarial attacks is an ongoing challenge that includes robust model training, monitoring and detection, ensemble methods, and regular updates. To achieve this, incorporate robustness as a design principle during AI model development. By training models on adversarial examples and employing techniques such as adversarial training, we can enhance their resilience. Anomaly detection systems help to identify unusual behavior that may indicate an attack in progress. Different architectures and defenses combine with multiple AI models, and ensemble methods make it more challenging for attackers to craft effective adversarial examples. The attackers constantly adapt their methods, so regular updates on AI models and defenses are critical to staying ahead.

2.8 DATA TAMPERING

AI can subtly alter data at scale, resulting in misinformation or undermining the reliability of AI-driven decision-making processes. Data tampering is a form of cyberattack that involves malicious alterations or manipulations of data, often with the aim of deceiving, causing damage, or gaining unaccredited access to systems or information. Data tampering poses significant threats in the context of AI and ML, as AI systems heavily rely on the quality and integrity of the data they train on. Data tampering can take various forms, and its impact can range from misleading AI models to causing serious security breaches [25].

Here are some key aspects to consider, such as training data alteration, poisoning attacks, model evasion, and data exfiltration. The attackers may modify or inject malicious data into the training datasets used to train AI models. This can lead to the model learning incorrect patterns, making it susceptible to generating erroneous results or making flawed decisions. The data poisoning attacks involve introducing malicious data points into a data set to corrupt the training process. For example, in a spam email classifier, the attackers might inject legitimate emails with spam-like features to confuse the model. The attackers may craft data inputs specifically designed to exploit vulnerabilities in AI models. These inputs can fool the model into making incorrect predictions or decisions. In cases where sensitive information is the target, data tampering attacks might involve altering data to subtly leak information to malicious actors.

Data tampering can have serious consequences, including misleading AI models, security breaches, and loss of trust. The tampered data leads to AI models making incorrect predictions or classifications, potentially affecting the quality and reliability of AI-powered systems. In cases where data tampering is part of a broader attack, such as a data breach, the attackers may use tampered data to gain unaccredited access to systems or steal sensitive information. The data tampering incidents erode trust in AI systems and the data they rely on, affecting user confidence and adoption.

The defense against data tampering is essential to ensure the reliability and security of AI systems with data quality assurance, access control, data monitoring, blockchain and immutable logs, robust model validation, and security awareness. The rigorous data quality assurance practices verify the integrity of training and input data. The data validation checks and anomaly detection help identify potential tampering. Access control restricts access to data repositories and implements strict access control measures to prevent unauthorized alterations. The data monitoring process continuously monitors data streams and repositories for any suspicious activities or unauthorized changes. The blockchain technology, or immutable logs, creates an unalterable record of data changes and transactions, providing transparency and traceability. The robust model validation techniques ensure that AI models can identify and reject tampered data inputs. Employees receive training on data security best practices, which includes identifying and reporting instances of data tampering.

2.9 SUPPLY CHAIN ATTACKS

AI helps attackers target supply chains by injecting malicious code or compromised components into software or hardware systems. The supply chain attacks represent a sophisticated and often stealthy form of cyberattack that targets vulnerabilities in the interconnected web of suppliers, vendors, and service providers that support an organization's operations. These attacks leverage the trust established within the supply chain to compromise the security of the targeted systems or organizations [26].

Supply chain attacks can take a variety of forms and exploit different aspects of the supply chain ecosystem. The key elements to consider include attack vectors, targeted organizations, infiltration and persistence, and payload delivery. The attackers may infiltrate a supply chain through multiple entry points, including compromised software updates, counterfeit components, vendor relationships, or subcontractors. These attacks often target organizations with access to valuable assets, sensitive data, or critical infrastructure. A supply chain attack, for example, targets a software vendor in order to compromise the software delivered to multiple customers. Once inside the supply chain, the attackers work to remain undetected, infiltrating networks and systems and establishing persistence to maintain access over an extended period. The ultimate objective is to deliver a malicious payload or manipulate data within the supply chain, which then propagates to end users or organizations downstream.

Supply chain attacks can have far-reaching consequences like widespread impact, reputation damage, and IP theft. The attacks on suppliers or vendors can have a cascading effect, affecting numerous downstream organizations and potentially causing data breaches, service disruptions, or financial losses. Customers and partners may perceive the targeted organizations as the source of the security breach, leading to damage to their reputation and loss of trust.

The defense against supply chain attacks requires a multifaceted approach that includes vendor risk assessment, secure software development, third-party audits, zero-trust architecture, supply chain visibility, and incident response planning. The vendor risk assessment is conducted through thorough assessments of the security practices of suppliers and vendors, including evaluating their cyber security measures and assessing their potential risk to your organization. Both in-house teams and external suppliers implement secure coding practices to minimize vulnerabilities in software and to continuously monitor for software updates and patches. The third-party cyber security auditors are engaged to assess the security posture of critical vendors or suppliers. A zero-trust security architecture is adopted, which assumes that threats may exist within the network. This approach requires verifying and authenticating all network requests and communications, even those originating from trusted sources. The supply chain maintains visibility into the supply chain, monitors network traffic, and implements intrusion detection systems to detect unusual or suspicious activities. The comprehensive incident response plans are developed that consider the possibility of a supply chain attack and define roles, responsibilities, and communication protocols to respond effectively.

2.10 INTELLECTUAL PROPERTY THEFT

Supply chain attacks could lead to the theft of IP, trade secrets, or sensitive customer data, potentially for exploitation or sale on the dark web. The advent of AI has ushered in an era of remarkable technological advancements, revolutionizing industries and simplifying tasks across the globe. However, great power also carries great responsibility, as some have harnessed AI's capabilities for nefarious purposes. The application of AI in IP theft is particularly concerning, as it allows for the theft, replication, and dissemination of proprietary information, creations, and innovations. In this section, we explore the insidious world of IP theft facilitated by AI and the profound implications it has for individuals, organizations, and society as a whole. IP encompasses a broad spectrum of creative and intellectual assets, including patents, trademarks, copyrights, trade secrets, and industrial designs. These rights protect the innovations and ideas of individuals and organizations, granting them exclusive rights to use, sell, or license their creations [27]. IP theft involves the unauthorized acquisition, replication, distribution, or utilization of IP assets. While it has existed for centuries, AI has significantly amplified the scale, speed, and sophistication of these thefts.

AI-driven bots scour the internet, databases, and internal systems for confidential documents, patents, and trade secrets, extracting information with unprecedented efficiency. AI can expedite the process of reverse engineering products and technologies, enabling malicious actors to replicate and market counterfeit goods or services. The NLP algorithms create convincing copies of written content, making it challenging to distinguish between original and plagiarized materials. AI-generated deepfake technology can manipulate audio and video to create convincing impersonations or fabrications, thereby infringing on copyright and tarnishing reputations. Organizations suffer substantial financial losses due to IP theft, affecting revenue, market share, and competitiveness. The fear of IP theft may discourage companies from investing in research and development, hampering innovation and technological progress. The incidents of IP theft, especially when involving deepfakes or counterfeit products, can irreparably harm a brand's reputation. The IP theft can extend to military technologies and critical infrastructure, posing significant national security risks. The governments and international organizations are working to strengthen IP protection laws and enforcement mechanisms. The organizations are investing in robust cyber security measures, including AI-driven threat detection and response systems. The blockchain can provide a secure immutable ledger for IP rights and transactions, reducing the risk of fraud. The developers must prioritize ethical considerations in AI design, reducing the likelihood of AI being weaponized for IP theft.

AI will continue to play a critical role in developing proactive defense mechanisms against IP theft, like predictive analytic and autonomous threat responses. The enhanced international cooperation is vital to combating cross-border IP theft, as malicious actors often exploit legal loopholes in different jurisdictions. The society must grapple with the ethical implications of AI, balancing innovation with responsible use and accountability.

2.11 CONCLUSION

The mechanics of deepfakes involve complex deep learning algorithms, neural networks, and meticulous data collection and preprocessing. As the technology advances, creating convincing synthetic content becomes increasingly accessible, posing ethical and security challenges. Understanding the mechanics of deepfakes is crucial for both recognizing their potential benefits and mitigating the risks they pose to society. Deepfakes are a powerful demonstration of AI's ability to manipulate reality in ways previously thought impossible. Despite their numerous positive applications, the potential for misuse is undeniable. The society faces the daunting task of striking a balance between harnessing the creative and transformative potential of deepfakes and safeguarding against their malicious use. As technology evolves, it becomes imperative to develop ethical guidelines, educational campaigns, and legal frameworks to navigate the increasingly complex landscape of manipulated reality.

The emergence of adaptive malware underscores the need for a dynamic and adaptive cyber security posture. As these threats continue to evolve, the organizations must embrace advanced techniques and technologies to detect, respond to, and mitigate the risks posed by this intelligent and resilient form of malware. The cyber security professionals and organizations must remain vigilant and proactive in the face of this evolving threat landscape. AI-powered phishing attacks represent an evolution of a common and persistent cyber security threat. As AI advances, so too will the sophistication and effectiveness of these attacks. To defend against them, the organizations must adopt a proactive stance combining advanced technology solutions with user education and robust incident response plans to mitigate the risks posed by AI-powered phishing attacks.

The automated attacks pose a significant challenge to cyber security due to their speed, scalability, and adaptability. As cyber threats continue to evolve, the organizations must adopt advanced technologies, proactive monitoring, and vigilant security practices to defend against automated attacks effectively. Combining technology with human expertise is essential in maintaining a strong defense posture in the face of these evolving threats. The emergence of AWS presents a complex and multifaceted challenge for the international community. Thoughtful deliberation, international cooperation, and the establishment of clear guidelines are necessary to balance the potential military advantages of AWS with ethical, legal, and security concerns, ensuring their consistent use with international humanitarian law and human rights standards. The ethical, legal, and security implications of AWS will continue to shape discussions on the future of warfare and international security.

The lethal AI represents a profound and complex ethical challenge that demands careful consideration, ethical development, and responsible governance. To ensure the use of lethal AI in ways that minimize harm to civilians, protect human rights, and uphold international humanitarian law, it is crucial to strike a balance between military advantages and ethical constraints. The ongoing discourse and global cooperation on lethal AI will shape the future of warfare and international security. The lack of accountability in AI and autonomous systems presents multifaceted challenges that require attention from policymakers, technologists, ethicists, and legal experts. Striking a balance between technological innovation and accountability is essential to harnessing the benefits of AI while mitigating its risks. Addressing this accountability gap is critical for ensuring that AI systems align with human values, uphold ethical standards, and remain within the bounds of legal and regulatory frameworks.

The proliferation of AI and autonomous systems is a defining feature of the modern technological landscape. While it offers numerous benefits, it also raises profound ethical, societal, and governance challenges. Addressing these challenges requires a concerted effort from stakeholders worldwide to ensure that AI proliferation aligns with human values, respects ethical principles, and contributes positively to the betterment of society and the global community.

AI-enabled impersonation attacks pose a serious threat to individuals, organizations, and the integrity of online content. As AI technology continues to advance, it is essential to remain vigilant and proactive in developing and deploying countermeasures to detect and prevent these deceptive impersonation attempts. Combining technological solutions, user education, and ethical considerations is crucial to defending against AI-enabled impersonation and preserving trust in the digital realm. AI-powered spear phishing attacks are a potent and evolving cyber security threat. As AI technology continues to advance, organizations and individuals must remain vigilant and proactive in adopting advanced technologies, educating users, and implementing ethical considerations to defend against these highly targeted and deceptive attacks. Combining technology with human expertise is essential in maintaining a strong defense posture against AI-powered spear phishing.

AI-powered manipulation of public opinion is a formidable challenge in the digital age. Countering the spread of disinformation and manipulation requires a multifaceted approach that combines technology, education, regulation, and ethical considerations. As AI continues to evolve, vigilance and proactive efforts are crucial in safeguarding the integrity of public discourse and democratic processes in an increasingly interconnected world. As AI continues to advance, its malicious usage poses a growing threat to individuals, organizations, and societies at large. To mitigate these risks, it is imperative to develop robust AI ethics frameworks, enhance cyber security measures, and establish international regulations governing the use of AI in areas such as autonomous weaponry. Educating the public about the potential for AI misuse and fostering responsible AI development and deployment are critical steps toward harnessing AI's benefits while minimizing its dark side.

The adaptive malware represents a significant and evolving threat in the world of cyber security. Defending against these threats requires a proactive approach that combines advanced technology, collaboration, and a commitment to staying informed about emerging threats and vulnerabilities. As adaptive malware continues to evolve, organizations and individuals must adapt their cyber security strategies to mitigate the risks posed by these dynamic and persistent threats. Phishing attacks continue to pose a significant cyber security threat, evolving in sophistication and tactics. The effective defense requires a combination of technology, user education, and vigilant monitoring to detect and mitigate these deceptive attacks. As phishing techniques become more advanced, organizations and individuals must adapt their cyber security strategies to safeguard sensitive information and maintain trust in digital communication. The automation of cybercrime tools and techniques drives the growing cyber security challenge of automated attacks. Effective defense against automated threats necessitates a combination of technology, proactive patch management, and vigilant monitoring to detect and mitigate these attacks. As automation in cybercrime continues to advance, organizations and individuals must adapt their cyber security strategies to safeguard against automated attacks and protect critical systems and data.

Data poisoning and manipulation pose serious threats to the integrity of data-driven decision-making and information systems. Defending against these threats requires a multifaceted approach that combines technology, data governance, ethical considerations, and user education to safeguard data integrity and maintain trust in digital information. As data continues to play a pivotal role in technology and society, vigilance against data poisoning and manipulation remains essential for preserving the reliability and security of data-driven systems. The adversarial attacks epitomize the dynamic nature of AI security, where attackers constantly seek to exploit vulnerabilities in ML models. Understanding the mechanics of these attacks and developing robust defenses are critical to maintaining the integrity, reliability, and security of AI systems in an increasingly AI-driven world. As the field of AI security evolves, the defenders must remain vigilant and proactive in countering adversarial threats. Data tampering is a potent threat that can undermine the integrity and security of AI systems. As AI continues to play a critical role in various domains, from finance to healthcare, protecting against data tampering becomes paramount. Effective defenses require a combination of technological solutions, robust data management practices, and a proactive approach to monitoring and securing data throughout its life cycle.

The supply chain attacks are a growing concern in the realm of cyber security due to their potential for widespread and cascading impacts. As the organizations become more interconnected, securing the supply chain becomes paramount. A proactive and vigilant approach to vendor risk management, supply chain transparency, and robust cyber security measures are essential to defending against these increasingly sophisticated threats. AI, with its immense potential for good, also harbors the capacity for harm when wielded unethically. AI facilitates IP theft, exemplifying the dual nature of technology. To mitigate the malicious use of AI in this context, it is imperative that governments, organizations, and individuals work together to develop robust legal frameworks, advanced cyber security measures, and ethical guidelines. Only through collective efforts can we protect IP, foster innovation, and secure a more promising future in the age of AI. For further details, the readers are encouraged to refer to the reference numbers [28–33].

REFERENCES

1. Floridi, L. (2018). "Artificial intelligence, deepfakes and a future of ectypes". Editor Letter, Philosophy of Technology, 31, 307–321. https://doi.org/10.1007/s1334701803253
2. Mo, H., Chen, B. and Luo, W., "Fake faces identification via convolutional neural network". In Proceedings of the 6th ACM Workshop on Information Hiding and Multimedia Security, ACM, 2018, pp. 43–47
3. Tolosana, R., VeraRodriguez, R., Fierrez, J., Morales, A. and OrtegaGarcia, J. (2020), "DeepFakes and beyond: A survey of face manipulation and fake detection". Information Fusion. 64. https://doi.org/10.1016/j.inffus.2020.06.014.
4. Pan, D., Sun, L., Wang, R., Zhang, X. and Sinnott, R. O., "Deepfake detection through deep learning". In 2020 IEEE/ACM International Conference on Big Data Computing, Applications and Technologies (BDCAT), Leicester, UK, 2020, pp. 134143, https://doi.org/10.1109/BDCAT50828.2020.00001.
5. Agarwal, S., Farid, H., Gu, Y., He, M., Nagano, K. and Li, H. (2019). "Protecting world leaders against deep fakes". Computer Vision and Pattern Recognition Workshops, 1, 38–45.
6. Westerlund, M. (2019). "The emergence of deepfake technology: A review". Technology Innovation Management Review, 9(11), 39–52. https://doi.org/10.22215/timreview/1282
7. Brundage, M., Avin, S., Clark, J., Toner, H., Eckersley, P., Garfinkel, B. and Anderson, H. (2018), "The malicious use of artificial intelligence: Forecasting, prevention, and mitigation". *arXiv*. https://arxiv.org/abs/1802.07228
8. Meskys, E., Liaudanskas, A., Kalpokiene, J. and Jurcys, P. (2020), "Regulating deep fakes: Legal and ethical considerations". Journal of Intellectual Property Law & Practice. 15, 2431.
9. Aslam, D. M., Dengpan, Y., Hanif, M. and Asad, M. (2020), "Adaptive machine learning: A framework for active malware detection". https://doi.org/10.1109/MSN50589.2020.00025.
10. Tabassum, I., Bazai, S. U., Zaland, Z., Marjan, S., Khan, M. Z. and Ghafoor, M. I., "Cyber security's silver bullet a systematic literature review of AI powered security". In 2022 3rd International Informatics and Software Engineering Conference (IISEC), Ankara, Turkey, 2022, pp. 17, https://doi.org/10.1109/IISEC56263.2022.9998305.
11. Pacheco, M. L., Hippel, M., Weintraub, B., Goldwasser, D. and NitaRotaru, C. "Automated attack synthesis by extracting finite state machines from protocol specification documents". In 2022 IEEE Symposium on Security and Privacy (SP), 2022, p. 5168.
12. Taddeo, M. and Blanchard, A. (2022). "A comparative analysis of the definitions of autonomous weapons systems". Science and Engineering Ethics, 28, 37.
13. Cath, corinne, 2018, "Governing artificial intelligence: Ethical, legal and technical opportunities and challenges. Philosophical Transactions of the Royal Society A: Mathematical, Physical and Engineering Sciences, 376(2133), 20180080
14. Hohma, E., Boch, A., Trauth, R. and Lütge, C. (2023). "Investigating accountability for artificial intelligence through risk governance: A workshop based exploratory study". Frontiers in Psychology, 14, 1073686. https://doi.org/10.3389/fpsyg.2023.1073686. PMID: 36760454; PMCID: PMC9905430.
15. Shahriari, K. and Shahriari, M., "Ethically aligned design: A vision for prioritizing human wellbeing with artificial intelligence and autonomous systems". In 2017 IEEE Canada International Humanitarian Technology Conference (IHTC), Toronto, ON, Canada, 2017, pp. 197201, https://doi.org/10.1109/IHTC.2017.8058187.

16. Gayan, A. (2021), "Impact of AI on Social Engineering". Medium. https://medium.com/unpackai/impactofaionsocialengineeringe9bd763a77db
17. Basit, A., Zafar, M. and Liu, X. et al. (2021)., "A comprehensive survey of AI enabled phishing attacks detection techniques". Telecommunication Systems 76, 139–154.
18. Ienca, M. (2023). "On artificial intelligence and manipulation". Topoi, 42, 833–842.
19. Guembe, B., Azeta, A., Misra, S., Osamor, V. C., FernandezSanz, L. and Pospelova, V. (2022). "The emerging threat of AI driven cyber attacks: A review". Applied Artificial Intelligence, 36(1), 2037254. https://doi.org/10.1080/08839514.2022.2037254.
20. Choi, J., Shin, D., Kim, H., Seotis, J. and Hong, J. B., "AMVG: Adaptive malware variant generation framework using machine learning". In 2019 IEEE 24th Pacific Rim International Symposium on Dependable Computing (PRDC), Kyoto, Japan, 2019, pp. 24624609, https://doi.org/10.1109/PRDC47002.2019.00055.
21. Ripa, S. P., Islam, F. and Arifuzzaman, M., "The emergence threat of phishing attack and the detection techniques using machine learning models". In 2021 International Conference on Automation, Control and Mechatronics for Industry 4.0 (ACMI), Rajshahi, Bangladesh, 2021, pp. 16, https://doi.org/10.1109/ACMI53878.2021.9528204.
22. Falco, G., Viswanathan, A., Caldera, C. and Shrobe, H. (2018). "A master attack methodology for an AI Based automated attack planner for smart cities". IEEE Access, 6, 48360–48373. https://doi.org/10.1109/ACCESS.2018.2867556.
23. Weerasinghe, S., Alpcan, T., Erfani, S. M. and Leckie, C. (2021). "Defending support vector machines against data poisoning attacks". IEEE Transactions on Information Forensics and Security, 16, 2566–2578. https://doi.org/10.1109/TIFS.2021.3058771.
24. Fan, M., Liu, Y., Chen, C., Yu, S., Guo, W. and Liu, X., "Combating false sense of security: Breaking the defense of adversarial training via nongradient adversarial attack". In ICASSP 2022 2022 IEEE International Conference on Acoustics, Speech and Signal Processing (ICASSP), Singapore, Singapore, 2022, pp. 3293–3297, https://doi.org/10.1109/ICASSP43922.2022.9746138.
25. Wang, Y. and Li, Y., "Research on digital media image data tampering forensics technology based on improved CNN algorithm." In 2021 5th Asian Conference on Artificial Intelligence Technology (ACAIT), Haikou, China, 2021, pp. 393397, https://doi.org/10.1109/ACAIT53529.2021.9731182.
26. Sani, A. S., Yuan, D., Meng, K. and Dong, Z. Y., "Idenx: A blockchain based identity management system for supply chain attacks mitigation in smart grids". In 2020 IEEE Power & Energy Society General Meeting (PESGM), Montreal, QC, Canada, 2020, pp. 15, https://doi.org/10.1109/PESGM41954.2020.9281929.
27. Xiong, Y., Ramachandran, G. K., Ganesan, R., Jajodia, S. and Subrahmanian, V. S. (2020). "Generating realistic fake equations in order to reduce intellectual property theft". IEEE Transactions on Dependable and Secure Computing, 19(3), 1434–1445. https://doi.org/10.1109/TDSC.2020.3038132.
28. Lalla, V., Mitrani, A. and Harned, Z., Fenwick, New York and Santa Monica, USA, (2022) "Artificial intelligence: Deepfakes in the entertainment industry", World Intellectual Property Organization, Wipo Magazine, https://www.wipo.int/wipo_magazine/en/2022/02/article_0003.html
29. Helmus, T. C.,(2022). Artificial Intelligence, Deepfakes, and Disinformation. *RAND Corporation*, 1–24., https://doi.org/10.7249/PEA1043-1, Document Number: PE-A1043-1, Year: 2022, Series: Expert Insights
30. Jones, N. (2023). "How to stop AI deepfakes from sinking society-and science". Nature, 621(7980), 676–679. https://doi.org/10.1038/d41586-023-02990-y

31. Satariano, A. and Mozur, P. (2023). The people onscreen are fake. The disinformation is real. *International New York Times*, NA-NA. https://www.nytimes.com/2023/02/07/technology/artificial-intelligence-training-deepfake.html

32. Gosain, M. T., Bhatia, M. K., Sharma, M. R., Bhanvra, M. S., Shaw, A., Singh, T. and Kashyap, B. H. (2024). "Artificial intelligence and the privacy paradox: Challenges and opportunities in legal adaptations". Educational Administration: Theory and Practice, 30(5), 10384–10394. https://www.livelaw.in/law-firms/law-firm-articles-/deepfakes-personal-data-artificial-intelligence-machine-learning-ministry-of-electronics-and-information-technology-information-technology-act-242916, https://www.livelaw.in/law-firms/law-firm-articles-/compliance-significant-beneficial-ownership-limited-liability-partnerships-zeus-law-associate-247825?infinitescroll=1 https://www.livelaw.in/law-firms/law-firm-articles-/compliance-significant-beneficial-ownership-limited-liability-partnerships-zeus-law-associate-247825?infinitescroll=1

33. Bhaumik, A., (2023) "Regulating deepfakes and generative AI in India", The Hindu Newspaper, https://www.thehindu.com/news/national/regulating-deepfakes-generative-ai-in-india-explained/article67591640.ece

Chapter 3

Classification of data anonymization techniques

Hakan Koyuncu and Raoof Altaher

3.1 INTRODUCTION

The advent of technology has brought about a shift in this era, where data is now considered as valuable as oil. As businesses and various industries increasingly rely on data to fuel their operations, protecting the privacy and security of this information has become important. Data anonymization, a method employed to protect sensitive data, plays a role in this ever-evolving landscape. This section aims to introduce the concept of data anonymization while emphasizing its importance in today's data-driven world.

3.1.1 Background and importance of data anonymization

Personal data is constantly being created, processed, and transmitted. From transactions to intricate machine-to-machine communications in Industry 4.0 environments, data flows are everywhere. However, as the volume of data grows, the likelihood of its misuse escalates. Unprotected personal data is susceptible to misuse. This results in breaches of privacy, identity theft, and other malicious activities. That's where data anonymization comes in as a ray of hope.

Data anonymization refers to the process of transforming data in a way that makes it nearly impossible or impractical to identify the individuals it originally belonged to. The main goal is to safeguard individuals' privacy and keep the usefulness of the data for analysis or research purposes. In essence, it allows organizations to leverage the power of data without compromising privacy. First, it serves as a barrier against breaches of sensitive information. Hackers and cybercriminals find the data less appealing when they eliminate or change identifiers. Even if they manage to access the information, it would be useless. Therefore, protect those involved.

Moreover, in today's era where regulations like the General Data Protection Regulation (GDPR) and the California Consumer Privacy Act (CCPA) impose guidelines on data handling, anonymization serves to ensure compliance (Tankard, 2016). Organizations can secure their data by anonymizing it. Analyze information without violating privacy laws, thereby avoiding legal consequences.

DOI: 10.1201/9781032657264-3

Additionally, data anonymization goes beyond minimizing risks. It also creates opportunities for collaboration and innovation. For example, sharing anonymized data sets can foster research partnerships and drive innovation in various fields. In sectors such as healthcare, anonymized patient data plays a role in advancing research while safeguarding patient confidentiality. The significance of data anonymization cannot be understated, as it protects privacy while promoting practices.

The upcoming sections of this chapter will provide an overview of the techniques and methodologies associated with data anonymization, delving deeper into this essential domain.

3.1.2 Role of data anonymization in cyber security and Industry 4.0

The rapid advancement of technology in this era has led to developments, particularly in cyber security and Industry 4.0. The interconnection of these domains creates a landscape filled with both opportunities and challenges. This connection centers on data, which can be valuable or risky. Data anonymization appears as a tool that ensures the effective use of data in these contexts. This section explores the role of data anonymization within cyber security and Industry 4.0, highlighting its importance, applications, and potential consequences.

3.1.2.1 Cyber security: the digital fortress

In today's world cyber security acts as a protective fortress against malicious intrusions, data breaches, and unauthorized access. With organizations accumulating amounts of data ranging from customer information to sensitive operational details the risks associated with data breaches have become more significant. The consequences of breaches are diverse and far reaching; they include losses, damage to reputation, legal complications, and compromised competitive advantage.

Data anonymization plays a role in strengthening this fortress. By converting data into a structure where sensitive information is either eliminated or obscured it guarantees that the data stays undecipherable even if someone gains access. This does not secure the privacy of individuals. Also shields organizations from the numerous dangers present in the digital world.

3.1.2.2 Industry 4.0: the dawn of a new industrial era

Industry 4.0, also known as the revolution, stands for the merging of traditional industrial methods with modern digital technologies. What sets this revolution apart is the integration of automation data exchanges, cloud computing, the Internet of Things (IoT), cognitive computing, and artificial intelligence (AI) into operations. This transformation gives rise to manufacturing facilities where machines, systems, and human operators communicate in time, resulting in improved productivity, efficiency, and adaptability (Ren et al., 2021).

However, data lies at the heart of Industry 4.0. Sensors, embedded systems, and interconnected devices continuously generate, process, and send data to enable real-time decision-making and optimization. Security and privacy challenges arise when dealing with the volume and speed of this data generation. This is where data anonymization becomes essential. As industries move toward manufacturing and interconnected operations, the complexity of data flows increases. By anonymizing this data while still harnessing its value for efficiency purposes, we ensure that associated risks are mitigated. It enables industries to strike a balance between utilizing data for advantage and maintaining security measures to protect privacy.

3.1.2.3 The convergence of cyber security and Industry 4.0

The convergence of cyber security and Industry 4.0 creates an environment where innovation intersects with security. As industries embrace technology and prioritize data-centric approaches, the cyber threat landscape evolves, becoming more complex and menacing. Within this context, data anonymization plays a role in ensuring that industries can innovate and expand while maintaining data security. Furthermore, the collaborative nature of Industry 4.0, characterized by cross-border data flows and interorganizational data sharing, necessitates data protection measures. Anonymization facilitates this by preserving the usefulness of information while eliminating any identifying markers. This does not promote collaboration and innovation. Additionally, it guarantees adherence to worldwide data protection regulations.

Data anonymization appears to be an actor in the relationship between cyber security and Industry 4.0, striking a balance between keeping the value of information while prioritizing security. As industries navigate the challenges and opportunities presented by the digital era, the significance of data anonymization will only become more pronounced. It will serve as a guiding light leading industries toward a future where data drives growth without compromising security and privacy.

3.1.3 Objectives of the chapter structure

The field of data anonymization is vast, complex, and constantly evolving in relation to the growing fields of cyber security and Industry 4.0. This chapter aims to supply an exploration of techniques used for data anonymization, explaining their importance, methodologies, and implications in the modern digital era. To ensure a coherent presentation on this multifaceted topic, this chapter has been carefully organized into distinct sections that delve into specific aspects of data anonymization (Jakob et al., 2020). This section offers an overview of the chapter's structure, giving readers a glimpse into the journey they're about to embark on.

a. **Introduction:** This first part sets the stage by introducing readers to the world of data anonymization. It supplies background information on why data anonymization is important, its role in cyber security and Industry 4.0, and gives a summary of how the chapter is structured.

b. **Basics of Data Anonymization:** Before diving into classifications in detail, it is essential to set up a foundation. This section clarifies concepts related to data anonymization highlights the pressing need, for it in today's data-driven world and addresses the challenges professionals face during the anonymization process (Smith & Agrawal, 2022).

c. **Classification of Data Anonymization Techniques:** In this section, we will thoroughly examine techniques used for data anonymization. We will distinguish between anonymization and pseudonymization. Explore both advanced methods of anonymization. Throughout the discussion, we will gain insights into how these techniques work their applications and the subtle differences between them (Fredj et al., 2015).

d. **Role of AI in Data Anonymization:** With the rise of AI data anonymization is experiencing a transformation. In this section, we will explore how AI is revolutionizing data anonymization, discuss the benefits and challenges it brings, and supply real-world case studies that prove AI-powered anonymization in action.

e. **Data Anonymization in Industry 4.0:** Industry 4.0, which combines technologies with industrial practices, presents unique opportunities and challenges for data management. This segment examines the importance of data privacy in this era of revolution and explores how data anonymization plays a crucial role in domains such as smart manufacturing, IoT, and supply chain (Zuo et al., 2021).

f. **Ethical and Regulatory Considerations:** Beyond aspects data anonymization also involves regulatory considerations. In this section, we delve into the implications surrounding data anonymization explore the frameworks governing it and highlight the delicate balance between preserving data utility and ensuring privacy (Rodriguez et al., 2022).

g. **Conclusion:** As we wrap up this chapter, we supply a recap of the points discussed so far. We also contemplate the future of data anonymization in the age of AI and Industry 4.0 while emphasizing the need for research and exploration in this field.

In essence, this chapter serves as an exploration of data anonymization. It aims to provide readers with an understanding, covering concepts, advanced methodologies, technical complexities, and ethical considerations. As we progress through each section, readers will gain a well-structured insight into the classification of data anonymization techniques and their profound significance in the fields of cyber security and Industry 4.0.

3.2 BASICS OF DATA ANONYMIZATION

3.2.1 Definition and key concepts

In the domain of data security and privacy, data anonymization acts as a guardian, safeguarding information against possible misuse while still maintaining its inherent worth for analytical and research objectives. This segment explores the principles that form the basis of data anonymization, providing readers with a grasp of its importance in the current data-centric environment.

Definition of Data Anonymization: Data anonymization involves a procedure of altering or encoding information in a way that makes it impossible to link it back to a particular individual. This process aims to minimize the chances of revealing details, thus safeguarding the privacy rights of individuals. By transforming the data, we can still fulfill its original purpose, such as research, analysis, or business operations, while maintaining the confidentiality of individuals (Sweeney, 2002). Here are a few important concepts that are crucial, for understanding data anonymization:

a. **Personal Data:** This refers to any information that can be used either alone or when combined with data to identify a person. Examples include names, addresses, phone numbers, and complex data like genetic information (Institute of Medicine (US) Committee on Health Research and the Privacy of Health Information: The HIPAA Privacy Rule, 2009).

b. **De-identification:** Often used interchangeably with anonymization, de-identification involves removing or modifying identifiers to prevent the association of data with specific individuals. However, it's important to note that depending on the technique used, identified data may still be re-identifiable under certain conditions or with additional data.

c. **Re-identification:** This term describes the process by which anonymized, or de-identified data is matched back to its source or linked to an individual. Effective anonymization strategies aim to minimize the risk of re-identification.

d. **Data Masking:** This technique involves obscuring data within a database to make it inaccessible, for users. It guarantees that confidential information stays private and can only be accessed by authorized individuals.

e. **Data Utility:** A concept, in the field of data anonymization data usefulness refers to finding a balance between safeguarding privacy and ensuring that the data still is functional and valuable for its intended purpose.

f. **Noise Addition:** This technique involves introducing data or "noise" to the original dataset. Doing it obscures the data, making it more difficult to decipher while still preserving the overall integrity and usefulness of the dataset for analysis.

The delicate equilibrium between keeping data private and retaining its usefulness is a theme in the realm of data anonymization. As we explore the techniques and methodologies used in the field of data anonymization in this chapter, these fundamental concepts will serve as guiding principles. Subsequent sections will further clarify why anonymizing data is crucial in today's age and shed light on the challenges faced in developing robust and effective anonymization strategies.

3.2.2 Challenges in data anonymization

Achieving data anonymization is a task that involves navigating various challenges related to balancing data utility and individual privacy. With the growth of data generation and usage worldwide, the intricacies surrounding successful anonymization have become even more pronounced. This section thoroughly explores the hurdles faced by researchers, data practitioners, and organizations in the field of data anonymization.

a. **Risk of Re-identification:** The concern regarding re-identification is widespread; despite implementing anonymization protocols and employing computational techniques there is still a possibility that anonymized data can be linked back to its sources or individuals through auxiliary data (Carvalho et al., 2021).

b. **Diminished Data Utility:** Overzealous anonymization efforts can unintentionally diminish the value of data making it less effective for research purposes. Striking a balance between ensuring data privacy and preserving its usefulness poses a challenge (Rafiq et al., 2022).

c. **Dynamic Data Landscape:** Anonymization challenges are further compounded by the changing nature of data ecosystems. As datasets evolve anonymized information may become vulnerable to re-identification requiring an adaptable approach to anonymization.

d. **Technological Proliferation:** The continuous progress of technology, in the fields of intelligence and data analysis, has both positive and negative consequences. On one hand, it allows for methods of anonymizing data. On the other hand, it can also be used to reverse engineer and de-anonymize data which raises concerns about privacy infringements.

e. **Regulatory Labyrinth:** Navigating through regulations such as the GDPR and the CCPA poses a challenge. This is particularly true when dealing with data that crosses authorities making the regulatory landscape more complicated.

f. **Ethical Quandaries:** In addition to regulatory obstacles ethical considerations are intertwined with data anonymization. Questions arise regarding the obligation to anonymize data, the potential consequences of re-identification, and overall ethical responsibilities of those managing the data (Saunders et al., 2015).

g. **Resource Constraints:** Implementing robust data anonymization techniques requires power, ability, and time investment. Smaller organizations or those with resources may face difficulties, in obtaining the means for effective anonymization (Pawar et al., 2018).

h. **Evolution of Attack Vectors:** Just as advancements are made in anonymization techniques malicious actors also evolve their strategies to undermine them. They continuously develop methods to exploit vulnerabilities to compromise anonymity. As technology advances, correlation, background knowledge, and inference attacks are constantly improving. This requires us to take an approach to ensure the anonymization of data.

3.3 CLASSIFICATION OF DATA ANONYMIZATION TECHNIQUES

Data anonymization is a field that offers techniques to address different challenges and needs in the digital world as shown in Figure 3.1. In this section, we will explore the methods used for data anonymization and discuss when they are most suitable.

3.3.1 Anonymization vs. pseudonymization

Differentiating between anonymization and pseudonymization plays a role in discussions about data privacy. While their ultimate goal is to protect data, the methods and consequences of each approach differ significantly.

3.3.1.1 Definition and core concepts

a. **Anonymization:** Anonymization involves making changes to data that cannot be reversed making it impossible to identify the individual

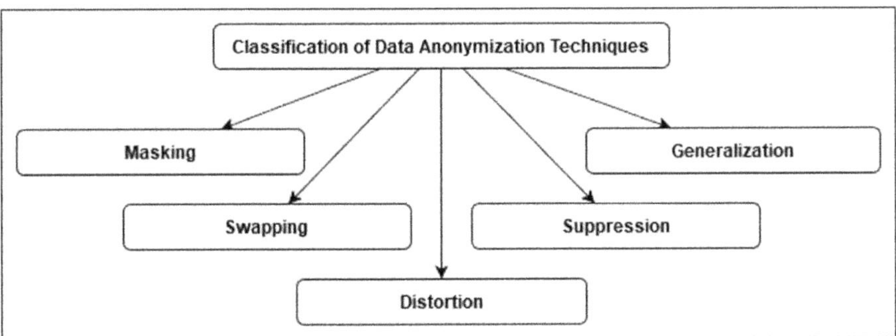

Figure 3.1 Classification of data anonymization techniques.

associated with the data (Sweeney, 2002). This allows for the sharing of information while still protecting privacy by changing the data.

b. **Pseudonymization:** On the other hand, pseudonymization is a process where personal data fields are replaced with pseudonyms or artificial identifiers. This allows for the retrieval of the data using an algorithm or key. Both pseudonymization and anonymization play roles in ensuring data processing, security, and access after the introduction of the GDPR.

3.3.1.2 Methodological differences

Anonymization techniques employ methods such as data masking, generalization, and noise addition to prevent reversal or individual identification of the data. On the other hand, pseudonymization techniques use encryption, tokenization, and shuffling of data to keep the information disguised (Pfitzmann & Hansen, 2010). Anonymization methods such as randomization and generalization alter the integrity of the data to break any connection between it and individual identities.

3.3.1.3 Implications for data protection

From a standpoint anonymized data is generally exempt from data protection regulations due to its irreversible nature. Conversely, pseudonymized data remains within the scope of these laws because it can be reversed back to its form if needed (Zuo et al., 2021). According to Recital 26 of GDPR, anonymized data falls outside the realm of regulation since it no longer holds information making it practically impossible to identify individuals.

3.3.1.4 Use cases and applications

Anonymization is commonly used in scenarios where preserving data utility is crucial, such as datasets or research studies. On the other hand, pseudonymization is suitable for situations where both protecting information and ensuring accessibility in its original form are necessary – for example, in clinical trials or financial transactions.

The contrast between anonymization and pseudonymization arises from their approaches, consequences, and uses. As data privacy becomes increasingly important, in our age, it is crucial to understand these concepts in a nuanced way.

3.3.2 Traditional anonymization techniques

Conventional techniques for preserving anonymity play a role in safeguarding data privacy while ensuring the usability of data for analysis and processing purposes. These techniques manipulate data to prevent the identification of individuals addressing concerns about privacy in scenarios involving sharing and publishing data (Monteiro et al., 2022).

3.3.2.1 Data masking

a. **Definition and Mechanism:** Data masking, also referred to as data obfuscation or data redaction, involves substituting data with modified content that still maintains a resemblance to the original (Fredj et al., 2015).

b. **Applications:** This technique is primarily employed in production environments like software testing, where it is necessary to mask real production data to protect sensitive information while keeping its value, for testing purposes.

c. **Strengths and Limitations:** The main advantage lies in generating a sanitized version of the data without altering its format. However, if not implemented correctly, there is a risk that the masked data can be reverse-engineered to reveal the information.

3.3.2.2 Data swapping (shuffling or permutation)

a. **Definition and Mechanism:** Data swapping involves exchanging values within a database to safeguard the privacy of data subjects. By shuffling values, between records individual records stay accurate. The dataset loses its integrity.

b. **Applications:** This technique is commonly used in census datasets to prevent the identification of individuals or households.

c. **Strengths and Limitations:** Data swapping effectively preserves the distribution and statistical properties of the dataset. However, it may result in lost relationships between records. There is a risk if the swapped data can be correlated with other datasets (Murthy et al., 2019).

3.3.2.3 Generalization

a. **Definition and Mechanism:** Generalization involves replacing data values with categories or ranges reducing the level of detail in order to make it more challenging to pinpoint individual records.

b. **Applications:** Generalization is widely used when releasing datasets where detailed information could lead to privacy breaches. For example, in health datasets specific ages might be generalized into age groups to prevent identification.

c. **Strengths and Limitations:** Generalization helps keep the usefulness of the dataset, for analyses. However excessive generalization can reduce its utility.

3.3.3 Advanced anonymization techniques

Techniques to prevent the compromise or linkage of sensitive information to individuals have become increasingly necessary in the evolving world of data privacy. Researchers have developed methodologies that surpass traditional approaches to address this issue. These advanced methods are based on computational principles aiming to strike a balance between safeguarding privacy and preserving the usefulness of data for legitimate purposes (Jiang et al., 2021).

The increasing concern over data privacy has led to the development of advanced anonymization techniques that effectively protect against data breaches and re-identification attempts. These innovative methods, grounded in rigorous mathematical and computational frameworks, aim to ensure a higher level of privacy while preserving the data's usefulness for legitimate purposes, not as extensions of traditional approaches.

3.3.3.1 Differential privacy

Differential privacy (DP) has become an aspect in the field of data privacy. It provides a framework to guarantee that the results of statistical analyses remain largely unchanged regardless of whether an individual's data is included or excluded (Ficek et al., 2021).

Definition: A mechanism is considered to achieve a level of privacy called ε-differential privacy when the likelihood of a result still is almost unchanged regardless of whether an individual's data is included. Here ε represents a small positive parameter that quantifies the amount of privacy loss.

Mechanism: DP uses calibrated noise to obfuscate the results of queries, making it difficult to identify data points. Distributions such as the Laplace or Gaussian ensure that this noise sufficiently obscures specific data points. As a result, it becomes computationally challenging to extract information about any individual (Kurz, 2021).

Applications: Leading tech companies have integrated DP (data privacy) principles into their data collection frameworks to safeguard user information. When it comes to making public data available, data privacy offers a framework for sharing statistics without putting individuals at risk of privacy breaches (Kenny et al., 2021).

3.3.3.2 k-Anonymity, l-Diversity, and t-Closeness

The concepts of k anonymity, l diversity, and t closeness are the foundations of advanced techniques for anonymizing data. These techniques offer levels of privacy protection based on the data involved and the potential risks it may face.

k-Anonymity: This principle mandates that any released data must be indistinguishable from at least k-1 other entities within the dataset. This guarantees the inability to identify individual data separately, thereby safeguarding against re-identification (Li et al., 2007).

l-Diversity: To enhance the privacy of sensitive released data l-diversity builds upon k-anonymity by requiring that each equivalence class contains, at least "l" distinct values in each equivalence class, This measure helps prevent attribute disclosure, where an attacker could infer an individual's attributes on other available data (Machanavajjhala et al., 2006).

t-Closeness: This principle ensures that the distribution of a sensitive attribute within any equivalence class closely matches the distribution of the attribute across the dataset. Closeness is measured using a distance metric like Earth Mover's Distance. This principle furnishes protection against sophisticated probabilistic attacks (Domingo-Ferrer et al., 2016).

These principles play a role in industries, especially in the medical field. It is essential to strike a balance between the usefulness of data and maintaining privacy (Rajendran et al., 2018).

3.3.3.3 Homomorphic encryption

Alloghani et al. (2019) mention homomorphic encryption (HE), a technique that enables computational operations on encrypted data without requiring its decryption. This characteristic ensures that data stays encrypted during processing, thereby enhancing its privacy and security.

Definition: According to Will and Ko (2015), a cryptographic scheme is homomorphic if it allows for operations on encrypted data that, upon decryption, yield results identical to those obtained from unencrypted data.

Applications: The emergence of HE brings forth promising opportunities in fields such as cloud computing and secure data analysis. For example, it allows researchers to explore encrypted records without accessing the actual data, thereby ensuring the protection of patient privacy (Munjal & Bhatia, 2022).

Challenges: Despite providing privacy protection for computational requirements, the processing burden associated with HE may render it unsuitable for certain applications. The demands associated with HE can present obstacles in situations that require real-time processing or have limited resources.

Advanced methods of anonymization present a range of approaches to safeguarding data privacy, each with its own strengths and challenges. As data continues to play a role in society, these techniques will play a vital part in upholding individuals privacy rights while still harnessing the valuable insights derived from data.

3.4 ROLE OF ARTIFICIAL INTELLIGENCE IN DATA ANONYMIZATION

The convergence of AI and data anonymization represents progress in data privacy. As the amount of data grows, it becomes increasingly complex to safeguard privacy. While traditional methods of anonymization have their advantages, they often face difficulties in achieving a balance between preserving data usefulness and ensuring privacy. AI, using its power and predictive capabilities, offers a solution to address these challenges by enabling strong data protection while keeping the inherent value of the information.

3.4.1 AI-driven data anonymization approaches

The incorporation of AI into data anonymization has fostered the growth of responsive techniques that can adapt to changing data environments and emerging security risks. These methods utilize machine learning (ML), deep learning, and other AI approaches to enhance the efficiency of anonymization procedures (Dwork & Roth, 2014).

a. **Generative Adversarial Networks (GANs) in Anonymization**: GANs, which are a type of network, have found applications in the field of data anonymization. The main idea behind this method is to use two networks: a generator and a discriminator, which compete with each other. The generator's goal is to generate data that closely resembles data, whereas the discriminators' task is to distinguish between real and synthetic data. Through training, the generator becomes skilled at generating data that maintains the properties of the original dataset without compromising individual identities (Abadi et al., 2016).

b. **Autoencoders for Data Reconstruction**: Autoencoders are a kind of neural network that is used to compress data and then reconstruct it. In the context of anonymization, noise is added during the compression stage to ensure that the reconstructed data, although structurally like the original, has been altered enough to avoid re-identification (McMahan et al., 2023).

c. **Feature Selection and Transformation**: AI algorithms have the capability to analyze and select the features or data attributes for a specific analysis while disregarding the rest. This does not reduce the complexity of the data. Additionally, it lessens the likelihood of inadvertently disclosing confidential data. Furthermore, we can use AI-powered techniques to alter data features, making them more challenging to interpret.

d. **Adaptive Anonymization**: Traditional methods of anonymization often rely on fixed thresholds and parameters which might not be suitable for all types of datasets. On the other hand, AI-powered techniques can dynamically identify the proper anonymization parameters by considering the unique characteristics of the data. This ensures a compromise between maintaining privacy and preserving usefulness (Yoon et al., 2020).

e. **Privacy-preserving Machine Learning**: AI can ensure privacy-preserving model training in situations where we need to analyze data using ML models. Federated learning techniques enable the training of models using data from sources without jeopardizing data privacy. The model only shares its updates, ensuring the data remains secure and untouched (Chaudhuri et al., 2011).

The use of AI-based techniques for anonymizing data signifies a shift in how we view and implement data privacy. By harnessing AI's adaptable capabilities, we can establish data protection that is well-suited to the intricate nature of contemporary datasets. As AI advances in its influence, shaping the future of data anonymization will only become more important.

3.4.2 Benefits and challenges of AI in data anonymization

The merging of AI with data anonymization has opened possibilities for data protection and privacy. This combination has resulted in benefits, making data anonymization more effective, efficient, and adaptable. However, with any progress, integrating AI into data anonymization comes with its own set of challenges. In this section, we carefully examine the advantages and difficulties associated with AI-powered data anonymization, offering an insight into its implications.

3.4.2.1 Benefits of AI in data anonymization

a. **Enhanced Accuracy:** By leveraging their capacity to analyze data sets and identify patterns AI algorithms can enhance the precision of the anonymization procedure thereby minimizing the chances of data breaches or re-identification (Raimundo & Rosário, 2021).
b. **Scalability:** Traditional methods of anonymization may face challenges when dealing with datasets, which can result in inefficiencies. On the other hand, approaches powered by AI are naturally scalable allowing for anonymization of massive datasets.
c. **Adaptive Anonymization:** AI models can learn and adapt from the data they analyze. This allows them to effectively anonymize data when there are data structures or changes, in the data landscape.
d. **Preservation of Data Utility:** Preserving the usefulness of data is a hurdle when it comes to anonymization. Utilizing AI-driven techniques can help keep the value of anonymized data ensuring its effectiveness, for analysis and gaining insights.
e. **Real-time Anonymization:** AI's computational capabilities allow for anonymization guaranteeing the preservation of privacy in constantly evolving settings where data is continuously produced.

3.4.2.2 Challenges of AI in data anonymization

a. **Complexity:** AI models, and learning models can be quite intricate. Grasping their complexities and ensuring implementation can pose a challenge, causing ability, in the field.
b. **Over-reliance on AI:** Although AI supplies benefits it is important to recognize that relying heavily on it can have negative consequences. It is crucial to understand that AI is merely a tool and like any tool it has its limitations.

c. **Ethical Concerns:** The ethical implications of using AI in the anonymization of data are significant particularly when it comes to the biases embedded within AI models. If left unattended these biases could distort the anonymization processes.

d. **Computational Costs:** Advanced AI models can require resources resulting in higher expenses particularly when managing large datasets.

e. **Data Dependency:** The effectiveness of AI models heavily relies on the quality and representativeness of the training data. When the data used for training is not truly reflective or contains biases it can lead to inaccuracies, in the anonymization process driven by AI.

There are advantages to incorporating AI into data anonymization. However, it is important to approach this convergence with a viewpoint acknowledging both the benefits and challenges it presents. As AI progresses and its use in data anonymization becomes more sophisticated, it has the potential to completely transform how we safeguard and keep data privacy. Nonetheless, a cautious approach that includes research and development is essential to fully exploit its capabilities while minimizing any associated risks.

3.5 DATA ANONYMIZATION IN INDUSTRY 4.0

3.5.1 Importance of data privacy in the fourth industrial revolution

The Fourth Industrial Revolution, also known as Industry 4.0, stands for a change in how industries operate. It is driven by the convergence of technologies, automation, AI, and the IoT. This transition goes beyond advancements and has a global impact on industries, economies, and societies. At the core of this transformation lies data referred to as the 'new oil' of the age. As industries leverage data to drive innovation improve operations. Developing business models ensuring data privacy becomes an essential concern (Alazab et al., 2022).

3.5.1.1 The pivotal role of data in Industry 4.0

Data serves as the foundation for Industry 4.0 supporting manufacturing processes that rely on real-time data to perfect production lines. Additionally, AI-powered supply chain management uses this data for inventory control. In Industry 4.0, devices and systems are interconnected, easing voluminous flows of data in directions. From sensors on production lines to AI-driven robots every connected device contributes to data generation, processing, and sharing (Farhan et al., 2018). However, managing this amount of data poses challenges since a huge part of it is sensitive in nature and includes business information or personal consumer details. Mishandling or misusing this data can result in consequences such as penalties, loss of consumer trust, and damage to the brand's reputation.

3.5.1.2 Why data privacy is paramount in the Fourth Industrial Revolution

a. **Stakeholder Trust:** Today as people become more aware of their digital footprints, businesses that prioritize data privacy will gain the trust of stakeholders. Trust is highly valued in this era. By showing a dedication to protecting customer data, companies can not only keep their existing customers but also attract new ones, which leads to sustainable business growth.

b. **Regulatory Landscape:** The regulations about data protection are constantly changing on a scale. Laws such, as the European Union's GDPR and the CCPA in the United States set forth rules for handling consumer data. Failure to comply can lead to fines, legal issues, and damage to one's reputation. By making data privacy a priority businesses can guarantee adherence to obligations.

c. **Operational Benefits:** In addition, to meeting regulatory requirements there are benefits, to safeguarding data privacy. When data is professionally managed and anonymized the likelihood of data breaches decreases, resulting in cybersecurity expenses. Furthermore, streamlined and anonymized datasets contribute to the effectiveness of AI algorithms enabling more insights (Jan et al., 2023).

d. **Innovation and Competitive Advantage:** Data is an asset when it comes to driving innovation. However, it's crucial to prioritize privacy alongside this pursuit. Businesses can achieve this balance by implementing data anonymization techniques. By doing they can harness the power of data for innovation while still safeguarding privacy. This approach does not promote innovation but also gives businesses a competitive advantage in the market (Castro et al., 2022).

e. **Ethical Responsibility:** Apart from the necessity for businesses to prioritize aims there is also an ethical responsibility to safeguard the privacy of individuals. In the landscape of Industry 4.0 where data sharing is constant and widespread, it becomes imperative for businesses to handle personal and sensitive information, with utmost care and respect (Girka et al., 2021).

3.5.1.3 The path forward

As we see the transformation of the industrial landscape, in the era of Industry 4.0, it becomes increasingly crucial to prioritize and safeguard data privacy. Enterprises that are taking the initiative in realizing this importance and taking measures will undoubtedly establish themselves as leaders in this era. By incorporating techniques, for anonymizing data harnessing AI-powered privacy solutions, and cultivating a culture of privacy within their organization's businesses can successfully navigate the hurdles presented by the

Fourth Industrial Revolution while upholding uncompromising standards of data privacy.

The advent of the Fourth Industrial Revolution offers prospects for development, ingenuity, and advancement. Nevertheless, with these prospects come difficulties and one of the challenges is the imperative to safeguard data privacy. As industries evolve and adjust to this era the protection of data privacy will become a cornerstone for ethical progress. Those businesses that acknowledge this reality and take a stance in safeguarding data privacy will appear as the champions of Industry 4.0.

3.5.2 Application of data anonymization in smart manufacturing, IoT, and supply chains

The advent of Industry 4.0 stands for a change toward an interconnected world, where devices, systems, and processes work together to achieve operational excellence and drive innovation. At the heart of this shift is the exchange of data, across networks enabling time decision-making and valuable insights. However, as the volume and complexity of data continue to grow, it becomes increasingly important to prioritize privacy protection and data security measures. In this section, we will explore how data anonymization techniques are implemented in three areas of Industry 4.0: Smart Manufacturing, the IoT, and Supply Chains.

3.5.2.1 Smart manufacturing

The transformation of the manufacturing industry through digitization, often referred to as Smart Manufacturing, uses technologies to enhance productivity, efficiency, and flexibility. The integration of data-driven insights is an element in this process.

a. **Process Optimization**: By using data collected from sensors and devices, we can carefully adjust and refine manufacturing processes. For instance, when we anonymize and analyze data from temperature sensors, we can ensure that each manufacturing process has the conditions for maintaining product quality (O'Donovan et al., 2015). This approach also helps protect information related to methods.

b. **Predictive Maintenance**: The careful examination of data collected from machines and equipment helps advanced predictive maintenance algorithms forecast failures or malfunctions. This allows for measures to be taken, reducing the amount of time, equipment is out of service (Sexton & Shao, n.d.).

c. **Quality Control**: By utilizing anonymized data collected from quality control sensors, we can ensure that only products meeting the standards move forward to the phase of production enabling real-time detection of anomalies (Lu, 2017).

3.5.2.2 Internet of things (IoT)

The world of IoT with its network of devices produces a significant amount of data. It is crucial to oversee this data to effectively address privacy risks.

a. **Smart Homes:** Data collected from home devices such as thermostats, security cameras, and smart refrigerators improves the user's experience while safeguarding their privacy (Al-Fuqaha et al., 2015). For example, intelligent thermostats powered by AI can adapt to a user's preferences over time while keeping their data anonymous and unidentifiable.
b. **Healthcare:** Wearable devices that check statistics generate data. By anonymizing this data, healthcare professionals can access information for diagnosis and treatment while still protecting the patient's identity (M. Li et al., 2010).
c. **Smart Cities:** Smart cities use data to improve living in aspects such as managing traffic and waste disposal. The collection of data from sensors helps in efficiently managing the city while also ensuring the privacy of its residents (Zanella et al., 2014).

3.5.2.3 Supply chains

In today's interconnected world, supply chains have become increasingly complex with networks that span across countries and even continents. To keep a flow of goods, it is crucial to check these supply chains in time and make decisions based on data-driven insights.

a. **Inventory Management:** Analyzing data from warehouses helps us predict demand patterns ensuring that we can replenish stock on time and reduce holding costs (Sivarajah et al., 2017).
b. **Logistics Optimization:** The use of data, from GPS and similar tracking devices, is extremely important for optimizing routes ensuring that deliveries are made on time and reducing transportation expenses (Ivanov et al., 2020).
c. **Vendor Management:** By analyzing data, from vendors, we can gain valuable insights into their performance and ensure that we only collaborate with the most dependable ones (Aronsson et al., 2011).

In the era of Industry 4.0 as the global industrial landscape undergoes changes, it becomes crucial to prioritize data privacy and security. To address this concern data anonymization methods, supply a solution that allows businesses to use data effectively while safeguarding privacy. Whether streamlining manufacturing operations, enhancing user experiences, with devices, or ensuring supply chain management data anonymization is set to play a vital role in the era of Industry 4.0.

3.5.3 Future trends and predictions

The future of data anonymization is closely tied to the advancements seen in Industry 4.0 and the broader field of AI. As we enter a decade, it becomes crucial to expect the paths that data anonymization will take. It is important to take initiative in preparing for changes while also using them to enhance data privacy and security. In this section, we will explore the expected trends and predictions in the field of data anonymization supplying insights into this domain's evolving landscape.

3.5.3.1 Evolution of quantum-safe anonymization techniques

The rise of quantum computing poses a threat to cryptographic foundations, which form the basis of various anonymization methods. Quantum machines, known for their power, challenge existing cryptographic algorithms. This shift in the paradigm emphasizes the need, for developing and implementing post-quantum anonymization techniques that can protect data privacy in the presence of quantum challenges.

The quantum realm operates based on the principles of quantum mechanics giving quantum computers the ability to perform computations faster than classical computers. This advantage is an edged sword; while it can lead to breakthroughs in fields, it also poses a threat to the security of current cryptographic algorithms that rely on the difficulty of certain mathematical problems (such as factoring large numbers and solving discrete logarithm problems). Quantum algorithms like Shor's algorithm and Grover's algorithm have shown the potential in solving these problems faster than the best-known algorithms used by classical computers (Bernstein et al., 2009).

Given the threat of an apocalypse, researchers and developers in the field of data anonymization are actively exploring quantum-safe cryptographic solutions. These techniques rely on problems that are believed to be resilient against attacks from quantum computers (Song, 2014). Lattice-based cryptography, hash-based cryptography, code-based cryptography, and multivariate polynomial cryptography are some of the leading approaches in this developing area (Peikert, 2016).

The journey toward developing anonymization techniques that can withstand quantum technology poses challenges. The early stage of quantum computing along with the need to establish algorithms for the post-quantum era presents a significant hurdle. Additionally, it is crucial to strike a balance during the transition phase ensuring compatibility with existing cryptographic systems while forging ahead with innovations that supply protection in the quantum era. Developing anonymization techniques that're resistant to quantum threats is an aspect of safeguarding data privacy and security in today's rapidly advancing technological landscape. As efforts in quantum cryptography research and standardization gain momentum the creation of resilient anonymization techniques capable of withstanding quantum attacks will play a pivotal role in discussions around data privacy. These techniques will serve as a haven for data, amidst the dominance of quantum capabilities.

3.5.3.2 Integration of AI and blockchain for enhanced data anonymization

The combination of technology and AI-powered anonymization methods signals an era of strengthened data privacy and security. Blockchain, known for its unchangeable nature, offers a foundation for storing and conducting data transactions. When integrated with AI, it has the potential to greatly enhance the layers of security and privacy. For example, smart contracts powered by AI can automate the process of anonymizing data to storing it on the blockchain guaranteeing that the data still is private and resistant, to tampering.

Integrating AI with technology can greatly strengthen data security and privacy. By using contracts based on blockchain along with data, AI has the potential to become a formidable tool in enhancing cybersecurity. This will introduce features like transparency and openness to the forefront (Charles et al., 2023).

There are projects in the field of scientific research that are investigating the intersection of blockchain and AI. The combination of these two technologies has been acknowledged for its ability to enhance the capabilities of blockchain by providing a range of data integration tools and supported by AI tools (Hechler et al., 2020). Furthermore, the combination of blockchain and AI allows for the use of hashing, digital identification, and encryption codes. After the data is processed, it is sent to the application layer that manages management. This integration results in a tool, for learning data analytics guaranteeing privacy and security of network data (Priya et al., 2023).

In industries, such as healthcare, there has been a lot of discussion about combining blockchain and AI to create a system. Various strategies and frameworks have been proposed to leverage blockchain and AI in order to establish an environment ensuring data privacy and secure management of healthcare information (Anoop & Asharaf, 2022).

The combination of AI and blockchain technologies presents an opportunity for changes in data anonymization. By using the advantages of both technologies, we have the potential to establish a reliable data management system that prioritizes security and privacy. Ongoing research in this field highlights the importance of integrating AI and blockchain to enhance data anonymization ensuring the protection of information in our digital world.

3.5.3.3 Rise of context-aware anonymization

Traditional approaches to data anonymization are being replaced by context techniques. These new methodologies recognize that the level and method of anonymization can vary depending on the context, such as records or online shopping history. This shift marks an era where techniques are customized to meet the requirements of diverse types of data and their intended use.

An interesting development in the field of context anonymization is the Sensitivity Context Aware Anonymization method. It introduces a mechanism specifically designed for Big Data situations (Ramya Shree et al., 2022). The key focus is to understand the sensitivity context in which data exists and customize the anonymization process accordingly. This ensures that the unique characteristics and needs of the data are considered.

In addition, there has been a discussion about adopting a method for protecting the privacy of public data. This highlights the necessity of ensuring that the process of sanitizing information aligns with ethical principles governing data handling (Zykov et al., 2019). This narrative emphasizes the significance of embracing a context-driven approach to anonymizing data rather than relying on a standardized solution that may not be suitable, in all cases.

Moreover the discussion on Context-Based Personal Data Discovery for Anonymization explains how taking context into account during data discovery procedures can greatly improve the anonymization methods used by companies. This approach is in line with the requirements of the GDPR which highlights the importance of considering context to ensure data privacy and compliance.

In a more expansive scope, researchers have investigated anonymization methods that take workload into account when dealing with datasets. These methods, which are based on decision trees and sampling techniques, demonstrate the ability to scale anonymization algorithms for datasets that exceed the memory. This represents an advancement toward context data anonymization that addresses the challenges posed by workload and data scale (LeFevre et al., 2008).

Furthermore, researchers have examined the use of clustering to assess the usefulness of anonymized data in the context of data anonymization. They have employed models such as k-anonymity and (ε, δ) DP to evaluate data utility. This approach emphasizes the significance of evaluating the usefulness of anonymized data, which aligns with the concept of context anonymization (Ferrão et al., 2022).

The burgeoning discourse and innovations in context-aware anonymization epitomize a significant trend in the domain of data privacy and security. These nuanced methodologies are emblematic of a broader shift toward more tailored, context-driven data anonymization frameworks that are attuned to the specific requisites of diverse data types and their intended use.

3.5.3.4 Enhanced focus on ethical anonymization

There is growing awareness about the considerations related to using and storing data, which has resulted in an emphasis on the need for anonymization techniques that are both technically dependable and ethically robust. The goal is to ensure that when data is anonymized it does not lead to biases or discrimination particularly when used in AI algorithms. There is debate, in the realm of data anonymization about whether anonymization alone is sufficient to ensure ethical research conduct. An analysis of documents reveals growing concerns about the effectiveness of anonymization methods in preventing re-identification. This concern is particularly prominent in research, where the identifying characteristics of genetic material combined with technological advancements have posed challenges to established identifiability standards. Even when individual identities are difficult to discern, there still is a potential for harm to groups that cannot be solely mitigated through anonymization. The analysis demonstrates that a significant portion of the literature supports the requirement for an ethics committee evaluation to effectively address concerns related to the use of anonymized samples and data, in research endeavors (Phillips et al., 2017).

Furthermore, ethical concerns about data anonymization also apply to the field of sciences. In this domain, we need to consider how we interact with participants, during fieldwork handle audio-video recordings transcribe them, and analyze the data. Ethical discussions, in this area usually revolve around guidelines and recommendations that emphasize practices and methodologies (Mondada, 2014).

Moreover, there is currently a lot of focus on examining the ethical components related to anonymizing or identifying health data. This scrutiny entails delving into the intricacies of anonymization techniques acknowledging their limitations and receiving recommendations, from researchers on how to enhance these techniques (Long et al., 2019)

The absence of regulations concerning accessible information also exposes individuals to a higher risk of being reidentified and facing other privacy concerns. This brings attention to the issues related to bias and fairness. When it comes to research, there has been an assessment of the legal and ethical aspects associated with anonymization. This evaluation combines insights, from sciences regarding opinions and technical approaches, for anonymization resulting in a framework that guides ethical practices for protecting identity (Houghton & Houghton, 2018).

The ethical foundations of data anonymization have been gaining more attention. There is a growing focus on developing anonymization techniques that're ethically robust and can effectively address biases, discrimination, and re-identification threats (Salajegheh et al., 2020). This shift in perspective shows a dedication to upholding standards in handling data ensuring that the pursuit of data privacy and security does not compromise moral and social values.

3.5.3.5 Dynamic anonymization in real-time environments

The increasing use of real-time data processing and analytics highlights the need for dynamic anonymization methods that can work in time without delays. This is especially crucial in industries such as driving and real-time health monitoring, where ensuring data privacy and efficiency are extremely important.

A research paper presented an approach to anonymizing data focusing on managing risks and finding the configuration for each dataset. The paper also introduced two privacy metrics (CAK and R-CAK) to gauge the risk of re-identification of the anonymized data (Adkinson Orellana et al., 2021). This effort highlights the importance of employing methods of data anonymization in situations where information is constantly changing. Relying solely on techniques could result in data analysis or increased risks of re-identification (Gupta & Bhatnagar, 2014).

In areas where data is constantly changing there have been suggestions for protocols that ensure privacy while still allowing for dynamic data handling. One protocol stands out as it effectively manages data holders by using an approach that minimizes privacy breaches and optimizes the usefulness of the data (Zykov et al., 2019). This accentuates the imperative for dynamic anonymization schemes in real-time environments where data holders might join or leave dynamically. The exigency for real-time processing in data streams challenges the efficacy of existing static k-anonymization algorithms, which necessitate repeated data scans during the anonymization procedure, a process deemed infeasible in real-time data stream processing (Jain et al., 2016). This elucidates the imperative for dynamic anonymization techniques adept at managing real-time data streams. Furthermore, the evaluation of data anonymization schemes in an IoT environment for big data has been underscored, where the authors assessed re-identification risks and evaluated the schemes predicated on privacy preserving-level and data utility metrics (Ni et al., 2022). The IoT environments, often characterized by real-time data transmission and processing, beckon for dynamic anonymization techniques to ensure data privacy while keeping data utility.

These deliberations signify a palpable shift toward dynamic anonymization techniques adept at navigating the intricacies of real-time environments. The journey toward achieving robust dynamic anonymization in real-time scenarios is emblematic of the broader evolution within the domain of data anonymization poised to redefine the paradigms of data privacy and security in the impending era of real-time data processing and analytics.

3.5.3.6 Regulatory evolution to address anonymization challenges

The escalating adoption of data anonymization techniques is prompting regulatory bodies globally to revise their frameworks to tackle the unique challenges posed by these methods. This encompasses updating compliance requirements, setting standards for anonymization, and ensuring organizational adherence to these standards.

At a supranational level, regulatory bodies such as the former Article 29 Working Party, now the European Data Protection Board, have shown a proclivity toward a relative approach to anonymization as opposed to an absolute one. This shift underscores the theoretical possibility of re-identification, indicating that an absolute approach to anonymization may not suffice to combat the challenges of re-identification (Scheibner et al., 2021).

Moreover, within the context of clinical trials, critical areas have appeared where regulatory evolution is clear. These areas include informed consent, the use of public data, anonymization, and the regulatory regime for international data transfer. This scenario elucidates the persistent complexity and uncertainty characterizing the regulatory landscape surrounding data anonymization and international data transfer (Minssen et al., 2020).

The GDPR of the European Union elucidates the definitions of pseudonymization and anonymization. Over time, the interpretation of anonymization in practice has emerged, including in the form of regulatory guidance, notably the 2014 Article 29 Data Protection Working Party opinion on anonymization techniques.

In Brazil, under the Legal General Data Protection Law, there is ambiguity about what constitutes reasonable anonymization. The law stipulates that if anonymization can be reversed with reasonable efforts, the data protection law will apply, indicating a need for clear regulatory guidelines on the standards of anonymization (Moraes et al., 2021).

The failure of data anonymization in protecting individual privacy has also prompted discourse in the legal and policy arena. Studies have demonstrated the ease with which individuals can be re-identified or deanonymized in anonymized data, urging for stronger regulatory measures to ensure the efficacy of anonymization techniques (Ohm, 2010).

The discussions reflect a growing recognition among regulatory bodies regarding the intricacies of data anonymization. The evolving regulatory frameworks aim to address the challenges posed by data anonymization techniques, ensuring that they uphold the privacy and security of individuals while enabling the beneficial use of data in a myriad of sectors.

3.5.3.7 Enhanced user control over personal data

The impending trends foretell a change in basic assumptions toward endowing users with greater autonomy over their personal data, encompassing its anonymization. Tools empowering users to delineate their predilections for data anonymization and even tailor the degree of anonymization contingent on their comfort, are poised to gain prominence. Anonymization of personal data has been traditionally orchestrated by data handlers with users still being largely oblivious to the intricacies involved. This has a long history stemming from the conventional practices of National Statistical Institutes that supply various data products. However, this traditional approach has received substantial criticism, particularly in the face of publicized failures of anonymization efforts (Elliot et al., 2018).

A novel paradigm, heralding enhanced user control, proposes privacy-preserving data collection protocols that anonymize data sans the intervention of a third-party anonymizer or a private channel for data transmission (Andrew et al., 2023), presented in Figure 3.2. This approach looks to mitigate risks associated with third-party data handlers and promote user-centric data governance. Further, there has been a discourse on the pivotal role of users in controlling access, use, and dissemination of their personal data, especially in the private sector.

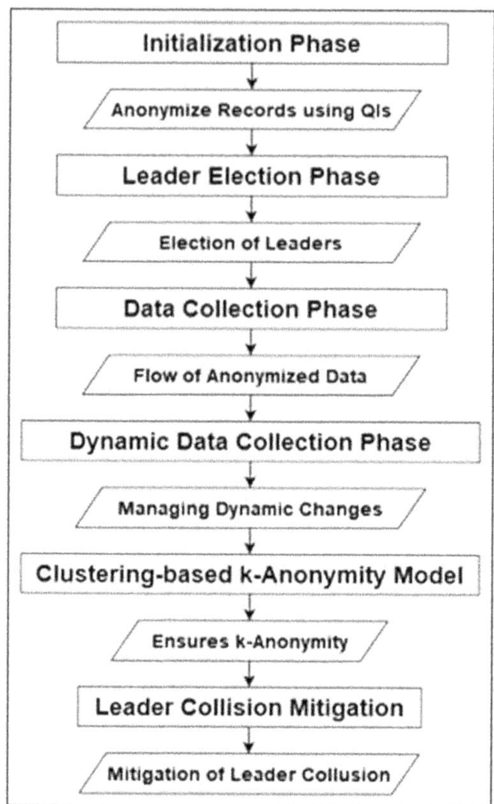

Figure 3.2 Privacy-preserving data collection protocol.

Noteworthy is the call for systemic oversight in big data health research, elucidating the need for a robust framework that accentuates user control amidst the implementation of AI technologies (Murdoch, 2021).

In the era of cloud computing, data anonymity is envisioned as a mechanism where the user's personal data is altered such that identification becomes implausible, and the cloud service provider is incapacitated from retrieving any personal information. This elucidates an emerging trend where data anonymization techniques are being honed to bolster user control and privacy in digital interactions (Hassan et al., 2022). Moreover, proposed privacy protection mechanisms entail comprehensive user and part identity authentication, fine-grained access control, and data desensitization to thwart unauthorized access and data leakage. These mechanisms are envisaged to be instrumental in fortifying user control over personal data and ensuring privacy at the mechanism level.

These deliberations underscore a palpable shift toward user-centric data anonymization paradigms. The burgeoning focus on user control augments the potential for more personalized and ethically aligned data anonymization strategies, thereby contributing to a robust data privacy and security ecosystem.

3.5.3.8 Proliferation of Anonymization-as-a-Service (AaaS)

In this era focused on data, the need for data anonymization solutions has become clearer. This growing demand has led to the emergence of platforms that supply Anonymization as a Service (AaaS) as shown in Figure 3.3. These platforms offer organizations tools and algorithms to anonymize their data dropping the requirement for expertise. The rise of this service-based model does not reflect the increasing significance given to data privacy. It also proves a broader trend toward outsourcing specialized technical services.

The inception of AaaS platforms is a response to the escalating concerns regarding data privacy and the concomitant regulatory landscape that mandates stringent data protection measures (Ciampi et al., 2021). By offering organizations an accessible avenue to anonymize their data, AaaS providers are playing a cardinal role in fostering a culture of data privacy and compliance. These platforms use innovative AI and ML algorithms to effectively anonymize data, ensuring that the anonymized data keeps its utility for analysis while following prevailing data protection laws.

Moreover, the AaaS model is poised to evolve in tandem with advancements in cryptographic techniques and privacy-preserving technologies. For instance, the integration of DP, a paradigm that quantifies privacy leakage

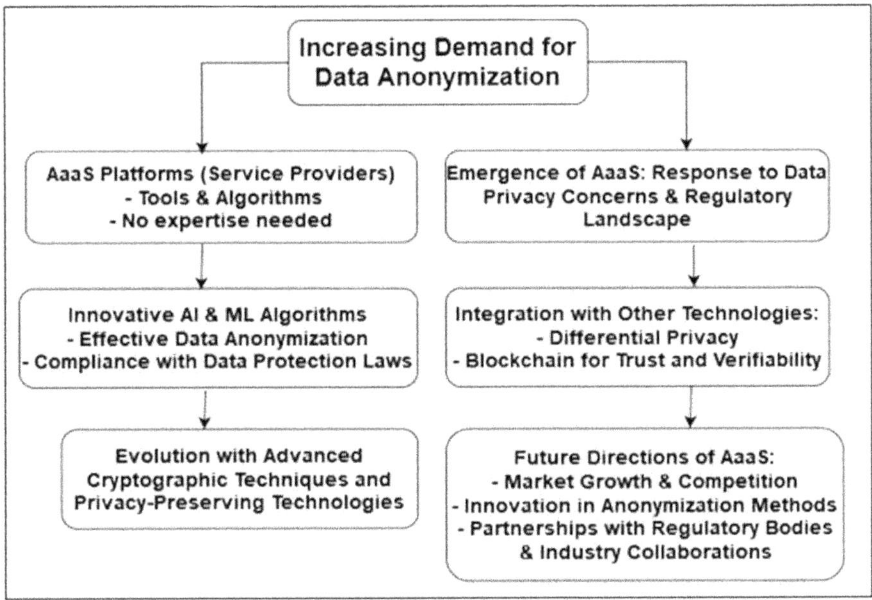

Figure 3.3 Anonymization-as-a-Service (AaaS).

in data analysis, into AaaS platforms is foreseen as a significant trend (Machanavajjhala et al., 2017). This amalgamation would empower organizations to not only anonymize their data but also to quantify the level of privacy preserved, thus supplying a more nuanced understanding of the trade-offs involved in data anonymization.

Furthermore, the confluence of blockchain technology with AaaS is expected to engender a new paradigm of trust and verifiability in data anonymization processes. By leveraging the immutable and transparent nature of blockchain, AaaS platforms can provide verifiable proofs of anonymization, thereby bolstering trust among stakeholders and regulators (de Haro Olmo et al., 2020).

As the market for AaaS (AI as a service) continues to grow, we can expect competition, which will drive innovation and the creation of more sophisticated methods of anonymization. The partnerships between AaaS providers and regulatory bodies along with collaborations across industries are expected to have an impact on shaping the future direction of the AaaS field.

3.6 ETHICAL AND REGULATORY CONSIDERATIONS

In today's era, data has become a resource that drives innovation, supplies valuable insights, and informs strategic decision-making. However, the responsible handling of data and information involves important ethical and regulatory obligations. As data anonymization methods grow more advanced and play a role in data management approaches, it is vital to grasp the implications and regulatory frameworks that govern their usage. This section delves into the considerations and regulatory landscapes surrounding the practice of data anonymization.

3.6.1 Ethical implications of data anonymization

The extensive collection, examination, and dissemination of data have prompted ethical apprehensions regarding appropriate handling protocols. The process of data anonymization, which serves as a means of safeguarding privacy, presents inherent ethical issues. This section examines the ramifications of data anonymization and investigates the intricate equilibrium between the utility of data, individual rights, and its societal repercussions.

a. **The Ethical Imperative of Privacy:** Privacy is more than an entitlement; it is an ethical obligation. The Universal Declaration of Human Rights acknowledges the significance of privacy as a right. Data anonymization is extremely important in ensuring the protection of privacy on a level. However, there can be dilemmas when privacy preservation techniques unintentionally compromise this goal. For instance, if anonymization is not properly implemented it can lead to data breaches where anonymous data can be connected to individuals thus undermining the essence of privacy that it was meant to protect (Rodriguez et al., 2022).

b. **Informed Consent and Autonomy**: Ensuring informed consent is a fundamental principle in the realm of data collection and processing. The awareness of data usage and the ability to exercise decision-making autonomy over personal information are crucial for individuals. Insufficient clarity in the communication of anonymization might undermine the establishment and maintenance of confidence. An instance that exemplifies ethical difficulties arises when data initially gathered for a specific aim is afterward anonymized and employed for an alternative goal, without the individual's awareness. The imperative to respect individuals' autonomy in relation to their anonymized data should not be disregarded (Rumbold & Pierscionek, 2018).

c. **Data Utility vs. Ethical Responsibility**: Anonymization often involves a trade-off between data utility and privacy. Highly anonymized data might lose its utility, making it less valuable for research or business insights. However, the ethical dilemma arises when the quest for utility compromises privacy. Is it ethically justifiable to slightly compromise privacy for significant societal benefits, such as medical research? This balance is a continuous ethical debate in the realm of data anonymization.

d. **Potential for Unintended Discrimination**: While anonymization aims to remove identifiers, it can sometimes lead to unintended consequences. For instance, if data from minority groups is anonymized to the extent that it becomes indistinguishable, it might lead to their underrepresentation in research or policy decisions. This inadvertent erasure can lead to a lack of inclusivity and potential discrimination (Mondada, 2014).

e. **Transparency and Accountability**: Ethical data practices demand transparency and accountability. Organizations must be forthright about their anonymization techniques, the potential risks associated, and the measures in place to mitigate those risks. Any breach of privacy, even unintentional, stemming from inadequate anonymization, places an ethical burden on the organization.

f. **The Ethical Challenges of Emerging Techniques**: With the advent of advanced anonymization techniques, such as DP, new ethical challenges have appeared. For instance, while DP offers robust privacy guarantees, it introduces "noise" into the data. If not communicated transparently, users of this data might make decisions based on 'noisy' or slightly altered data, leading to ethical concerns about the veracity and integrity of such data (Lowenberg & Puebla, 2022).

While data anonymization is a potent tool for preserving privacy, it is fraught with ethical challenges that evolve with changing technological landscapes and societal norms. Addressing these challenges requires a comprehensive approach, encompassing not only technical solutions but also ethical reflections, transparent communication, and a commitment to upholding the fundamental right to privacy. As data continues to shape our world, the ethical implications of its anonymization will remain at the forefront of discussions on digital ethics.

3.6.2 Regulatory frameworks and compliance (e.g., GDPR, CCPA)

The increasing prevalence of technology and the rapid proliferation of data have caused the implementation of legislation to safeguard individuals' privacy and promote responsible management of information. The regulations supply criteria for the safeguarding of data. Supply advice to corporations regarding compliance.

a. **GDPR:** The GDPR, which the European Union introduced in 2018, has influenced data protection around the world. Its main aim is to give individuals power over their data and make the regulatory landscape, for international businesses, more straightforward.
b. **Scope and Applicability:** The reach and application of the GDPR go beyond the borders of the EU covering any company that oversees the information of individuals living in the EU regardless of where the company is located geographically.
c. **Key Provisions:** The GDPR gives importance to the concept of minimizing data, which means that only necessary data should be processed and kept for the required period. It acknowledges anonymization and pseudonymization as methods of protecting data. However, it's important to note that while anonymized data is not subject, to GDPR pseudonymized data still falls within its jurisdiction (Yuan & Li, 2019).
d. **Rights of Data Subjects:** Individuals are endowed with rights like access, rectification, erasure of their data, and data portability, allowing them to use their data across numerous services.
e. **Compliance and Penalties:** Organizations must put in place measures for the protection of data both from an organizational standpoint. Failure to comply with these requirements may lead to penalties, which could amount to 4% of the company's annual turnover or €20 million, whichever is higher (Bukaty, 2019).
f. **CCPA:** The CCPA, enacted in 2018, is a pioneering data protection legislation in the United States, augmenting the privacy rights of California residents.
g. **Scope and Applicability:** The CCPA targets businesses that collect personal information of California residents and meet specific criteria concerning revenue, data volume, or profit from data sales (*California Consumer Privacy Act of 2018*, 2018).
h. **Key Provisions:** It introduces the notion of "sale" of personal information, encompassing a broad spectrum of data-sharing activities. Although

not explicitly mentioning anonymization, the CCPA supplies a carve-out for de-identified data, given it meets stringent criteria.

i. **Rights of Consumers:** Like the GDPR, the CCPA grants consumers rights to know about data collection, its purpose, and any third parties involved. They also have the right to opt-out of data sales and request data deletion. (Marks et al., 2020)

j. **Compliance and Penalties:** Businesses must implement reasonable security measures to safeguard data. Non-compliance can lead to civil penalties, and in case of data breaches, statutory damages may be imposed.

k. **Implications for Data Anonymization:** Regulatory frameworks like the GDPR and CCPA have far-reaching.

l. **Standardization of Techniques:** These regulations promote the adoption of standardized anonymization techniques, ensuring data protection is integral to data processing.

m. **Shift from Reactive to Initiative-taking Compliance:** Organizations are now proactively implementing data protection measures, including anonymization, to ensure compliance and avert penalties.

n. **Enhanced Trust:** Transparent data practices in line with regulatory requirements bolster consumer trust in digital services.

The emphasis on data protection has spurred innovation in anonymization techniques, with organizations exploring advanced methods that supply robust privacy guarantees without compromising data utility.

3.6.3 Balancing utility and privacy

The complexities of data anonymization revolve around the challenge of finding a balance between data usefulness and privacy. It is essential to ensure that while data fulfills its intended purpose, the privacy of individuals connected to that data is still protected. This delicate balance extends beyond aspects and delves into ethical considerations particularly regarding fundamental human rights and the broader social impacts of decision making driven by data.

a. **The Dual Imperative:** The primary goal of data anonymization is to change the data in such a way that it can no longer be used to identify individuals while remaining useful. This process, however, often involves a tradeoff. If we strengthen the anonymization, the data's usefulness may suffer. On the other hand, if we are less aggressive, anonymization may allow us to keep data utility. Additionally, there is an increased risk of sensitive information being exposed.

b. **Data Utility:** Data utility is essentially the degree to which data remains functional for its intended purpose. Within the context of data analysis, utility is vital as it ensures that the outcomes derived from anonymized data align with those gleaned from the original data. High utility is indispensable for precise data-driven decision-making, predictive modeling, and other analytical undertakings (Gangarde et al., 2022).

c. **Data Privacy:** This refers to the safeguarding of individual identities and their associated data. The aim of anonymization techniques is to ensure that even if adversaries manage to access the data, they are unable to deduce specific details pertaining to any individual (Zouinina et al., 2020).

d. The Trade-off Dilemma The tug-of-war between utility and privacy transcends being merely a technical challenge, embodying an ethical quandary as well. This is illustrated in the scenarios below:

e. **Medical Research:** Patient data is a linchpin for medical research, propelling advancements in treatment modalities and disease comprehension. However, the revelation of patient identities could have grave legal and ethical consequences. While stringent anonymization can cloak patient identities, it may also excise critical data nuances indispensable for research.

f. **Public Policy:** Governments often harness demographic data to craft policies. Anonymized data is pivotal as it ensures individuals remain unidentifiable, yet excessive anonymization could engender policies that misrepresent the actual needs of the populace (Esmeel et al., 2020).

g. **Navigating the Balance:** A spectrum of strategies and considerations can steer the process of reconciling utility and privacy:

h. **Granularity of Data:** Modulating the granularity of data is one avenue. For instance, aggregating ages into ranges as opposed to exact figures could bolster privacy albeit at the detriment of utility.

i. **Differential Privacy:** This avant-garde technique infuses controlled noise into the data, ensuring that the outcome of any analysis is nearly invariant, irrespective of the inclusion or exclusion of an individual's data. It extends a mathematical assurance of privacy while preserving a modicum of utility.

j. **Purpose Limitation:** A clear delineation of the purpose underlying data collection and processing can dictate the level of anonymization requisite. Broader aims such as general research may call for more stringent anonymization compared to specific, well-defined aims.

k. **Stakeholder Engagement:** Engaging stakeholders, inclusive of data subjects, can unearth insights into privacy expectations and acceptable trade-offs.

l. **Continuous Monitoring:** The equilibrium between utility and privacy is in flux. With the march of technology, data earlier deemed anonymized may become susceptible to re-identification, underscoring the need for ongoing monitoring and periodic reassessments (Rastogi et al., 2006).

3.7 CONCLUSION

The advent of the digital age, marked by a remarkable increase in data creation and consumption, has unveiled both prospects and dilemmas. As we traverse through this data-centric epoch, the importance of data privacy and safeguarding stays paramount. This chapter explored the domain of data anonymization, a crucial method that lives at the crossroads of data utility and privacy.

3.7.1 Recap of key points discussed

In the swiftly changing digital environment, protecting personal data has become a critical concern. Organizations and sectors face the dual challenge of leveraging data's potential while securing its privacy, bringing data anonymization techniques into the limelight as indispensable tools. This chapter embarked on an extensive examination of these techniques, shedding light on their importance, categories, and the hurdles they entail.

Introduction to Data Anonymization: The discourse began with an introduction, setting the scene for the later discussions. The core and importance of data anonymization were underscored, emphasizing its vital role in preserving individual privacy while helping data-driven insights. The link between data anonymization, cybersecurity, and Industry 4.0 was also examined, highlighting the technique's pertinence in modern industrial and technological scenarios.

Foundational Concepts: The chapter delved into the basics of data anonymization, illuminating its definition and pivotal concepts. The transformative aspect of data anonymization was emphasized, where raw data undergoes alterations to mask individual identities without giving up its analytical usefulness. The urgent need for data anonymization in today's data-centric world was discussed, spotlighting the dangers of data breaches and misuse.

Classifying Anonymization Techniques: A significant segment of the chapter was devoted to classifying data anonymization techniques. The difference between anonymization and pseudonymization was outlined, clarifying their respective scopes and applications. Conventional techniques like data masking, data swapping, and generalization were elaborated upon, supplying insights into their methodologies and use cases. The discussion transitioned to advanced anonymization techniques, delving into notions like DP, k-anonymity, l-diversity, t-closeness, and HE, with each technique thoroughly examined.

Interplay of AI and Data Anonymization: The narrative ventured into the sphere of AI, probing its burgeoning role in data anonymization. AI-driven approaches to data anonymization were discussed, accentuating their potential in augmenting traditional techniques, and ushering in novel methodologies. The advantages of incorporating AI into data anonymization processes, such as heightened accuracy and efficiency, were underscored, albeit with challenges like ethical dilemmas and AI biases also brought into discussion.

Data Anonymization in the Context of Industry 4.0: The relevance of data anonymization in Industry 4.0 was expounded, stressing its function in ensuring data privacy in the Fourth Industrial Revolution. The chapter explored the application of data anonymization techniques in various aspects of Industry 4.0, including smart manufacturing, IoT, and digital supply chains.

Ethical and Regulatory Landscape: The chapter also probed into the ethical ramifications of data anonymization, spotlighting the moral duties of organizations and industries in safeguarding data privacy. Regulatory frameworks governing data anonymization, like GDPR and CCPA, were dissected, emphasizing the necessity for compliance and the challenges it encompasses.

This chapter offered a comprehensive overview of data anonymization techniques, their significance, and the hurdles they pose in the modern digital era. As we navigate the intricacies of the data-driven universe, the principles and methodologies delineated here will serve as guiding torches, ensuring that data privacy and utility harmoniously coexist.

3.7.2　The future of data anonymization in the age of AI and Industry 4.0

The fusion of AI and Industry 4.0 heralds a new era of technological breakthroughs, redefining the facets of data management, analytics, and security. Amidst this transformative epoch, it's pivotal to envisage the evolution of data anonymization approaches. The intertwining of data anonymization with AI and Industry 4.0 forecasts a monumental shift in data feeling, processing, and safeguarding.

AI-Enhanced Anonymization Techniques: The amalgamation of AI with data anonymization stands to augment the efficacy and efficiency of prevailing techniques. ML algorithms, renowned for their pattern recognition and prognostic abilities, are instrumental in bolstering conventional anonymization methodologies. For instance, AI eases the discernment of best generalization tiers in datasets, striking a balance between data utility and privacy. Moreover, deep learning models are adept at identifying and obfuscating sensitive elements in unstructured data, transcending the limitations of traditional anonymization techniques.

GANs, a subset of neural networks, are a beacon of potential in the data anonymization domain. They can fabricate synthetic datasets that mirror the statistical attributes of the original data while obliterating the traceability of individual data points. This dual capability addresses privacy qualms and improves data augmentation, particularly in fields plagued by data paucity.

Data Anonymization in the Smart Ecosystems of Industry 4.0: Industry 4.0, typified by interlinked smart systems, unveils a distinct array of challenges and prospects for data anonymization. The surge of IoT gadgets engenders a torrent of data, a sizable part of which is personal and sensitive. The imperative of anonymizing this data, whilst keeping its analytic utility, ascends to prominence.

Advanced anonymization methodologies are set to be pivotal in intelligent manufacturing processes. In a smart factory scenario, sensor-generated data can be anonymized in real-time before transmission to a centralized nexus, preserving worker privacy alongside enabling proficient monitoring and quality assurance.

The decentralized essence of Industry 4.0 ecosystems, coupled with the rise of edge computing, calls for the concoction of distributed data anonymization strategies. Data could be anonymized at the source, that is, the device level, prior to transmission to a central server or cloud, diminishing data breach risks during transmission.

Challenges and Considerations: The path ahead, although laden with promise, is not bereft of hurdles. The intricate and adaptable nature of AI can engender challenges in ensuring uniform data anonymization. The enigmatic "black box" demeanor of certain AI models may obfuscate the evaluation of anonymization depth and efficacy. Furthermore, the escalation in AI model sophistication harbors the risk of "de-anonymization", undoing the anonymization processes to extract original data.

The dynamic and interlinked ecosystems of Industry 4.0 breed numerous vulnerability points. The quest for consistent and effective data anonymization across a plethora of devices, platforms, and systems will indeed be a daunting endeavor.

A Vision for the Future: Peering into the future, one can envisage the advent of adaptive anonymization frameworks, proficient in real-time modifications contingent on data nature and usage context. AI-driven systems might autonomously figure out the prime anonymization technique for a specific dataset, considering variables such as data sensitivity, intended usage, and legislative mandates.

Moreover, with the approach of quantum computing, its repercussions on data security and anonymization are bound to be significant. Quantum-resistant anonymization approaches may become indispensable, ensuring data privacy in an epoch where conventional encryption methodologies may be given obsolete.

In summation, the future of data anonymization amidst the age of AI and Industry 4.0 is set to be a dynamic and evolving terrain. While challenges are plenty, the scope for innovation and progress is boundless. A synergistic approach, encompassing technologists, ethicists, regulators, and end-users, will be pivotal in sculpting a future where data-driven insights harmoniously coalesce with privacy and security.

3.7.3 Encouraging further research and exploration

The rapidly advancing domain of data anonymization, especially within the realms of AI and Industry 4.0, underscores the importance of ongoing research and exploration. The complexity of ensuring data privacy while unlocking its potential for innovation mandates an initiative-taking stance toward research and development in this field. This section delineates the significance of nurturing a culture of inquiry and innovation, pinpointing potential avenues for exploration and the broader repercussions for society and industry.

The Imperative for Continuous Research: The fluid nature of AI and Industry 4.0 signifies that today's techniques and tools may become outdated or insufficient tomorrow. Data anonymization, situated at the nexus of these domains, is highly susceptible to such rapid transformations. As computational capabilities escalate, and cyber threats sophisticate, the methodologies employed to anonymize data must evolve correspondingly. This causes a commitment to perpetual research, ensuring data anonymization techniques stay robust, effective, and pertinent.

Potential Areas of Exploration: Several realms beckon researchers within the ambit of data anonymization:

a. **Quantum-Resistant Anonymization:** The dawn of quantum computing threatens current encryption and anonymization techniques, making research into quantum-resistant data anonymization crucial.
b. **Real-time Anonymization:** The global shift toward real-time data processing and analytics demands the capability to anonymize data instantaneously without sacrificing speed or utility.
c. **Context-aware Anonymization:** Considering the varied applications of data across healthcare, finance, and smart cities, the need for context-aware anonymization techniques, which tailor the level and method of anonymization to the specific use-case, is clear.
d. **Ethical AI and Anonymization:** As AI models become integral to data processing, ensuring that these models are ethical, unbiased, and transparent is crucial. Investigating the intersection of data anonymization and AI ethics is of paramount importance.

Collaborative Research and Open Innovation: The multifaceted challenges posed by data anonymization traverse the domains of technology, ethics, law, and sociology. Tackling these challenges calls for a collaborative research approach, amalgamating ability from diverse fields. Open innovation, where findings and methodologies are shared freely within the community, can accelerate progress, promoting a collective problem-solving approach.

Implications for Society and Industry: The ramifications of data anonymization research are profound for both society and industry. For society, efficacious data anonymization techniques uphold the right to privacy in an age of pervasive data collection. For industries, especially those spearheading Industry 4.0, robust anonymization is both a regulatory imperative and a competitive edge, easing data-driven innovation without compromising consumer trust.

Conclusion: Although substantial progress has been made in data anonymization, the journey is far from over. The convergence of AI and Industry 4.0 unveils both challenges and opportunities, making it a thrilling epoch for researchers. By cultivating a culture of inquiry, collaboration, and open innovation, we can traverse the intricacies of data privacy, ensuring a future where the immense potential of data is harnessed without infringing on the fundamental right to privacy. The encouragement of further research and exploration is not merely a scientific necessity but a societal one, harboring the prospect of a world where data-driven insights and human dignity coexist harmoniously.

REFERENCES

Abadi, M., Chu, A., Goodfellow, I., McMahan, H. B., Mironov, I., Talwar, K., & Zhang, L. (2016). Deep learning with differential privacy. In Proceedings of the 2016 ACM SIGSAC Conference on Computer and Communications Security (pp. 308–318). https://doi.org/10.1145/2976749.2978318

Adkinson Orellana, L., Dago Casas, P., Sestelo, M., & Pintos Castro, B. (2019, May). A new approach for dynamic and risk-based data anonymization. In Computational Intelligence in Security for Information Systems Conference (pp. 327–336). Springer International Publishing. https://doi.org/10.1007/978-3-030-57805-3_31

Alazab, M., Gadekallu, T. R., & Su, C. (2022). Guest editorial: Security and privacy issues in industry 4.0 applications. IEEE Transactions on Industrial Informatics, 18(9), 6326–6329. https://doi.org/10.1109/TII.2022.3164741

Al-Fuqaha, A., Guizani, M., Mohammadi, M., Aledhari, M., & Ayyash, M. (2015). Internet of things: A survey on enabling technologies, protocols, and applications. IEEE Communications Surveys & Tutorials, 17(4), 2347–2376. https://doi.org/10.1109/COMST.2015.2444095

Alloghani, M., M. Alani, M., Al-Jumeily, D., Baker, T., Mustafina, J., Hussain, A., & J. Aljaaf, A. (2019). A systematic review on the status and progress of homomorphic encryption technologies. Journal of Information Security and Applications, 48, 102362. https://doi.org/10.1016/j.jisa.2019.102362

Andrew, J., Eunice, R. J., & Karthikeyan, J. (2023). An anonymization-based privacy-preserving data collection protocol for digital health data. Frontiers in Public Health, 11, 1125011. https://doi.org/10.3389/fpubh.2023.1125011

Anoop, V. S., & Asharaf, S. (2022). Integrating artificial intelligence and blockchain for enabling a trusted ecosystem for healthcare sector. In C. Chakraborty; & M. R. Khosravi (Eds.), Intelligent Healthcare: Infrastructure, Algorithms and Management (pp. 281–295). Springer Nature. https://doi.org/10.1007/978-981-16-8150-9_13

Aronsson, H., Abrahamsson, M., & Spens, K. (2011). Developing lean and agile health care supply chains. Supply Chain Management: An International Journal, 16(3), 176–183. https://doi.org/10.1108/13598541111127164

D. J. Bernstein; J. Buchmann; & E. Dahmen(Eds.) (2009). Post-Quantum Cryptography. Springer. https://doi.org/10.1007/978-3-540-88702-7

Bukaty, P. (2019). The California Consumer Privacy Act (CCPA): An Implementation Guide. IT Governance Publishing. https://doi.org/10.2307/j.ctvjghvnn

California Consumer Privacy Act of 2018. (2018). Cyber/Data/Privacy Insights. https://cdp.cooley.com/ccpa-2018/

Carvalho, A. P., Canedo, E. D., Carvalho, F. P., & Carvalho, P. H. P. (2021). Anonimisation, impacts and challenges into big data: A case studies. In J. Filipe; M. Śmiałek; A. Brodsky; & S. Hammoudi (Eds.), Enterprise Information Systems (pp. 3–23). Springer International Publishing. https://doi.org/10.1007/978-3-030-75418-1_1

Castro, H., Costa, F., Ferreira, L., Ávila, P., Putnik, G. D., & Cruz-Cunha, M. (2022). Data science for Industry 4.0: A literature review on open design approach. Procedia Computer Science, 204, 877–884. https://doi.org/10.1016/j.procs.2022.08.106

Charles, V., Emrouznejad, A., & Gherman, T. (2023). A critical analysis of the integration of blockchain and artificial intelligence for supply chain. Annals of Operations Research, 327(1), 7–47. https://doi.org/10.1007/s10479-023-05169-w

Chaudhuri, K., Monteleoni, C., & Sarwate, A. D. (2011). Differentially private empirical risk minimization. Journal of Machine Learning Research: JMLR, 12, 1069–1109.

Ciampi, F., Demi, S., Magrini, A., Marzi, G., & Papa, A. (2021). Exploring the impact of big data analytics capabilities on business model innovation: The mediating role of entrepreneurial orientation. Journal of Business Research, 123, 1–13. https://doi.org/10.1016/j.jbusres.2020.09.023

de Haro Olmo, F., Varela Vaca, A., & Álvarez-Bermejo, J. (2020). Blockchain from the perspective of privacy and anonymisation: A systematic literature review. Sensors, 20. https://doi.org/10.3390/s20247171

Domingo-Ferrer, J., Sánchez, D., & Soria-Comas, J. (2016). Beyond k-anonymity: L-diversity and t-closeness. In J. Domingo-Ferrer; D. Sánchez; & J. Soria-Comas (Eds.), Database Anonymization: Privacy Models, Data Utility, and Microaggregation-Based Inter-Model Connections (pp. 47–51). Springer International Publishing. https://doi.org/10.1007/978-3-031-02347-7_6

Dwork, C., & Roth, A. (2014). The algorithmic foundations of differential privacy. Foundations and Trends® in Theoretical Computer Science, 9(3–4), 211–407. https://doi.org/10.1561/0400000042

Elliot, M., O'Hara, K., Raab, C., O'Keefe, C. M., Mackey, E., Dibben, C., Gowans, H., Purdam, K., & McCullagh, K. (2018). Functional anonymisation: Personal data and the data environment. Computer Law & Security Review, 34(2), 204–221. https://doi.org/10.1016/j.clsr.2018.02.001

Esmeel, T. K., Hasan, M. M., Kabir, M. N., & Firdaus, A. (2020). Balancing data utility versus information loss in data-privacy protection using k-anonymity. In Proceedings of the 2020 IEEE 8th Conference on Systems, Process and Control (ICSPC), 158–161. https://doi.org/10.1109/ICSPC50992.2020.9305776

Farhan, L., Kharel, R., Kaiwartya, O., Quiroz-Castellanos, M., Alissa, A., & Abdulsalam, M. (2018). A concise review on internet of things (IoT) -problems, challenges and opportunities. In Proceedings of the 2018 11th International Symposium on Communication Systems, Networks & Digital Signal Processing (CSNDSP), 1–6. https://doi.org/10.1109/CSNDSP.2018.8471762

Ferrão, M. E., Prata, P., & Fazendeiro, P. (2022). Utility-driven assessment of anonymized data via clustering. Scientific Data, 9(1), 456. https://doi.org/10.1038/s41597-022-01561-6

Ficek, J., Wang, W., Chen, H., Dagne, G., & Daley, E. (2021). Differential privacy in health research: A scoping review. Journal of the American Medical Informatics Association : JAMIA, 28(10), 2269–2276. https://doi.org/10.1093/jamia/ocab135

Fredj, F. B., Lammari, N., & Comyn-Wattiau, I. (2015). Abstracting anonymization techniques: A prerequisite for selecting a generalization algorithm. Procedia Computer Science, 60, 206–215. https://doi.org/10.1016/j.procs.2015.08.120

Gangarde, R., Shrivastava, D., Sharma, A., Tandon, T., Pawar, A., & Garg, R. (2022). Data anonymization to balance privacy and utility of online social media network data. Journal of Discrete Mathematical Sciences and Cryptography, 25(3), 829–838. https://doi.org/10.1080/09720529.2021.2016225

Girka, A., Terziyan, V., Gavriushenko, M., & Gontarenko, A. (2021). Anonymization as homeomorphic data space transformation for privacy-preserving deep learning. Procedia Computer Science, 180, 867–876. https://doi.org/10.1016/j.procs.2021.01.337

Gupta, P., & Bhatnagar, V. (2014). Dynamic anonymization techniques analogy for multiple releases of data. International Journal of Data Mining Modelling and Management, 6, 239–260. https://doi.org/10.1504/IJDMMM.2014.065147

Hassan, J., Shehzad, D., Habib, U., Aftab, M. U., Ahmad, M., Kuleev, R., & Mazzara, M. (2022). The rise of cloud computing: Data protection, privacy, and open research challenges—A systematic literature review (SLR). Computational Intelligence and Neuroscience, 2022, 8303504. https://doi.org/10.1155/2022/8303504

Hechler, E., Oberhofer, M., & Schaeck, T. (2020). AI and blockchain. In E. Hechler; M. Oberhofer; & T. Schaeck (Eds.), Deploying AI in the Enterprise: IT Approaches for Design, DevOps, Governance, Change Management, Blockchain, and Quantum Computing (pp. 253–271). Apress. https://doi.org/10.1007/978-1-4842-6206-1_11

Houghton, F., & Houghton, S. (2018). "Blacklists" and "whitelists": A salutary warning concerning the prevalence of racist language in discussions of predatory publishing. Journal of the Medical Library Association: JMLA, 106(4), 527–530. https://doi.org/10.5195/jmla.2018.490

Institute of Medicine (US) Committee on health research and the privacy of health information: The HIPAA privacy rule. (2009). Beyond the HIPAA Privacy Rule: Enhancing Privacy, Improving Health Through Research (S. J. Nass, L. A. Levit, & L. O. Gostin, Eds.). National Academies Press (US). http://www.ncbi.nlm.nih.gov/books/NBK9578/

Ivanov, D., Sokolov, B., & Dolgui, A. (2020). Introduction to scheduling in industry 4.0 and cloud manufacturing systems. In B. Sokolov; D. Ivanov; & A. Dolgui (Eds.), Scheduling in Industry 4.0 and Cloud Manufacturing (pp. 1–9). Springer International Publishing. https://doi.org/10.1007/978-3-030-43177-8_1

Jain, P., Gyanchandani, M., & Khare, N. (2016). Big data privacy: A technological perspective and review. Journal of Big Data, 3(1), 25. https://doi.org/10.1186/s40537-016-0059-y

Jakob, C. E. M., Kohlmayer, F., Meurers, T., Vehreschild, J. J., & Prasser, F. (2020). Design and evaluation of a data anonymization pipeline to promote open science on COVID-19. Scientific Data, 7(1), Article 1. https://doi.org/10.1038/s41597-020-00773-y

Jan, Z., Ahamed, F., Mayer, W., Patel, N., Grossmann, G., Stumptner, M., & Kuusk, A. (2023). Artificial intelligence for industry 4.0: Systematic review of applications, challenges, and opportunities. Expert Systems With Applications, 216, 119456. https://doi.org/10.1016/j.eswa.2022.119456

Jiang, H., Gao, Y., Sarwar, S. M., GarzaPerez, L., & Robin, M. (2021). Differential privacy in privacy-preserving big data and learning: Challenge and opportunity (arXiv:2112.01704; Version 1). arXiv. https://doi.org/10.48550/arXiv.2112.01704

Kenny, C. T., Kuriwaki, S., McCartan, C., Rosenman, E. T. R., Simko, T., & Imai, K. (2021). The use of differential privacy for census data and its impact on redistricting: The case of the 2020 U.S. Census. Science Advances, 7(41), eabk3283. https://doi.org/10.1126/sciadv.abk3283

Kurz, C. (2021). Understanding differential privacy. Significance, 18(3), 24–27. https://doi.org/10.1111/1740-9713.01528

LeFevre, K., DeWitt, D. J., & Ramakrishnan, R. (2008). Workload-aware anonymization techniques for large-scale datasets. ACM Transactions on Database Systems, 33(3), 17:1-17:47. https://doi.org/10.1145/1386118.1386123

Li, N., Li, T., & Venkatasubramanian, S. (2007). t-Closeness: Privacy Beyond k-Anonymity and l-Diversity. 2007 IEEE 23rd International Conference on Data Engineering, 106–115. https://doi.org/10.1109/ICDE.2007.367856

Li, M., Lou, W., & Ren, K. (2010). Data security and privacy in wireless body area networks. IEEE Wireless Communications, 17(1), 51–58. https://doi.org/10.1109/MWC.2010.5416350

Long, J., Liu, M., Liu, S., Tang, F., Tan, W., Xiao, T., Chu, C., & Yang, J. (2019). H2S attenuates the myocardial fibrosis in diabetic rats through modulating PKC-ERK1/2MAPK signaling pathway. Technology and Health Care, 27(Suppl 1), 307–316. https://doi.org/10.3233/THC-199029

Lowenberg, D., & Puebla, I. (2022). Responsible handling of ethics in data publication. PLOS Biology, 20(3), e3001606. https://doi.org/10.1371/journal.pbio.3001606

Lu, Y. (2017). Industry 4.0: A survey on technologies, applications and open research issues. Journal of Industrial Information Integration, 6, 1–10. https://doi.org/10.1016/j.jii.2017.04.005

Machanavajjhala, A., Gehrke, J., Kifer, D., & Venkitasubramaniam, M. (2006). L-diversity: Privacy beyond k-anonymity. In Proceedings of the 22nd International Conference on Data Engineering (ICDE'06), 24–24. https://doi.org/10.1109/ICDE.2006.1

Machanavajjhala, A., He, X., & Hay, M. (2017). Differential privacy in the wild: A tutorial on current practices & open challenges. 1727–1730. https://doi.org/10.1145/3035918.3054779

Marks, L., CISA, CRISC, CGEIT, CFE, CISSP, CSTE, ITIL, PMP (2020). The California Consumer Privacy Act and Encryption: Theory, Practice, Risk Assessment and Risk Mitigation. ISACA. https://www.isaca.org/resources/isaca-journal/issues/2020/volume-2/the-california-consumer-privacy-act-and-encryption

McMahan, H. B., Moore, E., Ramage, D., Hampson, S., & Arcas, B. A. y. (2023). Communication-efficient learning of deep networks from decentralized data (arXiv:1602.05629). arXiv. https://doi.org/10.48550/arXiv.1602.05629

Minssen, T., Rajam, N., & Bogers, M. (2020). Clinical trial data transparency and GDPR compliance: Implications for data sharing and open innovation. Science and Public Policy, 47(5), 616–626. https://doi.org/10.1093/scipol/scaa014

Mondada, L. (2014). Ethics in action: Anonymization as a Participant's concern and a Participant's practice. Human Studies, 37(2), 179–209. https://doi.org/10.1007/s10746-013-9286-9

Monteiro, S., Oliveira, D., António, J., Sá, F., Wanzeller, C., Martins, P., & Abbasi, M. (2022). Data anonymization: Techniques and models. In J. L. Reis; M. Del Rio Araujo; L. P. Reis; & J. P. M. dos Santos (Eds.), Marketing and Smart Technologies (pp. 73–84). Springer Nature. https://doi.org/10.1007/978-981-99-0333-7_6

Moraes, T. G., Lemos, A. N. L. E., Lopes, A. K., Moura, C., & de Pereira, J. R. L. (2021). Open data on the COVID-19 pandemic: Anonymisation as a technical solution for transparency, privacy, and data protection. International Data Privacy Law, 11(1), 32–47. https://doi.org/10.1093/idpl/ipaa025

Munjal, K., & Bhatia, R. (2022). A systematic review of homomorphic encryption and its contributions in healthcare industry. Complex & Intelligent Systems, 1–28. https://doi.org/10.1007/s40747-022-00756-z

Murdoch, B. (2021). Privacy and artificial intelligence: Challenges for protecting health information in a new era. BMC Medical Ethics, 22(1), 122. https://doi.org/10.1186/s12910-021-00687-3

Murthy, S., Abu Bakar, A., Abdul Rahim, F., & Ramli, R. (2019). A Comparative Study of Data Anonymization Techniques. In Proceedings of the 2019 IEEE 5th Intl Conference on Big Data Security on Cloud (BigDataSecurity), IEEE Intl Conference on High Performance and Smart Computing, (HPSC) and IEEE Intl Conference on Intelligent Data and Security (IDS), 306–309. https://doi.org/10.1109/BigDataSecurity-HPSC-IDS.2019.00063

Ni, C., Cang, L. S., Gope, P., & Min, G. (2022). Data anonymization evaluation for big data and IoT environment. Information Sciences, 605, 381–392. https://doi.org/10.1016/j.ins.2022.05.040

O'Donovan, P., Leahy, K., Bruton, K., & O'Sullivan, D. T. J. (2015). An industrial big data pipeline for data-driven analytics maintenance applications in large-scale smart manufacturing facilities. Journal of Big Data, 2(1), 25. https://doi.org/10.1186/s40537-015-0034-z

Ohm, P. (2010). What the surprising failure of data anonymization means for law and policy. High Tech Law Institute Events. https://digitalcommons.law.scu.edu/hightechevents/reidentification/paulohm/1

Pawar, A., Ahirrao, S., & Churi, P. P. (2018). Anonymization techniques for protecting privacy: A survey. In 2018 IEEE Punecon, 1–6. https://doi.org/10.1109/PUNECON.2018.8745425

Peikert, C. (2016). A decade of lattice cryptography. Foundations and Trends® in Theoretical Computer Science, 10(4), 283–424. https://doi.org/10.1561/0400000074

Pfitzmann, A., & Hansen, M. (2010). A terminology for talking about privacy by data minimization: Anonymity, Unlinkability, Undetectability, Unobservability, Pseudonymity, and Identity Management. https://www.semanticscholar.org/paper/A-terminology-for-talking-about-privacy-by-data-and-Pfitzmann-Hansen/ea6c46d8b7a1bc7fdc3b7c63b4888bd971200eff

Phillips, A., Borry, P., & Shabani, M. (2017). Research ethics review for the use of anonymized samples and data: A systematic review of normative documents. Accountability in Research, 24(8), 483–496. https://doi.org/10.1080/08989621.2017.1396896

Priya, S. K., Balaganesh, N., & Karthika, K. P. (2023). Integration of AI, blockchain, and IoT technologies for sustainable and secured Indian public distribution system. In B. Bhushan; A. K. Sangaiah; & T. N. Nguyen (Eds.), AI Models for Blockchain-Based Intelligent Networks in IoT Systems: Concepts, Methodologies, Tools, and Applications (pp. 347–371). Springer International Publishing. https://doi.org/10.1007/978-3-031-31952-5_15

Rafiq, F., Awan, M. J., Yasin, A., Nobanee, H., Zain, A. M., & Bahaj, S. A. (2022). Privacy prevention of big data applications: A systematic literature review. SAGE Open, 12(2), 21582440221096445. https://doi.org/10.1177/21582440221096445

Raimundo, R., & Rosário, A. (2021). The impact of artificial intelligence on data system security: A literature review. Sensors (Basel, Switzerland), 21(21), 7029. https://doi.org/10.3390/s21217029

Rajendran, K., Jayabalan, M., & Ehsan, M. (2018). A study on k-anonymity, l-diversity, and t-closeness Techniques focusing Medical Data. https://www.semanticscholar.org/paper/A-Study-on-k-anonymity%2C-l-diversity%2C-and-Techniques-Rajendran-Jayabalan/db076c5139cd352c726a5f63fb723bf86712d8da

Ramya Shree, A. N., Kiran, P., Rakshith, R., & Likhith, R. (2022). SCAA—Sensitivity context aware anonymization—An automated hybrid PPUDP technique for big data. In S. Aurelia; S. S. Hiremath; K. Subramanian; & S. Kr. Biswas (Eds.), Sustainable Advanced Computing (pp. 615–626). Springer. https://doi.org/10.1007/978-981-16-9012-9_49

Rastogi, V., Suciu, D., & Hong, S. (2006). The Boundary Between Privacy and Utility in Data Anonymization. arXiv E-Prints, cs/0612103. https://doi.org/10.48550/arXiv.cs/0612103

Ren, W., Tong, X., Du, J., Wang, N., Li, S., Min, G., & Zhao, Z. (2021). Privacy enhancing techniques in the internet of things using data anonymisation. Information Systems Frontiers. https://doi.org/10.1007/s10796-021-10116-w

Rodriguez, A., Tuck, C., Dozier, M. F., Lewis, S. C., Eldridge, S., Jackson, T., Murray, A., & Weir, C. J. (2022). Current recommendations/practices for anonymising data from clinical trials in order to make it available for sharing: A scoping review. Clinical Trials, 19(4), 452–463. https://doi.org/10.1177/17407745221087469

Rumbold, J., & Pierscionek, B. (2018). Contextual anonymization for secondary use of big data in biomedical research: Proposal for an anonymization matrix. JMIR Medical Informatics, 6(4), e47. https://doi.org/10.2196/medinform.7096

Salajegheh, R., Nemergut, E. C., Rice, T. M., Joseph, R., Tsang, S., Sarosiek, B. M., Muthusubramanian, C. P., Hipwell, K. M., Horton, K. B., & Naik, B. I. (2020). Impact of a perioperative oral opioid substitution protocol during the nationwide intravenous opioid shortage: A single center, interrupted time series with segmented regression analysis. PLoS One, 15(6), e0234199. https://doi.org/10.1371/journal.pone.0234199

Saunders, B., Kitzinger, J., & Kitzinger, C. (2015). Anonymising interview data: Challenges and compromise in practice. Qualitative Research, 15(5), 616–632. https://doi.org/10.1177/1468794114550439

Scheibner, J., Raisaro, J. L., Troncoso-Pastoriza, J. R., Ienca, M., Fellay, J., Vayena, E., & Hubaux, J.-P. (2021). Revolutionizing medical data sharing using advanced privacy-enhancing technologies: Technical, legal, and ethical synthesis. Journal of Medical Internet Research, 23(2), e25120. https://doi.org/10.2196/25120

Sexton, R., & Shao, G. (n.d.). Data analytics for smart manufacturing systems. NIST. Retrieved October 8, 2023, from https://www.nist.gov/programs-projects/data-analytics-smart-manufacturing-systems

Sivarajah, U., Kamal, M. M., Irani, Z., & Weerakkody, V. (2017). Critical analysis of big data challenges and analytical methods. Journal of Business Research, 70, 263–286. https://doi.org/10.1016/j.jbusres.2016.08.001

Smith, M., & Agrawal, R. (2022). Anonymization techniques. In L. A. Schintler; & C. L. McNeely (Eds.), Encyclopedia of Big Data (pp. 30–33). Springer International Publishing. https://doi.org/10.1007/978-3-319-32010-6_9

Song, F. (2014). A note on quantum security for post-quantum cryptography. In M. Mosca (Ed.), Post-Quantum Cryptography (pp. 246–265). Springer International Publishing. https://doi.org/10.1007/978-3-319-11659-4_15

Sweeney, L. (2002). k-anonymity: A model for protecting privacy. International Journal of Uncertainty, Fuzziness and Knowledge-Based Systems, 10(5), 557–570. https://doi.org/10.1142/S0218488502001648

Tankard, C. (2016). What the GDPR means for businesses. Network Security, 2016(6), 5–8. https://doi.org/10.1016/S1353-4858(16)30056-3

Will, M. A., & Ko, R. K. L. (2015). Chapter 5—A guide to homomorphic encryption. In R. Ko; & K.-K. R. Choo (Eds.), The Cloud Security Ecosystem (pp. 101–127). Syngress. https://doi.org/10.1016/B978-0-12-801595-7.00005-7

Yoon, J., Drumright, L. N., & van der Schaar, M. (2020). Anonymization through data synthesis using generative adversarial networks (ADS-GAN). IEEE Journal of Biomedical and Health Informatics, 24(8), 2378–2388. https://doi.org/10.1109/JBHI.2020.2980262

Yuan, B., & Li, J. (2019). The policy effect of the general data protection regulation (GDPR) on the digital public health sector in the European Union: An empirical investigation. International Journal of Environmental Research and Public Health, 16(6), 1070. https://doi.org/10.3390/ijerph16061070

Zanella, A., Bui, N., Castellani, A., Vangelista, L., & Zorzi, M. (2014). Internet of things for smart cities. IEEE Internet of Things Journal, 1(1), 22–32. https://doi.org/10.1109/JIOT.2014.2306328

Zouinina, S., Bennani, Y., Rogovschi, N., & Lyhyaoui, A. (2020). A two-levels data anonymization approach. Artificial Intelligence Applications and Innovations, 583, 85–95. https://doi.org/10.1007/978-3-030-49161-1_8

Zuo, Z., Watson, M., Budgen, D., Hall, R., Kennelly, C., & Al Moubayed, N. (2021). Data anonymization for pervasive health care: Systematic literature mapping study. JMIR Medical Informatics, 9(10), e29871. https://doi.org/10.2196/29871

Zykov, S. V., Temkin, I. O., & Deryabin, S. A. (2019). Tradeoff-based architecting of the software system for autonomous robotized open pit mining. Procedia Computer Science, 159, 1740–1746. https://doi.org/10.1016/j.procs.2019.09.345

Chapter 4

Privacy-centric software design for scalable big data analytics

Deepika Dubey, Devanshu Tiwari, and Atharva Jaiswal

4.1 INTRODUCTION: PRIVACY PRESERVATION IN BIG DATA

Big data encompasses datasets of such immense size and complexity that traditional data processing methods are inadequate. It stands for the enormous amounts of data—both unstructured and structured—that businesses deal with on a daily basis due to advances in technology, the rise in internet usage, social media, sensor networks, and other applications. A new era of opportunities and challenges is being brought about by this explosion in data generation [1]. As an example, think about a small online bookstore. A plethora of data is generated each time a customer browses or purchases data as book preferences, browsing history, and purchasing patterns. This, multiplied by the thousands of patrons that the bookstore serves on a daily basis, gives you an enormous volume of data to handle and examine. Big data is important because it can lead to well-informed decisions and offer priceless insights. Businesses can spot trends, forecast customer behavior, and streamline processes by evaluating this data. But among all of this information is one very important worry: privacy [2]. There are serious threats to personal privacy and data security because of the sheer amount and variety of data that is gathered. Big data repositories containing personal data may expose it to abuse, exploitation, and illegal access. Inappropriate use of big data can have serious repercussions for both individuals and society at large, ranging from identity theft to intrusive profiling. Therefore, it's critical to protect privacy when using big data. It entails putting strong safeguards in place to safeguard private data at every stage of its lifecycle, from data creation and storage to processing and analysis [3]. Figure 4.1 illustrates the life cycle stages of big data, encompassing data generation, storage, and processing.

Techniques such as encryption, anonymization, and access controls play a crucial role in safeguarding privacy and ensuring compliance with regulatory requirements.

DOI: 10.1201/9781032657264-4

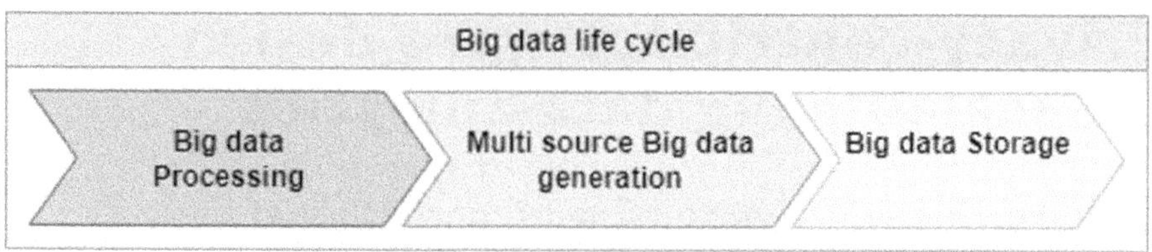

Figure 4.1 Data production, processing, and storage are the three steps of the big data life cycle that are displayed.

4.1.1 Objectives

1. **Understand Privacy Preservation Concepts**
   - To provide a thorough understanding of privacy preservation principles within the context of big data.
   - To discuss the fundamental importance and challenges of maintaining data privacy in large-scale data environments.
2. **Examine Social and Ethical Implications**
   - To analyze the social and ethical significance of privacy in big data applications.
   - To identify and explain the privacy requirements for handling big data responsibly.
   - To differentiate between privacy and security and to discuss their interplay in protecting data.
3. **Explore ETL Processes and Privacy**
   - To explain the ETL (Extract, Transform, Load) processes in big data systems.
   - To highlight methods for integrating privacy preservation measures during the ETL phases.
4. **Ensure Privacy During Data Creation**
   - To identify techniques for preserving privacy from the moment data is created.
   - To explore methods for anonymizing and protecting data at its inception.
5. **Implement Privacy Measures in Data Storage**
   - To discuss strategies for ensuring privacy during data storage.
   - To highlight best practices for maintaining data integrity and confidentiality in storage systems.
6. **Preserve Privacy in Cloud Storage**
   - To examine various privacy preservation methods for storing big data on cloud platforms.
   - To discuss encryption techniques and other tools to ensure data privacy in cloud environments.
   - To explore methods for verifying the integrity of stored big data.

7. **Maintain Privacy During Data Processing**
   - To identify methods for ensuring data privacy during processing and analysis.
   - To discuss techniques for maintaining privacy while performing computations on large datasets.
8. **Evaluate Privacy-Preserving Approaches**
   - To conduct a comparative analysis of de-identification techniques, including k-anonymity, l-diversity, and t-closeness.
   - To provide a comparative analysis of different privacy-preserving methods and their applications.
9. **Explore Hybrid Privacy Approaches**
   - To introduce hybrid methods like HybrEx that combine multiple privacy-preserving techniques.
   - To discuss the advantages and practical applications of hybrid approaches in big data environments.
10. **Implement Privacy-Preserving Aggregation**
    - To explore techniques for aggregating data in a privacy-preserving manner.
    - To explore techniques for securely aggregating data from diverse sources while safeguarding individual privacy.
11. **Operate on Encrypted Data**
    - To examine techniques for performing operations on encrypted data while preserving privacy.
    - To discuss the use of homomorphic encryption and other methods that allow computations on encrypted data without exposing the underlying information.

These objectives aim to provide a comprehensive guide on designing software that efficiently preserves privacy in big data environments, covering all aspects from data creation to storage and processing.

4.2 SOCIAL AND ETHICAL SIGNIFICANCE

Big data raises a lot of privacy concerns because of the enormous amounts of data that are gathered, stored, and processed—often without the knowledge or consent of the individuals involved. As Table 4.1 outlines, there are numerous privacy issues that aggravate this. Big data poses a wide range of risks to people's right to privacy, from data breaches and unauthorized use to discrimination, profiling, and a lack of transparency [4]. These worries are further exacerbated by the widespread monitoring and surveillance made possible by big data technologies, as well as the obtaining of needless data and the lack of adequate consent and control mechanisms. In order to ensure the ethical and responsible use of big data and protect people's right to privacy in a world

Table 4.1 Big-data privacy concerns

Privacy concern	*Description*
Data breaches	Unauthorized access to extensive collections of personal data can result in identity theft, financial fraud, and various other harmful activities.
Unintended data use	The inadvertent utilization of personal information in big data analytics, often without individuals' awareness or control over its secondary or tertiary uses.
Profiling and discrimination	Construction of detailed profiles of individuals based on online behaviors and demographics, potentially fueling discriminatory practices in various domains.
Lack of transparency	Obscure big data algorithms make it difficult for individuals to understand how their data is being used, undermining trust in data-driven decision-making processes.
Surveillance and monitoring	Extensive surveillance of individuals' activities, both online and offline, raising concerns about government overreach and intrusion into private lives. intrusion into private lives.
Data minimization	Collection of vast amounts of unnecessary data increases the risk of privacy violations, highlighting the need for strategies to limit data collection to essential information.
Consent and control	Limited control over the collection and utilization of personal information, posing challenges in obtaining meaningful consent for data collection and usage.

where data is becoming more and more important, it is imperative that these privacy issues are addressed.

4.2.1 Identifying the privacy requirements for big data

While big data analytics offers immense potential, a lack of standardized security and privacy tools hinders widespread adoption within organizations. There can be potential methods for enhancing big data platforms that support privacy protection [5]. The framework's plans for growth and foundations support:

1. Defining privacy policies that control access to data stored in big data platforms of interest.
2. Developing robust mechanisms to enforce data privacy policies.
3. Integrating the monitors that are created into the analytics platforms of interest. The enforcement methods that were developed for traditional database management systems don't seem to be suitable for the big data environment because of the large data volumes, heterogeneity, and the need for rapid data analysis.

4.2.2 Privacy v/s security

Although they are closely linked ideas, privacy and security are not the same. The term "privacy" refers to a person's right to manage their information and personal data, making sure that it is not accessed, utilized, or revealed without authorization. Confidentiality, anonymity, and the capacity to restrict access to sensitive data are all included. Security, on the other hand, is concerned with defending networks, systems, and data against intrusions, assaults, and unwanted access [6]. It consists of safeguards against data loss, theft, or damage, including access restrictions, firewalls, authentication, and encryption. Security deals with preventing external threats and vulnerabilities to the information, whereas privacy concerns an individual's rights to that information.

4.3 ETL (EXTRACT, TRANSFORM, AND LOAD)

Large volumes of data are being generated and gathered by businesses and government organizations, which presents more opportunities to comprehend data processing across several domains. However, there are dangers to user privacy associated with this increase in the usage of big data. Big data analytics and mining operations nowadays frequently struggle to maintain compliance with privacy laws and regulations. Even when apps and privacy laws change, developers must make sure their products respect privacy agreements and protect sensitive data [7]. In order to improve privacy protection in data-driven systems, formal methodologies and testing processes need to be advanced. These difficulties are brought to light by addressing them.

1. **Pre-Hadoop Data Validation and Privacy:** This stage encompasses data loading and initial privacy-related actions. Sensitive data, potentially used for individual identification, is explicitly defined within privacy requirements. Data retention policies and schema limitations are also established during this phase [8].
2. **MapReduce and Privacy Constraints:** The MapReduce process optimizes large datasets for efficient query processing. Privacy regulations may impose limitations on data sharing between processing stages and mandate minimum record counts for specific data values.
3. **Validation of ETL Processes:** The warehouse justification for compliance with privacy requirements should be confirmed at this point, just like in step (2). Certain data values may be omitted from the warehouse or aggregated anonymously if they show a high likelihood of identifying specific people.
4. **Privacy Considerations for Data Reporting:** Purpose-based privacy terms are essential for ensuring that sensitive information is not disclosed for purposes other than those for which it is intended [9].
5. **Privacy Conformance Testing in ETL:** Figure 4.2 illustrates novel privacy conformance testing methodologies aligned with the ETL process stages: Extract, Transform, and Load.

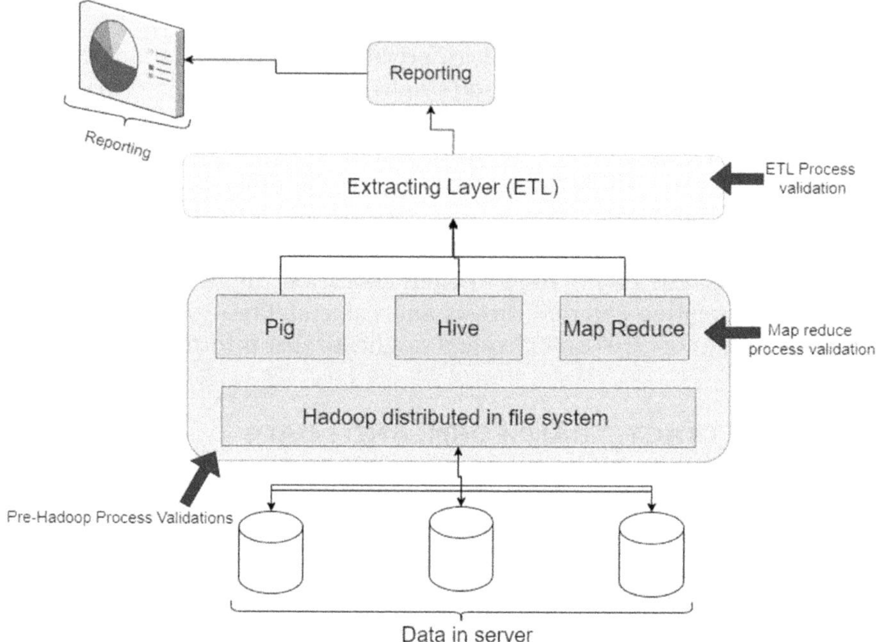

Figure 4.2 Novel privacy compliance testing frameworks aligned with the Extract, Transform, and Load (ETL) stages within the context of big data architecture and testing.

4.4 BIG DATA PRIVACY DURING THE DATA CREATION STAGE

Data generation can be categorized into active and passive methods. Passive data generation occurs unintentionally during online activities, often without user awareness, while active data generation involves deliberate data provision by the data owner to a third party [10]. Reduction of the possibility of privacy infringement during the collection of data can be achieved through fabrication or access restrictions.

1. Restrictions on access: The data owner will not supply such data if it believes that it may reveal private information that should not be disclosed. Several steps could be taken to protect privacy if the data owner is providing the information passively, including the use of encryption tools, ad or script blockers, and anti-tracking extensions [11].
2. Data Fabrication and Masking: In scenarios where sensitive data protection is compromised, data fabrication techniques can be employed to obscure original information before external access. This obfuscation renders the data less susceptible to exploitation. Tools like Socketpuppets further enhance data privacy by creating multiple digital identities for

a single individual, thereby dispersing personal information across various virtual personas and hindering identification [12]. Identity masking is possible with several security technologies, including Mask Me. This is particularly helpful when the data owner has to provide credit card information for online purchases.

4.5 BIG DATA PRIVACY IN DATA STORAGE PHASE

Advancements in data storage technologies, like cloud computing, have made storing large amounts of data easier. However, if a big data storage system is compromised, it can lead to serious consequences like exposing people's personal information. In distributed environments, applications often need data from different data centers, which can pose challenges for privacy protection [13].

To protect data, there are four main security categories: database-level security, file-level security, application-level encryption, and media-level security. Considering the characteristics of big data analytics (volume, velocity, and variety), storage systems should be scalable and able to adapt to different applications. Storage virtualization is a promising technology that combines multiple network storage devices into what seems like a single device, which helps meet these requirements.

One notable model for cloud data security is SecCloud, which focuses on both computation auditing security and data storage security. However, discussions about the privacy of data stored in the cloud are limited, indicating a need for more attention to this aspect [14].

4.6 METHODS FOR PRESERVING PRIVACY WHEN STORING DATA ON CLOUD

Cloud Data Security and Privacy: Safeguarding data in cloud storage necessitates a focus on confidentiality, integrity, and availability. The first two are paramount for privacy protection, as breaches can compromise sensitive information. Ensuring authorized access aligns with availability requirements [15]. To uphold individual privacy in big data storage systems, encryption techniques like public key encryption (PKE) are employed to restrict data access to authorized parties. Here are some approaches to safeguard user privacy in cloud-stored data:

- Attribute-based Encryption: It manages access by utilizing user identity, allowing full control over designated resources.
- Homomorphic Encryption: This technique allows computations to be performed on encrypted data without requiring decryption, enhancing its utility in identity-based (IBE) or attribute-based (ABE) encryption systems.

- Storage path encryption: It secures the storage of large data on cloud platforms.
- Hybrid clouds: It combines private and public cloud services, providing flexibility and security for organizations.

4.6.1 Integrity verification of big data storage

Cloud-based big data storage raises concerns about data control and security. Data owners may relinquish control over their data, necessitating trust in the cloud service provider to adhere to the agreed-upon storage conditions.

Verifying Data Integrity in the Cloud: To safeguard privacy and data integrity in cloud storage, robust verification mechanisms are essential [16]. Traditional methods like checksums, Reed-Solomon codes, trapdoor hash functions, digital signatures, and message authentication codes can be employed for this purpose.

One common approach is to retrieve all data from the cloud to verify its integrity. However, this can be inefficient. Alternatively, integrity verification schemes can be used where the cloud server provides evidence of data integrity only when all data are intact. Regular integrity verification is recommended for the highest level of data protection.

4.6.2 Big data privacy preserving in data processing

In big data processing, privacy protection is crucial and is approached in two distinct phases. The initial phase focuses on preventing unauthorized disclosure of sensitive information, employing robust measures like encryption, access control, and anonymization to safeguard data from unauthorized access or disclosure. This proactive approach ensures that data remains secure during storage, transmission, and processing, mitigating the risk of privacy breaches. In the subsequent phase, the emphasis shifts towards extracting meaningful insights and patterns from the data while preserving privacy [17]. Techniques such as data masking, differential privacy, and privacy-preserving algorithms are employed to ensure that analysis can be performed without compromising the confidentiality of individual data points. By integrating these privacy measures seamlessly into the data processing pipeline, organizations can harness the full potential of big data analytics while upholding privacy standards and regulatory compliance.

4.7 APPROACHES FOR BIG DATA THAT PROTECTS PRIVACY

An overview of a few classic techniques for protecting privacy in large data sets is given here. There is some privacy provided by these approaches that have been in use historically, but their drawbacks have prompted the development of other techniques.

4.7.1 De-identification

De-identification is a fundamental technique used in privacy-preserving data mining, where data is sanitized by generalizing quasi-identifiers and suppressing certain values before being released for analysis. This approach aims to protect individual privacy by making it harder to re-identify individuals from the data. Concepts like t-closeness, l-diversity, and k-anonymity have been introduced to enhance traditional de-identification methods and mitigate re-identification risks. However, in the context of big data analytics, de-identification faces challenges as attackers can leverage external information to re-identify individuals, increasing privacy risks [18]. Therefore, relying solely on de-identification may not be sufficient for safeguarding big data privacy.

Privacy-preserving big data analytics confronts significant challenges, including flexibility, effectiveness, and the risk of re-identification. Developing efficient privacy-preserving algorithms is crucial for mitigating re-identification risks and enhancing de-identification processes. Key concepts in this domain include identifier attributes, quasi-identifier attributes, sensitive attributes, equivalence classes, and insensitive attributes, which underpin the implementation of privacy protection measures.

4.7.2 k-Anonymity

A dataset is considered k-anonymous if it is impossible to uniquely identify an individual within the dataset based on the information contained therein [19]. A database, represented as a table with n rows and m columns, holds records pertaining to individual members of a population. In the context of k-anonymization challenges, the rows do not necessarily need to be unique. The columns represent the attributes associated with each population member. Table 4.2 shows a non-anonymized database containing patient records from a hypothetical hospital in Hyderabad. This dataset includes six attributes and ten records. There are two common methods for achieving k-anonymity for a given value of k.

1. Repression: This approach substitutes an asterisk (*) for some attribute values. A column's values can have all or part of its values replaced with '*'. All of the values in the "Name" and "Religion" attributes in the anonymized Table 4.3 should be changed to a "*"
2. Explanation: With this approach, a more inclusive category takes the place of an attribute's distinct values. For example, "≤20" may replace the value "19" in the attribute "Age," "20 < age ≤ 30" could replace the value "23" and so on.

When it comes to the attributes "Gender," "Age," and "State of domicile," Table 4.3 has 2-anonymity since there are always at least two rows in the table that have the precise combination of these qualities for every row in

Table 4.2 A database that is not anonymized and contains the patient records

Name	Age	Gender	State of domicile	Religion	Disease
Jhonny	20	Male	Andhra Pradesh	Christian	Viral Infection
Badri	24	Male	Karnataka	Buddhist	TB
Kishor	28	Male	Karnataka	Hindu	Heart-related
Salini	27	Female	Tamil Nadu	Muslim	TB
Yamini	23	Female	Karnataka	Hindu	Viral Infection
John	16	Male	Andhra Pradesh	Christian	Heart-related
Ramu	19	Male	Tamil Nadu	Hindu	Cancer
Sunny	28	Male	Andhra Pradesh	Parsi	No Illness
Joshna	23	Female	Andhra Pradesh	Christian	Heart-related
Ramya	28	Female	Andhra Pradesh	Hindu	Cancer

Table 4.3 Anonymity in regard to the characteristics "Age," "Gender," and "Domicile State"

Name	Age	Gender	State of domicile	Religion	Disease
*	≤20	Male	Andhra Pradesh	*	Viral Infection
*	≤20	Male	Andhra Pradesh	*	Cancer
*	≤20	Male	Andhra Pradesh	*	Heart-related
*	20–30	Male	Karnataka	*	TB
*	20–30	Female	Andhra Pradesh	*	Heart-related
*	20–30	Male	Karnataka	*	No Illness
*	20–30	Female	Andhra Pradesh	*	Viral Infection
*	20–30	Male	Karnataka	*	Heart-related
*	20–30	Female	Tamil Nadu	*	TB
*	20–30	Female	Tamil Nadu	*	Cancer

the database. "Quasi-identifiers" are the characteristics that an enemy can access. A dataset with k-anonymity has at least k entries containing each "quasi-identifier" tuple. Attacks such as the temporal assault, the unsorted matching attack, and the complementary release attack can still be applied on k-anonymous data.

To obtain this level of secrecy, a group of size k must withhold a minimal quantity of information, and the optimization issue is NP-hard. Similarly, suppressing attributes, rather than individual entries, is an NP-hard problem. As a result, we adopt the l-diversity technique for data anonymization.

4.7.3 l-Diversity

Data privacy, an important technique called l-diversity, emerges as a shield against intrusive data mining while preserving the usability of the data. Picture a scenario where a dataset contains sensitive information about individuals, such as their medical conditions or age groups. The k-anonymity model is a foundational concept in data anonymization, ensuring that groups of individuals with similar attributes are indistinguishable from one another. However, it has its limitations, particularly when it comes to protecting sensitive values within these groups [20].

This is where the l-diversity model steps in, extending the principles of k-anonymity but with a refined approach. Instead of merely grouping individuals based on similarity, l-diversity focuses on the diversity of sensitive attributes within these groups. For instance, consider a group of patients who share the same age range and medical condition. With l-diversity, it ensures that their medical conditions vary sufficiently to safeguard their identities effectively.

However, the implementation of l-diversity poses challenges. When the dataset lacks diversity in sensitive attributes, the model may introduce fabricated or fictitious data to enhance security. While this enhances privacy protection, it can complicate data analysis and interpretation. Moreover, l-diversity is susceptible to certain attacks, such as skewness and similarity attacks, which can compromise the confidentiality of attributes.

In essence, l-diversity stands as a crucial tool in the arsenal of data privacy techniques, yet it requires careful consideration and management to strike a balance between privacy protection and data utility, ensuring both the security of sensitive information and the efficacy of data analysis.

4.7.4 t-Closeness

The t-closeness model is a refined approach to group-based anonymization aimed at enhancing data privacy by reducing data granularity. This reduction involves a trade-off where some efficiency in data management or mining algorithms is sacrificed to enhance privacy [21].

Unlike its predecessor, l-diversity, t-closeness treats attribute values differently by considering their distribution within the dataset. An equivalence class is deemed to have t-closeness if the difference between the representation of a sensitive attribute within that class and its overall distribution across the dataset is below a threshold value, denoted as t. Achieving t-closeness across all equivalence classes in a dataset is the goal.

One key advantage of t-closeness is its ability to prevent attribute disclosure, a crucial aspect of data privacy protection. However, challenges arise as datasets grow in size and diversity, increasing the risk of re-identification.

From a computational standpoint, achieving t-closeness can be demanding. The brute-force approach, which exhaustively examines every possible dataset partition to find the optimal solution, incurs significant computational costs. Efforts to improve this have resulted in advancements, but achieving polynomial time complexity remains a challenging task.

t-Closeness represents a significant step forward in data privacy, albeit with computational complexities and trade-offs that must be carefully considered in its implementation within modern data anonymization strategies.

4.7.5 Comparative analysis of de-identification privacy methods

While advanced data analytics can extract useful information from large datasets, there is a significant risk to user privacy. Many strategies to protect privacy before, during, and after the big data analytics process have been put forth. Three privacy techniques—k-anonymity, l-diversity, and t-closeness—are covered in this chapter [22]. With the exponential growth of consumer data and the constant advancement of technologies, there will be an increasing trade-off between violating and protecting privacy. Table 4.4 lists current privacy-preserving de-identification techniques along with their shortcomings for large data.

Table 4.4 The limitations of the current de-identification privacy-preserving techniques in big data

Sr. no	Privacy measure	Definitions	Limitations	Computational complexity
1	k-anonymity	A framework for creating and assessing information-release algorithms and systems that minimize the disclosure of sensitive information about entities needing protection.	Susceptible to homogeneity attacks and the exploitation of background knowledge.	$O(k \log k)$
2	l-diversity	An equivalence class is deemed to have l-diversity if the sensitive attribute includes at least "well-represented" values. A table achieves l-diversity when each of its equivalence classes satisfies this requirement.	l-diversity might not completely prevent attribute disclosure and can be difficult or sometimes unnecessary to implement.	$O((n^2)/k)$
3	t-closeness	An equivalence class is considered t-close if the distribution of sensitive data within it closely resembles the overall distribution of that data in the entire dataset. This similarity is measured by a threshold value, T. A dataset is considered t-close if all its equivalence classes meet this t-closeness criterion.	t-closeness mandates that the distribution of sensitive data within each group of similar records closely aligns with the overall distribution of that data in the entire dataset.	$2O(n)O(m)$

4.7.6 HybrEx

In cloud computing, the hybrid execution paradigm is a model for privacy and confidentiality. It uses public clouds exclusively for safe operations while integrating an organization's private cloud; in other words, it only uses public clouds for an organization's non-sensitive data and computation that are classified as public, while the model uses an organization's private cloud for sensitive, private data and computation [23]. Prior to a job's execution, it takes data sensitivity into account. It offers safety integration. HybrEx MapReduce facilitates new application types that use both public and private clouds in the following four areas:

1. Hybrid map: As seen in Figure 4.3a, the reduce phase is only carried out in one of the clouds, but the map phase is carried out in both the public and private clouds.
2. Dividers that are vertical: It can be seen in Figure 4.3b. Using public data as the input, map and reduce activities are carried out on the public cloud. The outcome is stored there, and intermediate data is shuffled among the processes. With private data, the same tasks are completed in the private cloud. Each job is handled separately.
3. Dividends that are horizontal: As shown in Figure 4.3c, the reduction phase is carried out on a private cloud, while the map phase is only carried out on public clouds.
4. Mixture: Both the map phase and the reduction phase are carried out on public and private clouds, as seen in Figure 4.4. Additionally, data communication between clouds is feasible.

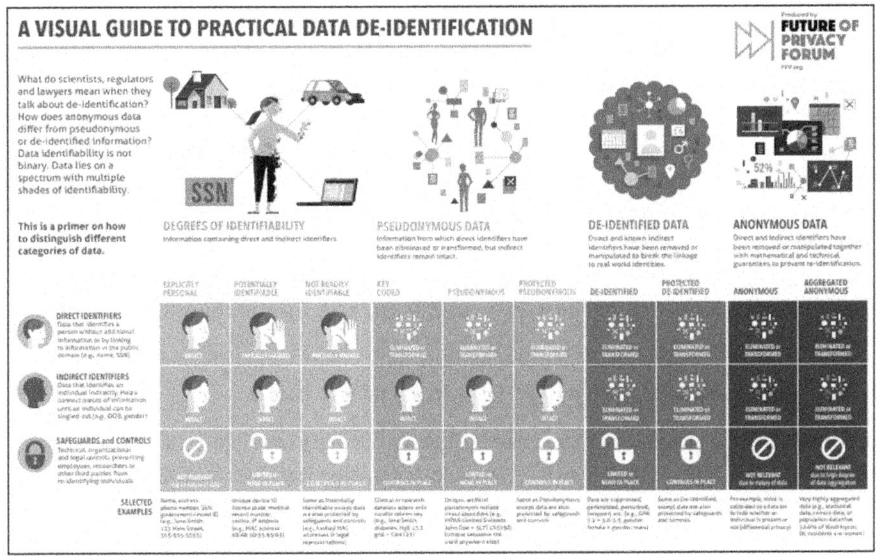

Figure 4.3 Practical data de-identification.

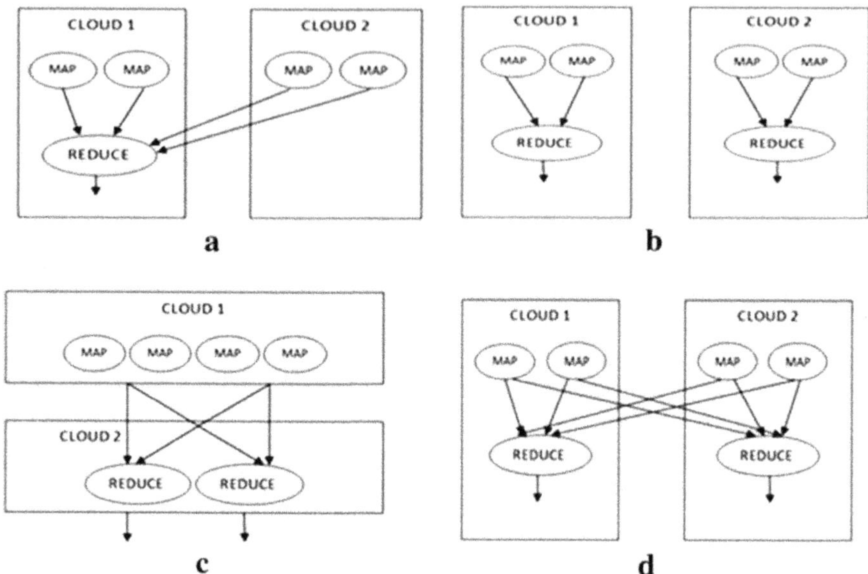

Figure 4.4 Methods for dividing data vertically, horizontally, and hybridly are called HybrEx. The following four categories illustrate how HybrEx MapReduce opens up new application types that make use of both public and private clouds.

4.7.7 Privacy-preserving aggregation

The foundation of privacy-preserving aggregation [24] is homomorphic encryption, which is a widely used method of gathering data for event statistics. Different sources can encrypt their unique data into cipher texts using the same public key if they have access to a homomorphic PKE technique. These encrypted texts can be combined, and the combined outcome can be obtained by using the matching private key. Aggregation, however, has a special purpose. Thus, during periods of massive data collection and storage, privacy-preserving aggregation can safeguard individual privacy. Considering its rigidity.

It can't do intricate data mining to take advantage of fresh insights. For massive data analytics, privacy-preserving aggregation is therefore insufficient.

4.7.8 Operations employing encrypted data

In modern big data analytics, the imperative to preserve individual privacy has catalyzed the exploration of operations conducted over encrypted data. This paradigm shift is exemplified by the ability to execute searches and analyses directly on encrypted datasets, a testament to the evolving landscape of privacy-preserving techniques.

Operating procedures over encrypted data entail performing computations or analyses on data that has been encrypted using cryptographic algorithms. This encryption shields the underlying data from unauthorized access or disclosure, ensuring confidentiality and integrity throughout processing and analysis. For instance, consider a financial institution encrypting transaction data to prevent unauthorized access to sensitive financial information.

However, despite the robust privacy protections afforded by encryption, the implementation of operating procedures over encrypted data is often perceived as inefficient within the context of big data analytics. This inefficiency arises due to several technical challenges inherent in working with encrypted data, particularly in high-volume and dynamic big-data environments.

One of the primary challenges is the computational overhead introduced by encryption and decryption operations. Encrypted data must be decrypted before performing operations such as searching, sorting, or aggregation, leading to increased processing time and resource utilization. This overhead becomes more pronounced as the volume of data and complexity of operations grow, impacting the scalability and responsiveness of analytics workflows.

Moreover, traditional encryption methods may not fully align with the distributed and parallel processing architectures commonly used in big data analytics platforms. Decrypting data in a distributed environment introduces coordination and synchronization overhead, further exacerbating performance issues.

To address these challenges, researchers have developed advanced cryptographic techniques such as homomorphic encryption and secure multiparty computation. These techniques enable computations to be performed directly on encrypted data without the need for decryption, preserving data privacy while mitigating the computational overhead associated with traditional encryption approaches.

Despite these advancements, optimizing the efficiency of operations over encrypted data in big data analytics remains an ongoing area of research and development. Balancing the imperatives of privacy protection and computational performance is crucial for realizing the full potential of privacy-preserving analytics in the era of big data.

4.8 EXISTING SOFTWARE AND THEIR DESIGN PITFALLS TO PRESERVE PRIVACY IN BIG DATA

Apache Hadoop

- Strengths: Popular for its scalability and ability to process large datasets.
- Pitfalls: Lacks built-in privacy-preserving mechanisms, making it challenging to ensure data privacy without significant customization. Security

measures focus more on access control rather than data anonymization or encryption during processing.

Google BigQuery

- Strengths: Offers robust data analysis capabilities with integrated machine learning.
- Pitfalls: While it provides data encryption, there are limitations in privacy-preserving data processing techniques like differential privacy. The reliance on user-defined configurations can lead to inconsistent privacy protection.

Amazon Redshift

- Strengths: Efficient for data warehousing with strong encryption mechanisms.
- Pitfalls: Privacy preservation during data processing is not inherently robust. The platform requires additional layers of privacy-preserving algorithms that are not native to its architecture, leading to potential vulnerabilities.

IBM Db2

- Strengths: Comprehensive database management with strong encryption.
- Pitfalls: Like other systems, it lacks built-in support for advanced privacy-preserving techniques like homomorphic encryption or federated learning, which are essential for modern big data privacy requirements.

Microsoft Azure SQL Database

- Strengths: High security with advanced threat protection and encryption.
- Pitfalls: While security features are advanced, privacy preservation, particularly in real-time data processing and anonymization, requires additional custom development and integration, increasing complexity and risk of errors

REFERENCES

1. Chamikara, M. A. P., Bertok, P., Liu, D., Camtepe, S., & Khalil, I. (2020). Efficient privacy preservation of big data for accurate data mining. Information Sciences, 527, 420–443.
2. Wang, T., Zheng, Z., Rehmani, M. H., Yao, S., & Huo, Z. (2018). Privacy preservation in big data from the communication perspective—a survey. IEEE Communications Surveys & Tutorials, 21(1), 753–778.
3. Pramanik, M. I., Lau, R. Y., Hossain, M. S., Rahoman, M. M., Debnath, S. K., Rashed, M. G., & Uddin, M. Z. (2021). Privacy preserving big data analytics:

a critical analysis of state-of-the-art. Wiley Interdisciplinary Reviews: Data Mining and Knowledge Discovery, 11(1), e1387.

4. Hosseini, M., Wieczorek, M., & Gordijn, B. (2022). Ethical issues in social science research employing big data. Science and Engineering Ethics, 28(3), 29.

5. Mantelero, A. (2018). AI and big data: a blueprint for a human rights, social and ethical impact assessment. Computer Law & Security Review, 34(4), 754–772.

6. Richardson, S. M., Petter, S., & Carter, M. (2021). Five ethical issues in the big data analytics age. Communications of the Association for Information Systems, (1), 18.

7. Nwokeji, J. C., & Matovu, R. (2021). A systematic literature review on big data extraction, transformation and loading (ETL). In Intelligent Computing: Proceedings of the 2021 Computing Conference, Volume 2 (pp. 308–324). Springer International Publishing.

8. Sabtu, A., Azmi, N. F. M., Sjarif, N. N. A., Ismail, S. A., Yusop, O. M., Sarkan, H., & Chuprat, S. (2017, July). The challenges of Extract, Transform and Loading (ETL) system implementation for near real-time environment. In 2017 International Conference on Research and Innovation in Information Systems (ICRIIS) (pp. 1–5). IEEE.

9. Vassiliadis, P., & Simitsis, A. (2009). Extraction, transformation, and loading. Encyclopedia of Database Systems, 10, 14.

10. Koo, J., Kang, G., & Kim, Y. G. (2020). Security and privacy in big data life cycle: a survey and open challenges. Sustainability, 12(24), 10571.

11. Xu, L., Jiang, C., Wang, J., Yuan, J., & Ren, Y. (2014). Information security in big data: Privacy and data mining. Ieee Access, 2, 1149–1176.

12. Archana, R. A., Hegadi, R. S., & Manjunath, T. N. (2018). A study on big data privacy protection models using data masking methods. International Journal of Electrical and Computer Engineering, 8(5), 3976.

13. Mazumdar, S., Seybold, D., Kritikos, K., & Verginadis, Y. (2019). A survey on data storage and placement methodologies for cloud-big data ecosystem. Journal of Big Data, 6(1), 1–37.

14. Goel, P., Patel, R., Garg, D., & Ganatra, A. (2021, May). A review on big data: privacy and security challenges. In 2021 3rd International Conference on Signal Processing and Communication (ICPSC) (pp. 705–709). IEEE.

15. Karthiban, K., & Smys, S. (2018, January). Privacy preserving approaches in cloud computing. In 2018 2nd International Conference on Inventive Systems and Control (ICISC) (pp. 462–467). IEEE.

16. Zhou, L., Fu, A., Yu, S., Su, M., & Kuang, B. (2018). Data integrity verification of the outsourced big data in the cloud environment: a survey. Journal of Network and Computer Applications, 122, 1–15.

17. Lu, R., Zhu, H., Liu, X., Liu, J. K., & Shao, J. (2014). Toward efficient and privacy-preserving computing in big data era. IEEE Network, 28(4), 46–50.

18. Lee, H. J., Cho, S. H., Seong, J. W., Lee, S., & Lee, W. (2020, February). De-identification and privacy issues on bigdata transformation. In 2020 IEEE International Conference on Big Data and Smart Computing (BigComp) (pp. 514–519). IEEE.

19. Andrew, J., & Karthikeyan, J. (2021). Privacy-preserving big data publication: (K, L) anonymity. In Intelligence in Big Data Technologies—Beyond the Hype: Proceedings of ICBDCC 2019 (pp. 77–88). Springer Singapore.

20. Muttoo, S. K., & Singhal, N. (2023). A novel privacy-preserving technique using steganography and l-diversity for multi-relational educational dataset. International Journal of Information Technology, 15(6), 3307–3325.

21. Li, N., Li, T., & Venkatasubramanian, S. (2006, April). t-closeness: Privacy beyond k-anonymity and l-diversity. In 2007 IEEE 23rd International Conference on Data Engineering (pp. 106–115). IEEE.
22. Begum, S. H., & Nausheen, F. (2018, January). A comparative analysis of differential privacy vs other privacy mechanisms for big data. In 2018 2nd International Conference on Inventive Systems and Control (ICISC) (pp. 512–516). IEEE.
23. Ko, S. Y., Jeon, K., & Morales, R. (2011). The {HybrEx} Model for Confidentiality and Privacy in Cloud Computing. In 3rd USENIX Workshop on Hot Topics in Cloud Computing (HotCloud 11).
24. Silva, L. V., Marinho, R., Vivas, J. L., & Brito, A. (2017, April). Security and privacy preserving data aggregation in cloud computing. In Proceedings of the Symposium on Applied Computing (pp. 1732–1738).

Chapter 5

Machine learning-based intrusion detection system

Chirag Joshi, Ranjeet K Ranjan,
and Abhishek Raghuvanshi

5.1 INTRODUCTION

An intrusion detection system (IDS), whether it's a hardware device or a software program, plays a crucial role in monitoring networks and systems for potentially harmful actions or breaches of established rules. Whenever such intrusive behavior or rule violations occur, these incidents are typically logged in a central repository through an Security information and event management (SIEM) system, and they may trigger notifications to an administrator, or sometimes both.

To distinguish genuine threats from false alarms, SIEM systems gather and analyze data from various sources and incorporate mechanisms for filtering out unnecessary alerts. While the primary function of an IDS is to detect and report anomalies in network or system activity, certain systems are also capable of taking action against suspicious activities.

An Intrusion Prevention System (IPS) also monitors network packets to identify potentially harmful network activities, much like an IDS does. However, its primary goal is to not only identify but also proactively prevent attacks once they have been recognized. While IDSs aim to spot irregularities to catch hackers before they inflict significant damage on a network, intrusion prevention systems focus on thwarting attacks.

IDSs come in two primary forms: host-based and network-based. Network-based IDSs are positioned on the network itself, while host-based IDSs are installed on individual client computers. Both of these IDSs function by scanning for signs of known attacks or deviations from typical behavior. They scrutinize alterations or anomalies higher up the protocol and application layers in the network stack. The implementation of an IDS can take the form of a network security appliance or a software program running on client hardware.

DOI: 10.1201/9781032657264-5

5.1.1 Types of intrusion detection system

Different types of IDSs are developed based on the way they are detecting any kind of intrusion. The following are the types of intrusion detection:

a. Network Intrusion Detection System: It is strategically positioned within the network to monitor both incoming and outgoing traffic from all network-connected devices. This ensures comprehensive surveillance of network activities.
b. Host Intrusion Detection System (HIDS): Each host linked to the network, providing direct internet access or internal network connectivity within a business, operates a HIDS. It is capable of identifying not only external threats but also internal risks, like instances where malware attempts to propagate from the host to other systems.
c. Signature-Based Intrusion Detection System (SIDS): Much like antivirus software, this system continuously monitors network traffic, scrutinizing each packet as it traverses the network. It then cross-references these packets with a database of attack signatures or attributes associated with previously identified and documented security threats.
d. Anomaly-Based Intrusion Detection System (AIDS): In response to the rapid generation of new malware, AIDS was developed to detect previously unknown and emerging threats. This approach utilizes machine learning (ML) to create a robust model of normal network and system activity. It then compares incoming data to this model, flagging anything that deviates from the expected behavior as suspicious. The strength of ML-based techniques lies in their adaptability, as these models can be trained to align with specific applications and hardware configurations, providing a more versatile defense mechanism compared to SIDS.

There are many methods available for the detection of Intrusion in a computer system. Starting from signature-based techniques to ML techniques are available for the detection of intrusion.

5.1.2 Motivation

A variety of methodologies, from signature-based to ML, have been put forward in recent years for the identification of intrusions. Because each intrusion dataset has a unique set of traits, signature-based approaches are ineffective when there are many such bots. However, some intrusion datasets began to expand and create several instances, making them harder to identify. As a result, approaches that can identify an intrusion during a zero-day assault are needed. Machine learning (ML) techniques have made major contributions to the identification of zero-day attacks. A dataset is typically used in ML methodologies to train and test the model, which is subsequently used to detect new attacks. Even though there are several approaches for intrusion identification, there is still much room for performance improvement. The improvements may be implemented by modifying a few hyperparameters in ML models and by using an effective feature selection method. In this chapter, we will be presenting an experimental analysis of three intrusion datasets using ML techniques.

5.1.3 Objective of the chapter

The main objective of this chapter is to provide an experimental analysis of various datasets and their behaviour after the application of ML techniques. The chapter will include:

1. Analysis of different Intrusion-based Datasets.
2. Experimental Analysis of ML Techniques.
3. Comparison of ML Techniques based on different datasets.
4. We will be using three different datasets, i.e., CTU-13, IoT-23, and CIC-IDS2017 for the experimental comparison in this chapter.

The rest of the chapter is organized as: In Section 5.2, we have added the existing work related to the chapter. Section 5.3 discusses the methodology used in this work. Section 5.4 discusses the information related to the datasets. Section 5.5 contains the results and analysis and Section 5.6 discusses the conclusion of the chapter.

5.2 EXISTING STUDIES

Cybersecurity has seen a surge in the use of ML in recent years. The ML-based algorithms have been used for a variety of tasks, including malware classification, intrusion detection, and anomaly detection. Dong et al. (2018) provided an overview of several ML methods for intrusion identification along with their benefits and drawbacks. Miller and Busby-Earle (2016 examined the function of several ML-based intrusion detection techniques. A thorough overview of the various botnet attack types, ML techniques for botnet and intrusion detection, and Software Defined Network (SDN) have all been proposed by Shinan et al. (2021). The authors identify the new problems and issues after carefully examining the various botnet detection approaches and SDN detection techniques. Hutchison and Mitchell (2007) employ intrusion detection algorithms using this newly developed capacity to stop Cross-Site Scripting (XSS)or denial of service assaults. The goal of this effort is to develop a technique that retains the majority of the distinctive data characteristics with the fewest possible attributes (Halim et al., 2021). The comprehensive experimental findings show that the suggested technique outperforms competing classifiers and creates interpretable fuzzy systems on the KDD-Cup99 intrusion detection benchmark data set (Tsang et al., 2007).

Joshi et al. (2022) presented a review of various Internet of Things (IoT) botnet detection techniques.

Joshi et al. (2023) implemented the ant colony-inspired feature selection for the detection of botnet intrusion and feature engineering. Hijawi et al. (2021) also provided a ML model for detecting Android botnets. Zhao et al. (2013), using botnet traffic, published a P2P botnet detection algorithm. The suggested method uses a correlation-based strategy for feature selection, as well as Bayesian and Decision Tree Classifiers. Joshi et al. (2021b) used fuzzy logic rules to generate new features, and then those features are used for intrusion bot detection. Joshi et al. (2021c) implemented techniques to detect different classes of bot intrusion using various ML and neural networks. In Graphic and Concept (2019), for the purpose of choosing crucial characteristics from the ISOT dataset, the authors exploited graph symmetry. For the purpose of detecting P2P botnets, Singh et al. (2014) proposed a methodology based on big data analytics. Hubballi and Suryanarayanan (2014) presented a survey on false alarm intrusion minimization. Kamesh and Sakthi Priya (2012) provided a thorough assessment in which they examined numerous botnet detection techniques. Joshi et al. (2021a) used ML techniques to detect botnets and then they compared all the results to find a conclusion that k-nearest neighbor (KNN) performs better than the dataset compared to other ML techniques.

Joshi et al. (2020) conducted a comprehensive analysis of different feature engineering algorithms for the detection of intrusion botnets. They applied the feature engineering techniques to identify the best feature set and then implemented the ML algorithms to identify the best techniques for those feature sets. Liu and Lang (2019) proposed a classification system for IDSs that revolves around data objects as the principal dimension for categorizing and enumerating the existing IDS literature based on deep learning and ML. This taxonomy structure is well-suited for researchers in the field of cybersecurity. Our survey begins by outlining the taxonomy and the fundamental concept of IDSs. Zamani and Movahedi (2013) examined some of these plans and evaluated their effectiveness. They have categorized the plans into two categories: computational intelligence (CI) approaches and traditional artificial intelligence (AI) methodologies. We describe how several CI method traits may be applied to create effective IDS. Additionally, Gao et al. (2019) used the ensembled ML techniques for intrusion detection. Singh et al. (2021) performed an analysis on the detection of credit card fraud using ML and DL methods. A multiclass classification of botnet is also performed by Joshi et al. (2021c).

5.3 METHODOLOGY

Computer science's ML field analyzes data to forecast outcomes based on past performance. It is a subfield of AI that uses data and algorithms to gradually learn from and outperform what people do. It uses mathematical

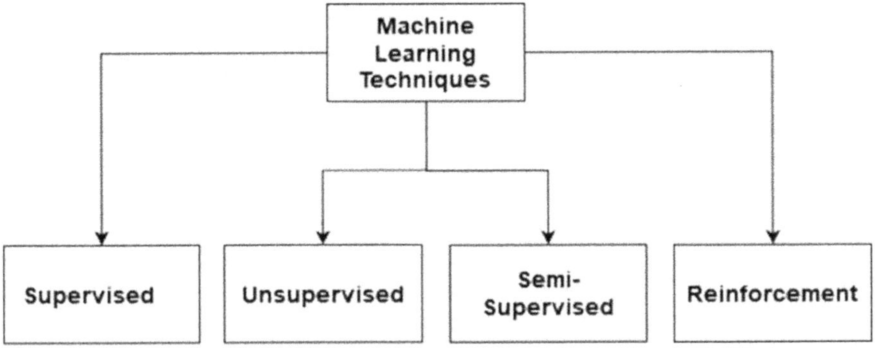

Figure 5.1 Classification of machine learning techniques.

models on a variety of data types to let computers learn without explicit instructions. Statistical methods for categorization or prediction are used to train ML algorithms, providing important data mining insights. Data patterns are discovered using ML algorithms, which are subsequently utilized to create prediction models. It is challenging to build a solution for an environment that is always changing when the data changes over time since the requirements and difficulties are constantly changing. The choice is obviously ML (Batta, 2020). Intrusion assaults may be used utilizing a variety of architectures, as was previously mentioned. Every intrusion has unique traits and an attack strategy that calls for extra care. Numerous dynamic techniques based on signature, behaviour, and traffic flow have been utilized in the past; however, they are not reliable for spotting hidden botnets. A categorization model based on ML can tackle the issue. An intrusion may be automatically detected using an ML-based model, which can evolve over time. We require more advanced approaches based on ML to identify a range of intrusion attacks. Generally speaking, there are four categories into which ML algorithms may be divided: supervised, unsupervised, semi-supervised, and reinforcement learning (RL) (Figure 5.1).

5.3.1 Supervised learning

The most popular learning method is supervised ML, which needs a labelled dataset with input and output variables. The supervised learning approach uses the labelled dataset with input and output variables. Both input and output variables are mapped together with a function. A learning technique then will be used on these variables for prediction. The training and testing will also be done on these two variables. Algorithms for supervised ML are further broken down into Regression and Classification (Burkart & Huber, 2021).

5.3.2 Unsupervised learning

Unsupervised learning is a method that uses data that has not been labeled. Recognizing previously unknown data and patterns is made easier via unsupervised learning. It is comparable to how the human brain learns. For instance, suppose an unsupervised learning task is given a dataset containing images of numerous animals. In this case, grouping the photographs based on attributes of the images, such as edge, texture, color shades, etc., may be accomplished by using an unsupervised learning technique. All the cat photographs in one group and the dog images in other groups should be able to be grouped together using a decent clustering approach (Saxena et al., 2017).

5.3.3 Semi-supervised learning

A semi-supervised approach is often trained using a mix of labeled and unlabeled data. Both huge numbers of unlabeled data and smaller clusters of labelled data may be handled by it. While unsupervised strategies deal with unlabeled data, supervised procedures work with labelled data. Semi-supervised learning techniques are particularly helpful when there is a dearth of labeled data and it may be difficult to create a reliable supervised classifier. On the other hand, knowing that specific data pieces belong to the same category may help the learning strategy for clustering approaches (van Engelen & Hoos, 2020).

5.3.4 Reinforcement learning

A technique called RL enables an agent to learn by doing experiments in a social setting and receiving feedback from its own actions and observations.

A learning agent must rely on its own experience because the learning algorithm is not fed with tagged data. The main objective of an agent in a standard RL algorithm that employs a reward function is to improve performance by maximizing positive rewards. The fundamental parts of a RL system include a policy, a reward signal, and a value function. A policy specifies how a learning agent will operate in a certain situation. The value that identifies an RL problem's objective is known as a reward signal. A value function calculates how beneficial it is for an agent to be in a certain state (Sutton & Barto, 2012).

Various ML techniques are available and out of those techniques we will be implementing below-described ML techniques on the three datasets.

5.3.5 Logistic regression

A logistic function (sometimes known as a sigmoid function) is used in logistic regression. A logistic regression model's output value is regarded as a sample's likelihood of belonging to a class. A kind of ML called logistic regression works with variables and a binomial outcome. Even though this model resembles a basic linear regression model in appearance, the functional form is binomial, making it impossible to assess the variables in the same way. The probability distribution of the number of survivors in a sample of a certain size is typically assumed to be binomial when the response variable is binary, given certain values of the explanatory variables (Bewick et al., 2005).

5.3.6 Support vector machine (SVM)

Popular supervised learning algorithms for tackling classification and regression issues include SVM. An SVM classifier is strengthened by the big margin and experiences fewer misclassifications as a result. The core elements of SVM are the separating hyperplane, maximum-margin hyperplane, soft margin, and kernel function. The amount of features in the dataset determines how difficult a linear SVM is. By employing a linear technique for classification and the kernel function, SVM may produce nonlinear hyperplanes (Pisner & Schnyer, 2019).

5.3.7 Decision tree

A non-parametric supervised learning technique called a decision tree is used to address classification and regression issues. Usually, training samples are iteratively partitioned to form a decision tree. Data splitting creates a hierarchical structure that may be seen as a tree. In a decision tree, the interior nodes of the tree reflect additional qualities while the leaf nodes indicate the class label. For classification and regression, a variety of methods are available. Some of the most well-known algorithms include CHID, CART, ID3, and C4.5, among others (Rana & Pandey, 2021).

5.3.8 K-nearest neighbor

A fresh sample is categorized into a class using the KNN method, a nonparametric supervised learning technique, depending on how closely it resembles training data samples. A test sample is categorized into a class based on which K training samples resemble each other more. The test sample is placed in the class that most closely resembles the majority of the class if there are two or more such classes. There are several distance measurement or similarity measurement approaches that may be used in KNN, depending on the type of features. The method will attempt to calculate the distance between a new data point, represented as a triangle, and its neighbors based on the value of K (Prasath et al., 2017).

In this chapter, we will be implementing the above-mentioned ML techniques on three different datasets and then will compare the results to understand the behaviour of each of the techniques on different datasets and their features.

5.4 DATASET

For our analysis, we have used three different datasets CTU-13, IoT-23, and CIC-IDS2017 for the implementation of ML techniques. All these datasets have contained malware, and all the datasets are updated regularly.

5.4.1 CTU-13

The botnet traffic dataset known as CTU-13 (Garcia et al., 2014) was recorded in 2011 at Czech Technical University in Prague, Czech Republic. The objective of the dataset was to compile a substantial collection of real botnet traffic, blending it with different types of network activity and background data. The dataset comprises 13 distinct captures, often referred to as scenarios, each representing various botnet samples. For each scenario, we executed a unique malware specimen that employed a range of protocols and carried out various operations. The details of each operation were documented in a PCAP file, encompassing all packets associated with the three distinct traffic types. To gain access to additional information such as NetFlows and WebLogs, these PCAP files underwent further analysis.

5.4.2 IoT-23

IoT-23 (Garcia et al., 2020) is a brand-new dataset of IoT device network traffic. The primary aim of this dataset was to create a comprehensive repository of genuine botnet traffic, interwoven with diverse network activities and background data. For each specific scenario, a distinct malware instance was executed, utilizing a range of protocols and executing various tasks. The particulars of each operation were meticulously documented within a PCAP file, encompassing all packets associated with the three distinct types of traffic. To access additional insights, such as NetFlows and WebLogs, in-depth analysis was performed on these PCAP files.

5.4.3 CIC-IDS2017

The CIC-IDS2017 (Sharafaldin et al., 2018) dataset comprises real traffic of various networks and the most recent and benign prevalent assaults (PCAPs). The dataset has various experimental results from network traffic analysis. The flows are categorized into timestamps, source and destination IP, and source and destination ports. The main focus of the generation of this dataset was to add and use some real background traffic which will make it an authenticate dataset.

5.5 RESULT DISCUSSION

We have implemented different ML techniques such as LR, KNN, SVM, and DT on three different datasets. The results are compared using the various performance metrics.

5.5.1 Implementation environment

In this experiment, we have used the Google Colab, an online cloud-based platform for the implementation of different ML techniques. RAM used for the experiment is 6GB and disk space is 50GB. We have also used various Python libraries like SkLearn, Matplotlib, Seaborn, NumPy, and Pandas.

5.5.2 Performance metrics

In this implementation, we have used performance metrics such as confusion matrix, precision, recall, accuracy score, and F1 score for the comparison of various ML techniques' performance on different datasets.

5.5.3 Result analysis

We have implemented Logistic Regression, KNN, SVM, and DT on IoT-23, CTU-13, and CIC-IDS2017:

In Tables 5.1–5.3, we have shown the comparison of different ML techniques.

Table 5.1 Comparison of different machine learning techniques on IoT-23 dataset

ML technique	Dataset	Recall	Precision	F1 score	Accuracy
KNN	IoT-23	99%	98.80%	99.39%	99.34%
Logistic regression		99.61%	97.08%	98.33%	98.19%
Decision tree		Overfit	Overfit	Overfit	Overfit
SVM		92.85%	98.26%	95.4%	95.45%

Table 5.2 Comparison of different machine learning techniques on CTU-13 dataset

ML technique	Dataset	Recall (%)	Precision	F1 score (%)	Accuracy (%)
KNN	CTU-13	99	99%	99.00	99.00
Logistic regression		99.03	98.68%	98.86	98.73
Decision tree		99.91	Overfit	99.95	99.95
SVM		99.89	99.67%	99.78	99.75

Table 5.3 Comparison of different machine learning techniques on CIC-IDS2017 dataset

ML technique	Dataset	Recall (%)	Precision (%)	F1 score (%)	Accuracy (%)
KNN	CIC-IDS2017	96	95.78	97.34	96.38
Logistic regression		94.59	94.32	96.77	97
Decision tree		98	98.67	98.22	98
SVM		95.46	95.45	95.88	97

The result analysis presented in three tables shows that most of the techniques performed well on all three datasets. Decision tree techniques are overfitted on the IoT-23 dataset and KNN achieves the highest accuracy in all the datasets.

We have also used the confusion matrix for analyzing the performance of various ML techniques. In Figures 5.2–Figure 5.5, we have shown some of the confusion matrix.

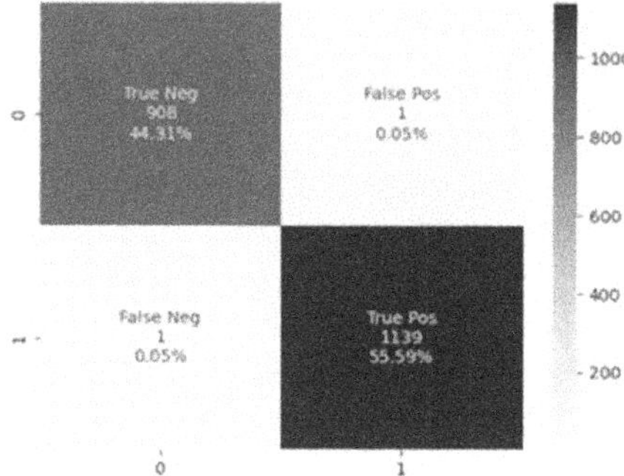

Figure 5.2 CTU-13 confusion Matrix-KNN.

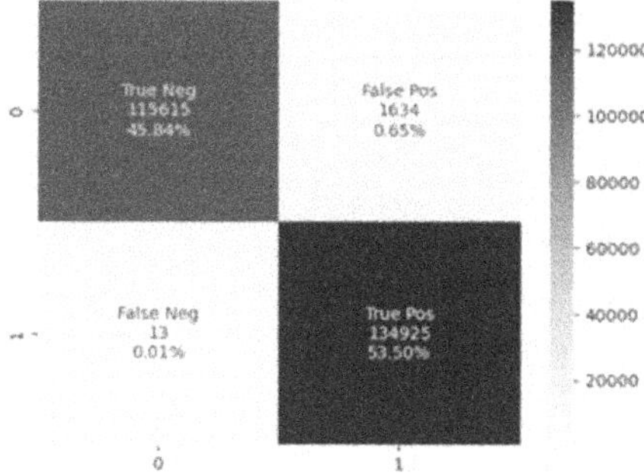

Figure 5.3 IoT-23 confusion Matrix-KNN.

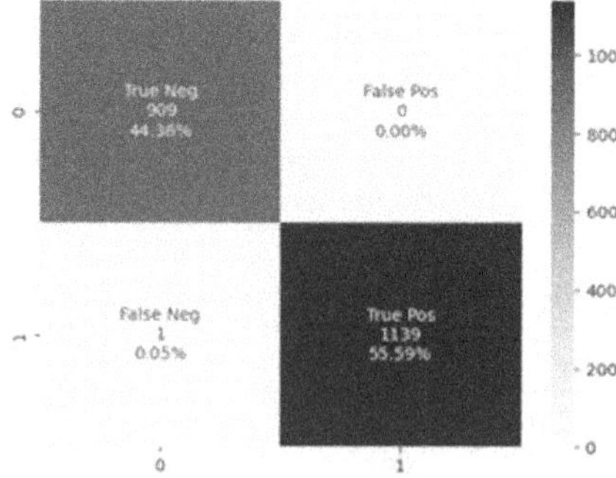

Figure 5.4 CTU-13 confusion Matrix-DT.

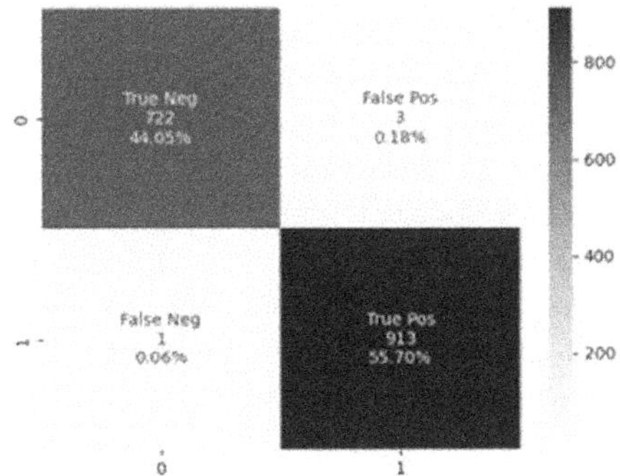

Figure 5.5 CTU-13 confusion Matrix-SVM.

Additionally, comparisons are made between each of the ML approaches employed in this work. The comparison we did in this work has shown that most of the ML techniques performed well for all the datasets used in this work. But KNN, performance best out of all the techniques, decision tree techniques overfitted on two datasets. The experimental comparison will help in understanding the nature of ML techniques on this type of dataset.

5.6 CONCLUSION

In this chapter, we delved into the realm of intrusion detection, employing a diverse array of ML techniques, including Logistic Regression, SVM, KNN, and decision tree models. Our focus was on doing an experimental comparison of three distinct datasets: IoT-23, CTU-13, and CIC-IDS2017 in terms of the accuracy of ML algorithms. The first step in our process involved rigorous data preprocessing, ensuring that the datasets were cleaned and transformed to make them suitable for ML applications.

Subsequently, we applied each of the aforementioned ML techniques to these datasets, seeking to discern their relative performance. The results obtained in our investigation unveiled a compelling outcome: KNN emerged as the standout performer, surpassing its counterparts on all three datasets. This was highlighted by its remarkable achievement of an accuracy rate reaching an impressive 99% on nearly all three datasets.

However, it is essential to note some noteworthy observations regarding the decision tree model. This technique exhibited signs of overfitting when applied to the IoT-23 and CTU-13 datasets. Overfitting occurs when the model learns the training data too well to the point where it struggles to generalize effectively to unseen data. This observation underscores the importance of tuning hyperparameters and possibly applying regularization techniques to mitigate overfitting when employing decision trees.

As we consider the future of this research, several avenues for improvement and exploration come to the fore. First, it would be advantageous to explore different hyperparameter configurations for each of the ML techniques. This could potentially lead to even better results and more robust models. The importance of hyperparameter tuning cannot be overstated, as it can significantly impact the accuracy of ML models.

Furthermore, the realm of feature engineering holds immense potential for enhancing the performance of IDSs. By carefully selecting, transforming, or creating new features from the existing data, we can equip the ML models with more discriminative information. This, in turn, may lead to better detection accuracy and generalization to new, unseen instances of intrusion attempts.

In conclusion, our exploration of ML techniques for intrusion detection, applied to diverse datasets, has yielded valuable insights. KNN stands out as a powerful method for this purpose, but there is still much room for refinement and improvement. The future of this field is ripe with opportunities for experimentation, parameter tuning, and feature engineering, all of which promise to enhance the efficacy of IDSs and fortify our digital defenses.

REFERENCES

Batta, M. (2020). Machine learning algorithms - A review. International Journal of Science and Research, 9(1), 381-undefined. https://doi.org/10.21275/ART20203995

Bewick, V., Cheek, L., & Ball, J. (2005). Statistics review 14: Logistic regression. Critical Care, 9(1), 112–118. https://doi.org/10.1186/cc3045

Burkart, N., & Huber, M. F. (2021). A survey on the explainability of supervised machine learning. Journal of Artificial Intelligence Research, 70, 245–317. https://doi.org/10.1613/JAIR.1.12228

Dong, X., Hu, J., & Cui, Y. (2018). Overview of botnet detection based on machine learning. In Proceedings of the 2018 3rd International Conference on Mechanical, Control and Computer Engineering, ICMCCE 2018, 476–479. https://doi.org/10.1109/ICMCCE.2018.00106

Gao, X., Shan, C., Hu, C., Niu, Z., & Liu, Z. (2019). An adaptive ensemble machine learning model for intrusion detection. IEEE Access, 7, 82512–82521. https://doi.org/10.1109/ACCESS.2019.2923640

Garcia, S., Grill, M., Stiborek, J., & Zunino, A. (2014). An empirical comparison of botnet detection methods. Computers & Security, 45, 100–123.

Garcia, S., Parmisano, A., & Erquiaga, M. J. (2020). IoT-23: A Labeled Dataset With Malicious and Benign IoT Network Traffic. Stratosphere Lab., Praha, Czech Republic, Tech. Rep.

Graphic, U., & Concept, S. (2019). SS symmetry A Feature Extraction Method for P2P Botnet Detection. https://doi.org/10.3390/sym11030326

Halim, Z., Yousaf, M. N., Waqas, M., Sulaiman, M., Abbas, G., Hussain, M., Ahmad, I., & Hanif, M. (2021). An effective genetic algorithm-based feature selection method for intrusion detection systems. Computers & Security, 110, 102448. https://doi.org/10.1016/j.cose.2021.102448

Hijawi, W., Alqatawna, J., Al-Zoubi, A. M., Hassonah, M. A., & Faris, H. (2021). Android botnet detection using machine learning models based on a comprehensive static analysis approach. Journal of Information Security and Applications, 58(February), 102735. https://doi.org/10.1016/j.jisa.2020.102735

Hubballi, N., & Suryanarayanan, V. (2014). False alarm minimization techniques in signature-based intrusion detection systems: A survey. Computer Communications, 49, 1–17. https://doi.org/10.1016/j.comcom.2014.04.012

Hutchison, D., & Mitchell, J. C. (2007). Detection of intrusions and malware. In 4th International Conference, DIMVA 2007 Lucerne, Switzerland, July 12–13: Vol. 4579 LNCS.

Joshi, C., Bharti, V., & Ranjan, R. K. (2020). Analysis of feature selection methods for p2p botnet detection. In Advances in Computing and Data Sciences: 4th International Conference, ICACDS 2020, Valletta, Malta, April 24–25, 2020, Revised Selected Papers 4 (pp. 272–282). Springer Singapore.

Joshi, C., Bharti, V., & Ranjan, R. K. (2021a). Botnet detection using machine learning algorithms. In Proceedings of the International Conference on Paradigms of Computing, Communication and Data Sciences: PCCDS 2020, 717–727.

Joshi, C., Ranjan, R. K., & Bharti, V. (2021b). A fuzzy logic based feature engineering approach for botnet detection using ANN. Journal of King Saud University - Computer and Information Sciences, 34(9), 6872–6882. https://doi.org/10.1016/j.jksuci.2021.06.018

Joshi, C., Ranjan, R. K., & Bharti, V. (2021c). ANN based multi-class classification of P2P botnet. International Journal Of Computing and Digital System, 1(1).

Joshi, C., Ranjan, R. K., & Bharti, V. (2023). ACNN-BOT: An ant colony inspired feature selection approach for ANN based botnet detection. Wireless Personal Communications, 0123456789. https://doi.org/10.1007/s11277-023-10695-8

Joshi, C., Sharma, R., Khyati, & Singh, H. (2022). A study on IoT - botnet detection techniques. A study on IoT -. Botnet Detection Techniques, 5(1), 26.

Kamesh, S. P. N., & Sakthi Priya, N. (2012). A survey of cyber crimes Yanping. Security and Communication Networks, 5, 422–437.

Liu, H., & Lang, B. (2019). Machine learning and deep learning methods for intrusion detection systems: A survey. Applied Sciences, 9(20), 4396. https://doi.org/10.3390/app9204396

Miller, S., & Busby-Earle, C. (2016). The role of machine learning in botnet detection. 359–364.

Pisner, D. A., & Schnyer, D. M. (2019). Support vector machine. In Machine Learning: Methods and Applications to Brain Disorders. Elsevier Inc. https://doi.org/10.1016/B978-0-12-815739-8.00006-7

Prasath, V. B., Alfeilat, H. A. A., Hassanat, A., Lasassmeh, O., Tarawneh, A. S., Alhasanat, M. B., & Salman, H. S. E. (2017). Distance and similarity measures effect on the performance of K-nearest neighbor classifier--a review. *arXiv preprint arXiv:1708.04321*. 1–39. https://doi.org/10.1089/big.2018.0175

Rana, A., & Pandey, R. (2021). A review of popular decision tree algorithms in data mining. Asian Journal of Multidimensional Research, 10(10), 230–237. https://doi.org/10.5958/2278-4853.2021.00837.5

Saxena, A., Prasad, M., Gupta, A., Bharill, N., Patel, O. P., Tiwari, A., Er, M. J., Ding, W., & Lin, C. T. (2017). A review of clustering techniques and developments. Neurocomputing, 267, 664–681. https://doi.org/10.1016/j.neucom.2017.06.053

Sharafaldin, I., Lashkari, A. H., & Ghorbani, A. A. (2018). Toward generating a new intrusion detection dataset and intrusion traffic characterization. In ICISSP 2018 - Proceedings of the 4th International Conference on Information Systems Security and Privacy, 2018-Janua(Cic), 108–116. https://doi.org/10.5220/0006639801080116

Shinan, K., Alsubhi, K., Alzahrani, A., & Ashraf, M. U. (2021). Machine learning-based botnet detection in software-defined network: A systematic review. Symmetry, 13(5), 1–28. https://doi.org/10.3390/sym13050866

Singh, K., Guntuku, S. C., Thakur, A., & Hota, C. (2014). Big data analytics framework for peer-to-peer botnet detection using random forests. Information Sciences, 278, 488–497. https://doi.org/10.1016/j.ins.2014.03.066

Singh, A., Ranjan, R. K., & Tiwari, A. (2021). Credit card fraud detection under extreme imbalanced data: A comparative study of data-level algorithms. Journal of Experimental\& Theoretical Artificial Intelligence, 0(0), 1–28. https://doi.org/10.1080/0952813X.2021.1907795

Sutton, R. S., & Barto, A. G. (2012). Reinforcement learning: An introduction second edition. Learning, 3(9), 322. https://books.google.com/books?id=CAFR6IBF4xYC&pgis=1%5Cnhttp://incompleteideas.net/sutton/book/the-book.html%5Cnhttps://www.dropbox.com/s/f4tnuhipchpkgoj/book2012.pdf

Tsang, C. H., Kwong, S., & Wang, H. (2007). Genetic-fuzzy rule mining approach and evaluation of feature selection techniques for anomaly intrusion detection. Pattern Recognition, 40(9), 2373–2391. https://doi.org/10.1016/j.patcog.2006.12.009

van Engelen, J. E., & Hoos, H. H. (2020). A survey on semi-supervised learning. Machine Learning, 109(2), 373–440. https://doi.org/10.1007/s10994-019-05855-6

Zamani, M., & Movahedi, M. (2013). Machine learning techniques for intrusion detection. ArXiv Preprint ArXiv:1312.2177.

Zhao, D., Traore, I., Sayed, B., Lu, W., Saad, S., Ghorbani, A., & Garant, D. (2013). Botnet detection based on traffic behavior analysis and flow intervals. Computers and Security, 39, 2–16. https://doi.org/10.1016/j.cose.2013.04.007

Chapter 6

Border surveillance

Prominent usage scenarios

*Kirti Raj Bhatele, Devanshu Tiwari,
and Anand Singh Bisen*

6.1 INTRODUCTION

One of the most important aspects of the nation's security is border security. An effective and advanced border security system does the following:

- Protects over the border from the illegal task of weapons, drugs, and people
- Encourages lawful entry and exit from the border.
- Prevents illegal immigration
- Avoids insurgency
- Avoids trans-border crime

The use of artificial intelligence (AI) can greatly enhance border control efforts and mitigate security threats associated with cross-border illegal activities and terrorism. This is part of a larger trend of making a nation's borders more intelligent, which involves creating interconnected and extensive information systems and implementing decentralized methods of information exchange for border and security purposes. At first, these systems may focus on essential functions, but they can be progressively expanded and improved to include a wider range of individuals and process more diverse data, such as biometric information.

Throughout history, countries have readily adopted "new" technologies in order to address the common issue of accurately identifying individuals for the purpose of regulating movement and addressing criminal activity. The use of various identification methods such as passports, body measurements, fingerprints, photography, lie detectors, and face recognition systems has been influenced by the prevailing scientific, social, and political ideologies and worries of the time and place.

There are four major types of AI applications that can be considered in the context of border security as well as border control:

1. Algorithmic risk assessment
2. Detection of emotion.
3. Biometric identification, i.e. faces recognition and automated fingerprint
4. AI tools for migration monitoring, analysis, and forecasting.

DOI: 10.1201/9781032657264-6

The implementation of AI technologies in border control can bring about numerous advantages, including the ability to identify fraud and misconduct, faster access to crucial information for decision-making, and improved protection for vulnerable individuals. However, it is important to consider the potential threats to basic rights that these technologies may present [1]. Biometric technology has varying accuracy across different methods and situations. Even fingerprint identification, which is relatively well-established, faces challenges when collecting biometric data. Factors such as a person's age and the environment can affect the accuracy of the data. Face recognition technology works well in real life when the images are clear and the matching algorithms are good. The quality of these algorithms depends on training data (like images) and techniques used to improve them. Emotion-detection algorithms may not be reliable or scientifically proven. People are worried that some systems for sharing information about borders and security might not have accurate data. People are especially worried about face recognition technology because it could lead to problems with basic rights, like bias and discrimination, data protection, and mass surveillance. Even if AI systems are accurate and fair, they can still have other big risks, like data protection and privacy [2]. The growing reliance on biometric information in information systems heightens the potential for illegal profiling, such as the identification of ethnicity through facial images. Additionally, even if profiling does not rely on biometric or personal information, other forms of data or a combination of different data types used in algorithmic profiling may result in discrimination based on protected characteristics. The current measures in place, such as human-in-the-loop protection (which involves human intervention) and the right to an explanation, may not effectively address these potential dangers.

Improving visibility in critical domains like borders and security is necessary for the development of emotion-detection technologies. Ultimately, a comprehensive comprehension and contemplation of wider considerations such as the origins of technology and societal and political perspectives and expectations would greatly contribute to the advancement and acceptance of powerful AI advancements. Incorporating new technologies without addressing issues like technological determinism and the belief in technological neutrality would erode basic rights, openness, and responsibility.

6.2 BACKDROP OF BORDER SURVEILLANCE AND PEOPLE IDENTIFICATION

Clearly defined physical boundaries, marked by man-made structures such as barriers or fencing, serve as a powerful symbol of a government's authority over a particular area. Although there have been persistent attempts to build and reinforce physical borders like walls and fences, border control is primarily focused on the inside rather than a specific geographical area. The expansion of authority beyond physical boundaries is achieved through various remote control techniques, such as pre-departure enrolment, remote inspections, and digital monitoring. Technologies such as biometric identification, polling, and risk analysis can facilitate the movement of borders, whether they are expanding or contracting [3].

6.3 REASONS FOR IDENTIFYING RISKY AND MIGRANT PEOPLE

Throughout history, the methods used to determine the inhabitants of a region have arisen as a result of the modern state's increasing demand to understand its citizens in order to effectively govern larger and expanding populations [4].

The issue of identifying individuals became more pressing after the Industrial Revolution, as the rise in both domestic and international movements led to the development of unfamiliar and potentially perilous communities.

6.4 IDENTIFICATION TECHNOLOGIES

Over time, innovative methods were developed to distinguish individuals using technology in a distinct and unmistakable manner. These practices were an evolution of earlier methods, like travel passes, which were designed to facilitate the movement of a select few, such as merchants and diplomats, while also regulating the movements of the less privileged and perceived as a threat to society. The introduction of passports after the French Revolution aimed to mainly manage the domestic travel of individuals. Nevertheless, the progressive implementation of passports helped improve border control and to clearly differentiate between citizens (who have the right to enter a country) and non-citizens (who need special permission) [5].

In addition, passports served as a means of regulating departure by either permitting or preventing migration. Due to their simplistic nature, passports and other forms of documentation were prone to misuse. Prior to the creation of dedicated border agencies and infrastructure, passports could potentially be counterfeited or exchanged between individuals, making it difficult to consistently enforce passport inspections and border control. This posed a significant challenge in both verifying the authenticity of documents and confirming the identity of the traveller. While some documents contained physical descriptions of individuals, they were often vague and based on personal interpretation. Before World War 1, different countries had their own passport designs. But after World War 2, countries made a bigger effort to agree on a standard passport format. Since its establishment in 1944, the International Civil Aviation Organization (ICAO) has been instrumental in setting guidelines for travel documents. As an illustration, in 2005, the member countries of ICAO agreed upon a new regulation mandating all nations to provide machine-readable travel documents (MRTDs) by 2020 [6].

6.4.1 Introduction to the MTRD

In 1968, the ICAO of the United Nations began developing MRTD. The year 1980 marked the introduction of a standard that requires a distinct machine-readable zone (MRZ) to be scanned using optical technology. The document includes fundamental details such as the document ID, its expiry date, the individual's name, gender, date of birth, and nationality. Since 1997, ICAO is working on incorporating additional information into various identity documents, such as passports, utilizing biometric technology [6]. To accomplish this, optical storage devices do not have enough space. In order to reduce physical touch and improve convenience, a noncontact integrated circuit (IC) chip was chosen. Its storage capacity enables it to store digital images and biometric patterns and provide authentication through robust encryption methods. The initial edition of the standard was published in 2004 due to the urgency to enhance border security following the 9/11 terrorist attacks. The minimum criteria for MRTDs include a digital signature from the issuing country, a facial photograph, and a digital version of the MRZ. A Public Key Infrastructure is always equipped with a digital signature. The ICAO standard provides:

1. An authentication protocol that relies on passivity to verify the data.
2. A potential basic access control (BAC) measure to safeguard the chip's contents from unauthorized entry.
3. An active authentication (AA) process to demonstrate the authenticity of the chip.

Although BAC does not provide much protection, the standard allows for the advancement of Extended Access Control (EAC). This is currently being developed and standardized in Europe. Likewise, there have been reported vulnerabilities in AA.

Identification and verification at ABC e-gates are dependent on travel documents, making them a crucial aspect of the border crossing procedure for travellers. There is currently no single travel document accepted at all borders globally, including within the EU. Available options include e-passports, Schengen visas, national ID cards, and residence permits, as outlined by ICAO (2006a), ICAO (2015), Visa (2007–2016), BSI (2011), and ICAO (2011). The e-reader's features have been previously outlined by ICAO (2014a). Besides the Schengen agreement, the Council Regulation 2252/2004, dated December 13, 2004, mandates the inclusion of the document holder's facial image and two fingerprints in compatible formats (Article 1). Within a period of 18 months, member states were directed to implement the regulations regarding facial images, and within 36 months for fingerprints, as stated in Article 6. According to this rule, biometrics data is strictly intended for verifying the legitimacy of documents and the identity of the owner (Article 4) and may not be utilized for any other reason. The following section provides a breakdown of the various types of e-passports used in Schengen countries and third country nationals (TCN) borders, as they are the most frequently utilized travel documents in ABC systems. An e-passport is an identification document that includes an RFID chip, holding important biographical and biometric data of the holder. Based on the type of travellers, who will soon undergo verification through RTP or EES (PWC, 2014), and the security authentication procedures, various e-passports are utilized by individuals from different countries. Since August 2006, it has been the responsibility of EU member states to provide their citizens with biometric passports, as mandated by the EU Council's regulation No. 2252/2004 (Council of the European Union, 2004). The initial version of these e-passports, which only included one biometric feature, specifically facial recognition, is now widely accessible. Since June 2009, the e-passport with two biometric techniques has been in use as the second generation. The ICAO has established the specifications for second-generation passports in ID-3 size, which comply with standard ICAO guidelines. The requirements mentioned are outlined in ICAO Document 9303, part one (ICAO, 2006a, 2006b). These specifications are also relevant for TCNs. The Federal Office for Information Security (BSI) specifications TR-03110, parts 1 and 3 (BSI, 2012, 2013) apply to the determination of EAC for e-passports issued by Schengen member states. In accordance with the ICAO guidelines of 2010, upcoming third-generation passports will implement a novel security measure known as supplemental access control (SAC). This advanced system builds upon the existing BAC and defends the contactless chip through encrypted data exchange with the reading device. The upcoming SAC is set to gradually take over for the BAC due to its enhanced security measures. Additionally, the SAC permits the use of a six-digit password and the Card Access Number to access the data saved on the card's RFID chip. The document usually displays the controller area network, which can potentially serve as an alternative to the MRZ in order to access information from the chip. Some airports have started implementing the transition from second to third-generation passports. BAC will continue to be utilized for worldwide compatibility and for the sake of compatibility with older systems. BAC and SAC, the two security measures, will coexist for a period of time [7].

Efforts were made in the criminal justice system to address the issue of identification by identifying and monitoring repeat offenders (recidivists). Anthropometry, the study of measuring human bodies, provided a potential means to establish a consistent identity for individuals. During the 1870s, Alphonse Bertillon, a police officer from France, created the most well-known anthropometric system. Bertillon aimed to gather a set of accurate measurements and details about a prisoner's physical features in order to document their identity, organize the data, and keep it on record for future comparisons. By the late 1800s, numerous nations in Europe and other regions had implemented the "Bertillonage system" [8].

6.4.2 Biometric identification

6.4.2.1 The discovery of fingerprints

Although there is evidence to suggest that fingerprints were used to identify individuals as far back as 4000 years ago, it wasn't until the late 19th century that the first identification systems were created. These systems, developed by William Herschel and Henry Faulds, involved a structured and consistent method of collecting and analysing fingerprints. Faulds, who was a British colonial, also proposed a system for organizing and comparing fingerprints, which could be used in criminal investigations. Despite being a scientist and the cousin of Charles Darwin, Francis Galton was the one who created a method of analysing fingerprints using specific features called "Galton points." This technique laid the foundation for fingerprint recognition as a scientific procedure. As the 1900s commenced, fingerprinting gradually became the preferred method for determining personal identities, replacing anthropometry. Even Bertillon, who utilized anthropometry for identification, eventually integrated fingerprints into his method. As fingerprinting gained popularity in criminal investigations, it also caught the attention of authorities involved in managing mobility and migration. For instance, in 1912, France passed a law requiring nomadic individuals to possess a "carnet des nomades" containing their fingerprints and pictures. During the late 1800s in the United States, as limitations on Chinese immigration were increasing, the idea of using fingerprints to identify Asian individuals was suggested. However, ultimately it was determined that photographs would be used instead. The utilization of fingerprints for identifying non-citizens gained popularity in Argentina when criminologist Juan Vucetich created a new fingerprinting system called "dactyloscopy" at the same time that the public expressed worries about the large number of immigrants from a perceived inferior race. Traditionally, fingerprints were taken in sets of 10, using ink and paper. The effectiveness and usefulness of pre-digital biometric identification techniques relied heavily on the quantity of stored information and the operator's ability to manually access the necessary records and compare the data. The effectiveness of fingerprints was also dependent on the presence of networks to share and connect fingerprint data at both domestic and global levels. In the beginning, attempts were made to create a centralized system for international criminal records. This led to the formation of the Bureau for Distant Identification in Copenhagen in 1922, followed by the International Criminal Police Congress in 1923, which eventually became Interpol [9].

Fingerprint identification has been in use since ancient times, although DNA evidence is now regarded as more precise. During the time of ancient Babylonia and China, thumbprints and fingerprints were commonly utilized as signatures on clay tablets and seals. The concept of fingerprints being distinct to each person has been around since the 14th century. In 1686, Marcello Malpighi, a physiologist, used a microscope to study fingerprints and observed a pattern of ridges and loops. In 1823, Jan Purkinje, a physiologist, observed a minimum of nine distinct fingerprint designs. Sir Francis Galton, an anthropologist, is considered the foremost expert in fingerprint recognition and was the first to demonstrate the scientific application of using fingerprints to distinguish people. Beginning in the 1880s, Galton (a cousin of Charles Darwin) studied fingerprints to seek out hereditary traits. Through his research, he concluded that every fingerprint is unique and does not change over time. In 1892, Galton released a book detailing his discoveries and identifying the top three fingerprint patterns: loop, whorl, and arch. These classifications are still used today. Law enforcement officials quickly realized the potential significance of fingerprint evidence. Sir Edward Richard Henry, a British official stationed in India, initiated the development of a fingerprint identification system for Indian criminals, consisting of 1,024 primary classifications. Similarly, in Argentina, police official Juan Vucetich utilized Galton's research to establish a fingerprint system and successfully identified a murderer in 1892.

6.4.2.2 Use of fingerprints today

At the start of the 1900s, Scotland Yard had initiated the collection of fingerprints, utilizing Henry's methods and establishing a Central Fingerprint Bureau. In 1903, a fingerprint system was implemented by the New York Police Department, the New York State Prison System, and the Federal Bureau of Prisons in the United States. The US Army also adopted fingerprint identification in 1905. In 1910, Thomas Jennings was accused of killing Clarence Hiller in Illinois. This was the first case in the United States where fingerprint evidence was effectively utilized to prove the accused's guilt. In 1911, incriminating fingerprints were found at Hiller's residence, leading to Jennings' conviction. Although Jennings appealed, the Supreme Court of Illinois upheld the evidence and he was executed in February 1912. This case, known as *People v. Jennings*, solidified the use of fingerprint evidence as a credible measure.

6.4.2.3 Dactyloscopy

The process of Dactyloscopy involves examining and categorizing the patterns found in single fingerprints. Fingerprints consist of a sequence of raised and indented lines on the skin of a finger, which typically create unique patterns in the form of loops, whorls, and arches. In addition, fingerprints possess unique attributes known as "minutiae," which refer to the quantity and arrangement of ridges that cannot be seen by the unaided eye. People can leave fingerprints on objects they touch, which can be seen or invisible. Visible fingerprints can come from substances that stick to the fingers, like dirt or blood, or they can be imprinted in a soft material like clay. Latent fingerprints are traces of sweat, oil, or other natural secretions on the skin, and they are not ordinarily visible. When the surface is solid, dusting techniques can reveal latent fingerprints, while chemical techniques can be used on porous surfaces. Before the computerization of fingerprint records, latent fingerprints left at a crime scene were not a reliable means of identifying a suspect because police did not have a way of determining which set of prints on file (if any) belonged to the suspect. However, fingerprints remain a highly valuable form of physical evidence for police in linking suspects to crimes. In the 1980s, there was a significant shift when the Japanese National Police Agency successfully implemented the first functional electronic fingerprint-matching system. Nowadays, most law enforcement agencies around the world rely on similar systems known as automated fingerprint identification systems (AFISs) to quickly search through millions of digitized fingerprint records. Prior to confirming a match, fingerprints identified by AFIS are reviewed by a fingerprint expert [11].

6.4.2.4 The face of crime

Since the creation of photography, pictures have been utilized to recognize individuals (crude facsimile technology was employed as far back as 1843). As early as the 1850s, pictures of recognized lawbreakers (referred to as "rogues" galleries') began to be exhibited openly or distributed among officials, initially in the United States and subsequently in Europe. Although pictures of faces may seem like a natural way to depict individuals, their effectiveness and usefulness in identifying people were restricted. One major challenge was the lack of a method for organizing and categorizing images for future identification and retrieval. Although Bertillon did incorporate facial pictures into his cards, this was not his main focus and he essentially created the concept of a "mug shot" – a police photo of an individual captured following their arrest. Efforts to use anthropometry to identify individuals have been intertwined with endeavours to attribute criminal behaviour to physical factors, establish racial hierarchies, and support eugenic measures in society. To illustrate, followers of physiognomy (which gained widespread popularity in 18th century Europe thanks to Johann Kaspar Lavater, a Swiss clergyman) attempted to determine a person's personality and traits based on their outward appearance, specifically their facial features. Despite being scientifically disproven, the allure of physiognomy endured for a considerable amount of time. It was believed that photographs could offer insight into a person's inner thoughts and character. The implementation of this innovative technology revived the examination of criminal physiognomy. For instance, Francis Galton utilized composite images created by overlapping photos of male convicts in order to find a visual depiction of a typical criminal visage. According to Lavater and Galton, physiognomy could be used to improve society, which helped justify the eugenics movement in the 1900s.

6.4.3 Mind-reading machines (border management and emotion detection)

The current excitement surrounding highly advanced AI has revived long-held aspirations and assertions of machines capable of identifying falsehoods, reading human thoughts, and uncovering criminal intentions from over a hundred years ago. At present, there is ongoing development and use of facial expression recognition technologies to identify human emotions and mental conditions. This includes their application in determining if individuals are being deceitful or honest. Similar to other biometric technologies, emotion detection is being tested in border management as a significant area of focus. In the 1880s, initial efforts were made to create machines that could detect criminal intent, also known as "criminal dangerousness." Efforts were made to measure variations in respiration through the use of "pneumographs" and blood pressure through the use of "sphygmographs." By the 1920s, all of these efforts resulted in the creation of the initial polygraph, which could detect lies by measuring and analysing various physical responses like blood pressure, heart rate, breathing, and skin sensitivity. Despite its inconsistent outcomes and the significant risks at play, the lie detector continued to be a widely used method of questioning in criminal inquiries until the conclusion of the 1900s. The FBI and the military utilized it in the United States until 1998 when the US Supreme Court determined that there was not an agreement on the reliability of polygraph evidence. The polygraph is still utilized in the present day for various objectives, including the supervision of sexual and aggressive offenders in the United States and United Kingdom after they have committed a crime. Despite scepticism surrounding the accuracy of the polygraph, emerging technologies are reviving researchers' optimism for the possibility of mind reading through body measurements. They are not only utilizing face recognition technologies but also implementing other methods such as electroencephalography (EEG), functional magnetic resonance imaging technology, thermal facial analysis, eye-tracking, and voice analysis [12].

6.4.3.1 Mind-reading project utilizing fMRI

The findings of a study conducted by Radboud University in the Netherlands were recently disclosed. The experiment involved displaying photos of faces to two participants while they were inside a high-resolution brain imaging machine called functional magnetic resonance imaging (fMRI). The fMRI scanner is a form of brain imaging technology that is non-invasive and tracks changes in blood flow to detect brain activity. While the participants viewed facial images, the fMRI monitored the neuron activity in the visual areas of their brains. After collecting the data, the scientists inputted it into a computer's AI algorithm, which was able to create a precise visual representation using the fMRI scan data.

According to the findings of the study, the fMRI/AI technology successfully replicated the original images shown to the participants with almost identical accuracy. According to the Mail Online, Thirza Dado, a cognitive neuroscientist and AI researcher who headed the research, is confident that the remarkable outcomes showcase the ability of fMRI/AI systems to accurately decipher thoughts in the future.

"I believe we can train the algorithm not only to picture accurately a face you're looking at but also any face you imagine vividly, such as your mother's," explains Dado.

> "By developing this technology, it would be fascinating to decode and recreate subjective experiences, perhaps even your dreams," Dado says. "Such technological knowledge could also be incorporated into clinical applications such as communicating with patients who are locked within deep comas."

According to the Mail Online, Dado's main objective is to utilize technology in aiding individuals who have lost their vision due to illness or injury, in order to restore their sight. "We are already developing brain-implant cameras that will stimulate people's brains so they can see again," Dado says [13].

The volunteers were exposed to a series of unfamiliar faces while undergoing brain scans in order to "train" the AI system. Based on the study, it was discovered that the images they viewed were not actual individuals, but rather computer-generated paint-by-numbers artwork, where every small dot of light or shade is assigned a specific computer code. Using fMRI scans, researchers were able to observe the neural activity of volunteers when they viewed the "training" images. Afterwards, an AI system translated this neural response into computer code, allowing for the accurate recreation of a photographic portrait for each volunteer. During the test, both the volunteers and the AI system were presented with faces that had never been seen before, yet were still decoded and recreated with remarkable accuracy.

6.4.3.2 Mind-reading project based on electroencephalography

Telepathy, also known as mind reading, involves the transfer of thoughts between individuals without the use of traditional methods of communication like speaking. While it has long been a popular topic in science fiction, recent scientific developments indicate that it may soon become a tangible phenomenon.

Dr. Hans Berger (1873–1941), a German psychiatrist, was the first to attempt the creation of a synthetic mind-reading system using electrical brain activity. While his efforts were unsuccessful, they paved the way for the development of EEG. Today, EEG is commonly used in the medical field to record brain activity and is also a popular tool for attempting to decode speech imagery in order to achieve mind reading.

As individuals, we often engage in internal dialogue. We repeatedly practice how to handle a challenging scenario, what to say to a potential client, and how to respond to tough questions in an interview. This process is known as speech imagery or covert speech. Unlike overt speech, which involves physically speaking to someone, speech imagery does not require movement of the articulators. Even if a person is unable to use their muscles and cannot physically move their articulators, they can still mentally imagine speaking or actively think about it [14].

6.4.3.3 Why EEG instead of fMRI, fNIRS, or other methods?

While there are other modalities like fMRI and functional near-infrared spectroscopy (fNIRS) available for decoding imagined speech, the EEG remains the most widely used method. However, when compared to these other modalities, EEG has a few limitations, such as susceptibility to noise and a lack of structure. Despite these challenges, EEG is favoured for its excellent temporal resolution, non-invasiveness, and cost-effectiveness. For example, the EEG signals recorded from a participant while imagining saying the words "in" and "cooperate" are displayed below. Despite the words having varying lengths, this is not apparent in the EEG data. The absence of any discernible "structure" in the signals makes it difficult to interpret the word envisioned by the participant [14].

An EEG-based method for deciphering imagined speech has a wide range of potential uses. For example, individuals with total muscle paralysis (such as those with locked-in syndrome) can still generate thoughts of speaking, which can be translated into audible signals to facilitate communication with others. Additionally, the system can aid individuals with significant speech disabilities. A potential military use for this technology is in combat situations where ambient noise hinders traditional verbal interactions. In this situation, the soldiers' ideas could be conveyed instead of their voices being hidden in the background noise. However, we do not believe that we have achieved this capability. The systems we currently possess for extracting imagined speech from EEG signals have significant drawbacks. To begin with, the systems do not operate in real-time, resulting in a notable lag between the EEG data collection and the ability to forecast the person's thoughts. Moreover, it is not possible for us to promise uniform precision levels across different individuals, implying that a system that functions well for one individual may not produce comparable outcomes for another. Furthermore, the existing systems have a restricted range of words, constraining the variety of words that can be imagined. Our paper extensively covers the factors to be taken into account while creating a feasible system for interpreting imagined speech from EEG.

Based on our current understanding, using words of varying lengths is a favourable option. Additionally, incorporating words from diverse languages can be beneficial if the participant is proficient in multiple languages.

Our chapter presents a surprising finding from a study involving five bilingual individuals, whose primary language is Hindi and secondary language is English. The results showed that the system for interpreting imagined speech had a higher success rate when the participants imagined in Hindi rather than English. Additional experiments are necessary to determine the cause of this observation. Our chapter has addressed the factors that must be taken into account when selecting the prompts.

In the past, the Audrey was the first speech recognition system that filled an entire room and was only capable of recognizing numbers from familiar individuals. Now, we have Google Voice, Siri, and Alexa that can fit in our pockets. Although the ability to decipher imagined speech is still developing, we cannot predict how advanced and influential it will become in the next ten years. Eventually, there may be a flawless mind-reading technology available to us.

6.5 AUTOMATED IDENTIFICATION AND ARTIFICIAL INTELLIGENCE OR AI-ENABLED AUTOMATED IDENTIFICATION

Before the rise of digital technology, humans were responsible for collecting, verifying, and organizing biometric data with the help of non-digital tools. This frequently meant a time-consuming procedure, and the precision of the outcomes depended on the proficiency and diligence of the individuals in charge and the organizational mechanisms implemented. The advancement of digital technologies has enabled the automation of capturing images and analysing biometric markers for enrolment. This has resulted in the creation of specialized IT systems, such as databases, and a network infrastructure. The use of IBM punch cards to access fingerprint records marked the initial attempts to automate fingerprint identification in the 1940s, but it was not until the 1960s that fully automated systems known as AFISs were developed. The late 1980s saw the emergence of optical scanners, enabling the direct scanning of fingerprint images into computers. Although AFISs are now commonly used in border control, there are ongoing efforts to develop and evaluate other biometric technologies such as iris recognition, facial image recognition, and DNA profiling. Advancements in AI have resulted in a rise in face recognition applications. These technologies not only offer convenience for individuals, such as tagging photos on social media, unlocking devices, and accessing homes and bank accounts but also provide valuable tools for criminal identification and public surveillance. When discussing borders, some believe that facial recognition technology is changing ports of entry and exit into complete panopticon. This means that it is constantly monitoring and recognizing travellers at multiple checkpoints during their border control process and connecting these checkpoints that used to be separate. In addition to automated recognition, the utilization of AI technology is growing in the fields of borders, migration, and security for various objectives such as identifying emotions, evaluating individual risks, tracking and analysing migration, and aiding decision-making in immigration processes.

6.5.1 Automated biometric identification systems

The main purposes of using automated biometrics are:

1. The process of biometric identity verification involves capturing an individual's biometric image and comparing it with an image stored on a biometric passport or in a database.
2. Biometric identification requires comparing a biometric image with multiple images in a database (referred to as one-to-many matching). Closed-set identification takes place when the subject's data is already stored in the database, while open-set identification takes place when there is uncertainty about the presence of the subject's data in the database.

6.5.1.1 Automated fingerprint identification

The process of an AFIS involves obtaining mathematical representations (templates) from fingerprint images (samples) in order to effectively compare fingerprints.

The fingerprint matching techniques are of two types:

1. Pattern matching.
2. Minutiae matching.

It is possible to obtain finger impressions from an object touched by an individual (known as latent fingerprints) and convert them into a digital format. However, these typically have poor-quality details. The identification process varies depending on the number and source of the fingerprints. A standard process of identification entails matching the fingerprints of a person undergoing a police or border inspection with a central repository of fingerprint collections as depicted in Figure 6.1.

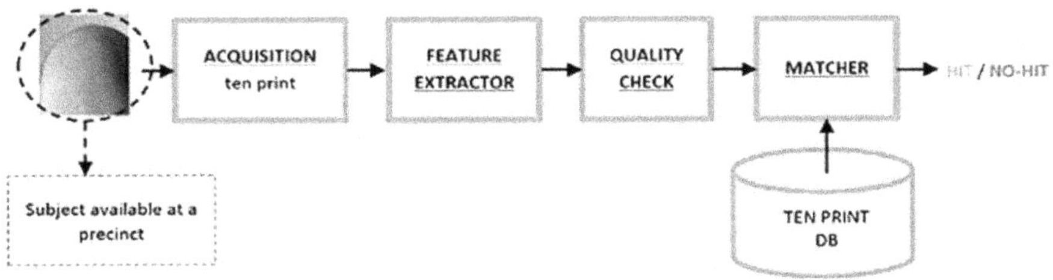

Figure 6.1 The primary steps involved in the process of fingerprint identification.

6.5.1.2 Face recognition

Facial recognition is the deed of verifying an individual's identity through their facial structures. It involves collecting and evaluating patterns from a person's face and then comparing them. The initial step in identifying and pinpointing human faces in images and videos is the use of face detection techniques. The face capture process transforms physical features (a face) into a series of digital details (data or vectors) based on the subject's facial characteristics. You can identify whether two faces belong to the same person by conducting a face match. To better understand this process, let's examine a recent instance using three steps. A student residing in the wider Washington DC region utilized a freely available facial recognition application to identify and remove duplicate images of faces from 827 videos shared on Parler during January 6, 2021 event at the Capitol building. They have curated a website named Faces of the Riot, showcasing these portraits.

1. Protesters, troublemakers, and reporters have utilized their smartphones to complete a portion of the process of capturing facial features (from a physical form to a digital image).
2. He employed facial recognition to isolate faces from 200,000 pictures.
3. The responsibility of investigating, converting digital pixels to vectors, and potentially matching faces with existing databases to identify individuals lies with the FBI, using an AFIS/ABIS system [15].

6.5.1.3 Face recognition data to identify and verify

Biometrics refers to the process of verifying and confirming a person's identity by utilizing distinct and verifiable information that is exclusive to that individual. To learn more about the definition of biometrics, please visit our web collection on the topic.

The question such as "Who are you?," is answered by Identification.

Whereas the question: "Are you really who you say you are?," is answered by Authentication.

A few more examples of this are:

1. Facial biometrics involves using a 2D or 3D sensor to "capture" a face and converting it into digital data through an algorithm. This data is then compared to images stored in a database.
2. These automated systems have the capability to quickly verify a person's identity by analysing their facial characteristics, such as the distance between their eyes, the shape of their nose, the outline of their lips, the ears, and the chin. They are also effective in crowded and ever-changing settings.
3. The iPhone X owners have already been familiarized with the technology of facial recognition.

6.5.1.4 Why face recognition is preferred nowadays?

1. The primary biometric benchmark remains to be facial biometrics.
2. This is because it is simple to deploy and apply. The interaction is not taking place physically.
3. The verification/identification processes for face match and face detection are also quick.

The field of deep learning has greatly contributed to the progress of computer vision. This type of AI, which is a subdivision of machine learning, utilizes artificial neural networks (NN) to detect patterns in a given set of data. Deep learning processes can be utilized for both supervised learning, where data is pre-labelled by humans, and unsupervised learning, where data is not labelled [15]. A face recognition system that uses deep learning methods usually involves identifying a face, adjusting the image to a standard format, and analysing facial characteristics to match them with one or more known faces is illustrated with the aid Figure 6.2.

Face recognition technology is primarily employed in automated border control (ABC) systems at airports. These systems use electronic travel documents or tokens to verify a passenger's identity and eligibility to cross the border, based on predetermined rules and by checking border control records. Presently, ABC systems are capable of accommodating various types of biometric technology, such as facial and iris recognition. The facial recognition method employed in ABC utilizes a comparison between the real-time image of the traveller's face and the image stored on their travel document's chip. By utilizing biometric passports, it is now feasible to confirm a traveller's identity without having to save their recent facial images in a database. As more countries adopt biometric passports, ABC gates equipped with facial recognition technology are becoming a standard element in airport security screenings.

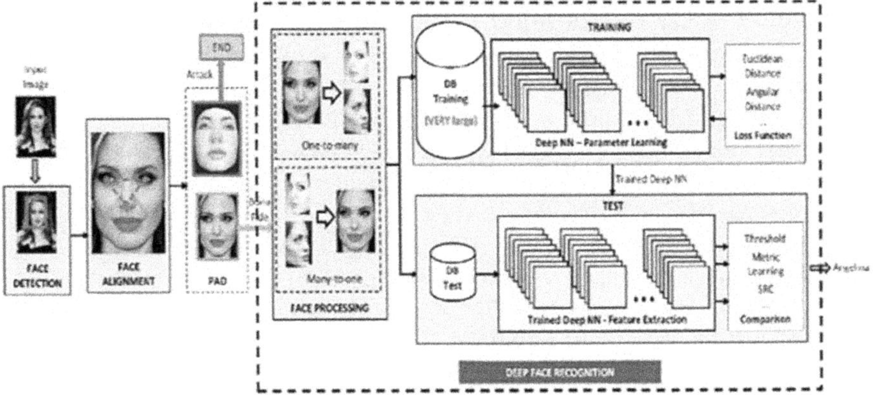

Figure 6.2 Illustration of a deep learning-based face recognition system works.

Live stream video from CCTV cameras can be analysed to use face recognition technology for real-time surveillance of public areas as well as airports. This includes analysing facial photographs obtained from cameras and matching them against pictures of people on surveillance lists. In 2020, Europe has seen a rapid increase in the deployment of face recognition systems for public surveillance, not just in airports but also in schools, stadiums, and event venues. The United States has already been a leader in implementing this technology at airports, and ABC gates that utilize face recognition have been installed in airports in Italy and Portugal.

6.5.1.5 *The present status of research and advancement in the domain of facial recognition around the globe*

Several projects are competing for the topmost spot in the biometric innovation competition. Among the major contenders are Microsoft, Apple, Facebook, Amazon, Google (collectively known as GAFAM). In order to accelerate our comprehension, the leading internet companies regularly exchange their discoveries in AI, visual identification, and facial analysis. A few of the recent undertakings include:

1. Academia: The Chinese University of Hong Kong researchers developed the GaussianFace algorithm in 2014, which achieved a facial identification score of 98.52%, surpassing the human performance of 97.53%. This impressive result was achieved despite the algorithm's requirement for a large amount of RAM and slower calculation speed.
2. Facebook and Google: In 2014, Facebook introduced DeepFace, a software with a 97.25% success rate in determining if two images of a face belong to the same individual. When compared to humans, the Facebook computer only performs 0.28% better, accurately identifying matches in 97.53% of cases.
3. With FaceNet, Google improved in June 2015: The accuracy achieved by FaceNet on the popular Labelled Faces in the Wild (LFW) dataset, with a score of 99.63% (0.9963 ± 0.0009), has set a new standard. Utilizing an artificial NN and a novel algorithm, the Mountain View start-up has effectively connected a face with its corresponding owner. This technology enables Google Photos to automatically organize and categorize pictures based on the individuals they identify. Proving its importance in the biometrics landscape, it was quickly followed by the online release of an unofficial open-source version known as OpenFace.
4. Microsoft, IBM, and Megvii: In a study conducted by MIT researchers in February 2018, technologies from Microsoft, IBM, and China's Megvii (FACE++) showed higher error rates when recognizing women with darker skin tones, compared to lighter-skinned men. Microsoft published a blog post at the end of June 2018 stating that its biased facial recognition algorithm had been greatly enhanced.

5. Amazon: As reported by Ars Technica in May 2018, Amazon is actively promoting Recognition, a face recognition technology that operates through the cloud, to law enforcement agencies. According to Newsweek's July 2018 report, Amazon's facial recognition software mistakenly identified 28 members of the US Congress as suspects in criminal activity, despite its ability to match faces against databases with tens of millions of entries and identify up to 100 individuals in one image.

6.5.1.6 *Major technology providers for biometric matching*

In late May 2018, the US Homeland Security Science and Technology Directorate revealed the results of sponsored experiments conducted at the Maryland Test Facility. In practical tests, 12 facial recognition systems were tested in a 2-m by 2.5-m corridor. Thales' solution, using LFIS software, showed great results: 99.44% face acquisition rate in under five seconds and 98% true identification rate in under five seconds. Furthermore, it boasted a 1% error rate, which is significantly lower than the accepted industry standard of 32%. According to a November 2018 report from the National Institute of Standards and Technology (NIST), the accuracy of 127 algorithms was evaluated and linked to the names of the participants. In November 2018, the NIST released a study that includes data on the accuracy of 127 algorithms and associates their performance with the names of the participants. Furthermore, the NIST Ongoing Face Recognition Vendor Test (FRVT) 3, which was finalized at the end of 2019, also reveals additional findings. NIST has a report that can be obtained by request. In January 2020, Information Technology and Innovation Foundation (ITIF) research indicated that NIST's findings demonstrate that the top facial recognition algorithms do not exhibit bias against any specific race or gender, disproving critics. NIST's publications on "Face recognition accuracy with face masks using post-COVID-19 algorithms" from August 2020 and March 2021 [16] further illustrate the continuous improvement of algorithm performance in less than a year.

6.5.1.7 *Facial emotion recognition (FER)*

FER is the process of using image processing software to identify emotions like contempt, joy, anger, surprise, fear, or sadness by mapping them onto a person's facial expressions in real-time or from still photographs. Understanding human emotions involves three steps: identification, interpretation, and analysis.

1. Identification of facial features
2. Recognition of facial expressions
3. Categorization of expressions into distinct emotional states.

The broad spectrum of potential uses for detecting facial expressions adds to its attractiveness. Unlike facial recognition, which focuses on identifying a person rather than their emotions, it is not synonymous. Facial expressions can be described using various methods such as geometric or physical features, Eigenfaces derived from edited images, dynamic models, and 3D models. Some companies that offer these services for brand marketing purposes are Noldus, Affectiva, Sightcorp, and Kairos (facial and emotion recognition) [17].

6.6 AI-ENABLED EMOTION DETECTION

The purpose of emotion detection technologies is to identify mental states and emotions by analysing facial expressions, often in combination with other physical characteristics like eye contact, body movements, tone of voice, heart rate, body temperature, skin conductivity, etc. Emotion detection systems that utilize AI are heavily influenced by the work of American psychologist Paul Ekman. Ekman asserts that lying produces non-verbal cues in facial expressions that can be identified through analysis. These systems are based on Ekman's belief that human emotions can be categorized into six universally recognized types: anger, disgust, fear, happiness, sadness, and surprise.

Researchers in the field of AI have attempted to develop technology that can automatically identify these fundamental emotions through facial expressions, although Ekman has expressed scepticism regarding the accuracy of such detection methods. AI-powered emotion sensing systems have already been implemented in various fields and situations, where they are utilized for tracking psychological well-being, identifying deceitful insurance reports, tracking students' involvement (including aiding children with autism in improving social and emotional abilities), evaluating job seekers, and identifying potential shoplifters. Law enforcement agencies also promote these applications for the purpose of preventing crime, conducting security screenings, and controlling border activities. In 2016, a facial expression analysis algorithm was reportedly created by researchers at Shanghai Jiao Tong University, capable of differentiating between portraits of convicted criminals and non-criminals. Two Stanford University researchers stated in 2017 that they had created an AI "gaydar" that could accurately identify gay men by analysing photographs with an 81% success rate. According to reports, the researchers asserted that their algorithms had the ability to be trained to assess an individual's intelligence and potential for criminal behaviour. In 2020, scientists at Harrisburg University revealed the creation of a facial analysis program that could forecast an individual's likelihood of being a criminal. Meanwhile, numerous companies have begun implementing these innovations in real-world situations. For instance, an Israeli company asserts that they have the capability to "identify, target, and capture potential terrorists or criminals before they can cause harm". In 2007, the US Transportation Security Administration introduced the Screening of Passengers by Observation Techniques system to identify potential terrorists by monitoring air travellers' expressions of fear and stress at borders. The effectiveness and fairness of the system were called into question. In 2021, the UK Border Agency experimented with a method to measure stress, anxiety, and deceit at immigration checkpoints, using facial and thermal analysis [18].

6.7 ALGORITHMIC PROFILING (ALGORITHMIC RISK ASSESSMENT)

Profiling is a commonly and justifiably used practice by law enforcement officers and border guards to identify and stop illegal immigration, as well as to deter, scrutinize, and charge criminal acts. In this scenario, profiling refers to "any type of computerized handling of personal information that uses said data to assess specific personal aspects of an individual, including but not limited to their work performance, financial status, health, personal choices, interests, trustworthiness, behavior, whereabouts, or movements." The practice of profiling is employed to direct border management and law enforcement tactics such as stop and search, restricting entry to specific regions, detentions, or referring individuals for more thorough screening at the border. Profiling serves two primary purposes.

1. The process of identifying people by their unique intelligence involves creating a profile that lists the distinguishing traits of potential suspects. This profile is based on factual evidence and information related to a specific event or events.
2. One approach to identifying individuals who may be of interest to law enforcement and border management is through predictive methods. These methods aim to analyse behaviour, but in reality, they often rely on visible physical traits such as age, gender, or ethnicity.

The use of algorithmic profiling poses significant concerns regarding fundamental rights, such as the possibility of discrimination and infringement on the rights to privacy and data protection. If not carried out with appropriate measures in place, both forms of profiling can be deemed illegal. These measures include having a valid and justifiable reason for conducting the profiling. This discussion focuses on the technical aspect of profiling, rather than its impact on human rights or legal implications [8].

The use of algorithmic profiling relies on various methods to create profiles of individuals based on correlations and patterns found in data. In this book, algorithmic profiling enables law enforcement and border management officials to closely examine individuals or specific groups that pose potential risks, determined through data analysis.

6.7.1 Why algorithm profiling: to support decision-making (automated decision-making)

The rise of predictive modelling techniques, made possible by the abundance of big data, is increasingly aiding or even replacing human decision making. The use of predictive modelling techniques involves the use of algorithms to aid in decision-making. An algorithm is a set of instructions for a computer to convert an input into an output. Many statistical methods are utilized in a wide range of algorithms, particularly those that handle large datasets. These algorithms employ methods to compute connections between various variables or characteristics within the dataset. One could potentially combine information about alcohol consumption among a particular group and their corresponding life expectancy data to determine the overall impact of alcohol on life expectancy. The result of these algorithms is consistently a probability, indicating that there is a level of doubt surrounding the output, such as the connections or categorizations that these algorithms have determined. To illustrate, email providers use algorithms to sort through emails, distinguishing between spam and legitimate messages. While these algorithms are effective, they are not flawless. In certain cases, spam email might not be recognized as such and could end up in the inbox instead. This is referred to as a "false negative" in the language of machine learning and AI. Occasionally, a genuine email (potential inbox material) may be mistakenly marked as spam and could be found in the junk folder; this is known as a false positive in machine learning systems.

6.7.2 Algorithmic profiling and automated decision system or algorithmic profiling and algorithmic govern mentality

The rise of ubiquitous computing sparked a data revolution and the development of algorithmic profiling, resulting in the emergence of a "surveillance society" or "information civilization." Algorithmic profiling is the practice of utilizing algorithms to examine surveillance data, with the goal of identifying patterns and forecasting decisions and anticipated behaviours. Algorithmic profiling is a technique that uses inferential analysis to detect connections or trends in datasets. These results can then be utilized as cues to categorize an individual as part of a certain group in an automated decision-making system.

Algorithmic profiling takes place in various settings such as insurance, financial services, variable pricing, advertising, job recruitment, schooling, government, and law enforcement. For instance, when deciding on a loan application, various factors such as personal risk of non-payment, local area, and socio-economic makeup may be taken into consideration. Algorithmic profiling is applied within the context of law enforcement activities at borders, specifically for border surveillance, management, and immigration control.

Risk evaluation algorithms and decision-making systems for immigration and border control are being implemented more frequently on a global scale. AI algorithms can also be used to sift through data of both kinds: (1) non-personal and (2) personal, in order to identify unknown persons who may be of interest to the authorities. Algorithmic governmentality is a term used to describe the incorporation of algorithmic profiling into government decision-making.

The rise of algorithmic governmentality relies on the extraction, consolidation, and computerized interpretation of large sets of data, with the goal of predicting, influencing, and taking action against potential behaviours before they occur. The goal of algorithmic profiling is to anticipate and proactively address individual actions. Using smart border technologies and algorithmic profiling, the goal is to constantly monitor and analyse data in order to create a pre-emptive intervention system for regulating migrant movement at the border. The collection of intelligent border technologies that can be utilized at intelligent borders includes fibre optic and ground sensors for detecting surface crossings, as well as radar, infrared, thermal imaging, and various biometric technologies.

Advanced border technologies detect and gather information from individuals in motion. The recorded biological identifiers, such as skin features, fingerprints, facial characteristics, retinas, irises, body movements, and types of motion (human or animal), are transformed into a usable format and saved in databases. This is done to create profiles, monitor movements, and conduct surveillance on individuals who are on the move [8].

REFERENCES

1. Tyler H. The Increasing Use of Artificial Intelligence in Border Zones Prompts Privacy Questions. MPI. https://www.migrationpolicy.org/article/artificial-intelligence-border-zones-privacy. (accessed on 1 Jan. 2024).
2. Molner P EU's AI Act Falls Short on Protecting Rights at Borders. Just security. https://www.justsecurity.org/90763/eus-ai-act-falls-short-on-protecting-rights-at-borders/. (accessed on 1 Jan. 2024).
3. Mahajan N, Chauhan A, Kajal M. An introduction to deep learning-based object recognition and tracking for enabling defense applications. Advances in Aerial Sensing and Imaging. 2024;5:109–127.
4. Causevic S, Salazar M, Orsini N, Kågesten A, Ekström AM. Sexual risk-taking behaviors among young migrant population in Sweden. BMC Public Health. 2022;22(1):625.
5. Madiega T. Artificial Intelligence Act. European Parliament: European Parliamentary Research Service. 2021.
6. International Civil Aviation Organization. Machine Readable Travel Documents. International Civil Aviation Organization. 2002.
7. Ramiro Troitiño D. The European Commission, the Council, and the European Parliament: Differentiated Theoretical Frame for the Digital Revolution. In Digital Development of the European Union: An Interdisciplinary Perspective 2023 (pp. 349–361). Cham: Springer International Publishing.
8. Vavoula N. Artificial intelligence (AI) at Schengen borders: Automated processing, algorithmic profiling and facial recognition in the era of techno-solutionism. European Journal of Migration and Law. 2021;23(4):457–484.
9. Hidayat F, Elviani U, Situmorang GBG, Ramadhan MZ, Alunjati FA, and Sucipto RF, Face recognition for automatic border control: A systematic literature review, IEEE Access. 2024;12:37288–37309, doi: 10.1109/ACCESS.2024.

10. Labati RD, Genovese A, Muñoz E, Piuri V, Scotti F, Sforza G. Biometric recognition in automated border control: A survey. ACM Computing Surveys (CSUR). 2016;49(2):1–39.
11. Damanik EL. Laborer identification and monitoring system: Dactyloscopy on the plantations of East Sumatra, 1926-1980. Jurnal Sejarah Citra Lekha. 2021;6(1):44–56.
12. Rockwell T. Dynamic empathy: A new formulation for the simulation theory of mind reading. Cognitive Systems Research. 2008;9(1-2):52–63.
13. Adams RB Jr, Rule NO, Franklin RG Jr, Wang E, Stevenson MT, Yoshikawa S, Nomura M, Sato W, Kveraga K, Ambady N. Cross-cultural reading the mind in the eyes: An fMRI investigation. Journal of Cognitive Neuroscience. 2010; 22(1):97–108.
14. Shah U Mind Reading! Decoding Imagined Speech From Brain Signals (Doctoral dissertation, Hamad Bin Khalifa University (Qatar)) 2022.
15. Carlos-Roca LR, Torres IH, Tena CF. Facial recognition application for border control. In 2018 International Joint Conference on Neural Networks (IJCNN) 2018 Jul 8 (pp. 1–7). IEEE.
16. Patrick G. Top 10 Biometric Technology Companies Revealing New Heights of Security and Surveillance. Verified Market research. https://www.verified-marketresearch.com/blog/top-biometric-technology-companies/. (accessed on 1 Jan. 2024).
17. Ko BC. A brief review of facial emotion recognition based on visual information. Sensors. 2018;18(2):401.
18. Li Y, Jiang Y, Tian D, Hu L, Lu H, Yuan Z. AI-enabled emotion communication. IEEE Network. 2019;33(6):15–21.

Chapter 7

Cyber security attacks and internet of things

Role of artificial intelligence

Shitanshu Jain and Varun Kumar Mishra

7.1 INTRODUCTION

7.1.1 Define the term Internet of Things (IoT)

The IoT encompasses a network of physical devices embedded with sensors, software, and connectivity, enabling them to collect and exchange data. This network includes a vast array of objects, from everyday items like lightbulbs and thermostats to complex machinery such as industrial robots and autonomous vehicles [1]. These devices' data can be utilized to enable new services and business models, increase productivity, and automate operations.

IoT devices are frequently distinguished by their compact size, restricted power supplies, and restricted computational and storage capacities. Given that they might not have the security measures seen in more conventional computing equipment, this makes them appealing targets for cybercriminals. Furthermore, a large number of IoT devices are frequently installed, expanding the possible attack surface.

The enormous number and variety of devices present one of the biggest obstacles to IoT security, making it challenging to continuously implement conventional security procedures. In addition, a lot of IoT devices interact via wireless networks, which are easily intercepted or interfered with. IoT security is therefore a serious issue that demands a new perspective on cybersecurity [9].

Artificial intelligence (AI) and machine learning (ML) offer promising solutions to the challenges posed by IoT security. By analyzing vast datasets from IoT devices, AI algorithms can effectively identify patterns indicative of cyberattacks [3]. Furthermore, AI-driven automation can streamline security responses, reducing human intervention and accelerating reaction times.

The IoT is a rapidly expanding sector with many benefits, but there are also significant security dangers. AI and ML are two promising emerging technologies that can help solve these issues, but in order to fully realize their potential, further research and development is needed.

The IoT has rapidly expanded in recent years, with millions of interconnected devices collecting and sharing data across the globe. IoT presents serious security issues in addition to its many advantages, like more automation, efficiency, and convenience [17].

DOI: 10.1201/9781032657264-7

IoT device cybersecurity vulnerabilities can have significant consequences such as physical harm, invasions of privacy, and data breaches. To address these issues, AI has come to light as a promising technology that offers fresh methods for identifying and preventing cybersecurity threats on IoT devices.

7.1.2 Importance of cybersecurity in the context of IoT

It is impossible to exaggerate the significance of cybersecurity in relation to IoT. The attack surface for cybercriminals expands along with the number of connected devices. High-security measures are required, as there were reportedly 1.5 billion cyberattacks on IoT devices in 2020 [4].

IoT ecosystems are made up of several components that are interrelated, including the cloud, communication networks, actuators, and sensors, all of which provide hackers with numerous entry points. For example, obsolete components, insecure update systems, and default settings in IoT devices are all susceptible to exploitation. Additionally, an effective method for breaching the integrity of the device is Bluetooth network hacking. When a device is compromised, it can act as a gateway to the entire network, compromising systems and confidential data.

IoT networks and devices are proliferating in the manufacturing, smart city solutions, retail, and healthcare industries. Their expansion into new industries and the way society depends on them will grow along with their efficacy and efficiency. Maintaining their integrity is essential to preventing shutdowns and guaranteeing the ongoing operations of businesses.

A variety of techniques, including the establishment of operational technology frameworks to protect and maintain a check on the availability, dependability, productivity, and safety of IoT devices, can be used to address these issues and improve IoT cybersecurity. Another promising approach to protecting IoT networks is blockchain technology, which provides transparent accessibility and a decentralized storage space resistant to data manipulation [12].

The increasing variety of interconnected devices and increasing reliance on IoT systems make cybersecurity a serious issue that needs to be addressed immediately. Businesses may safeguard their networks, devices, and data from cyberattacks by implementing strong security measures. This will guarantee business continuity and preserve stakeholder and consumer trust [33].

7.1.3 Role of AI in IoT cybersecurity

In several ways, AI is important to IoT cybersecurity. Real-time threat detection and response, vulnerability identification, and security protocol enhancement are all possible with AI-powered security solutions. Large volumes of data produced by IoT devices can be analyzed by AI techniques, such as ML, to find anomalies and possible security risks [28]. AI can also be used to automate security procedures, which can speed up response times and eliminate the need for manual intervention.

Cybersecurity for the IoT is greatly influenced by AI in various aspects. AI-driven security systems can find flaws, improve security procedures, and offer real-time threat detection and response [11]. To identify irregularities and possible security risks, AI techniques like ML can analyze vast volumes of data produced by IoT devices. Furthermore, security procedures can be automated with AI, which lowers the need for human interaction and speeds up response times.

The potential of AI to adapt and learn from new threats is one of the main advantages of employing it in IoT cybersecurity. AI-powered security solutions can be trained to identify and react to new attack techniques that cybercriminals develop [13]. AI may also be utilized to develop stronger access control and authentication systems, such as biometric authentication, to safeguard IoT networks and devices.

IoT security enabled by AI is not without its difficulties, though. An important worry is the possibility of hostile AI, in which cybercriminals employ AI to design and execute attacks. Attackers may utilize AI to produce deepfakes, or synthetic media, that mimic reputable people or gadgets. Furthermore, attacks such as adversarial ML attacks, in which attackers alter input data to trick AI algorithms into returning false positives, can target AI algorithms directly [8].

Organizations must establish strong security measures into effect for their IoT devices, like access control, encryption, and secure software development methods, to meet these difficulties. These precautions can be strengthened by AI-powered security solutions, but they shouldn't be their sole means of defense. In order to make sure that their AI-powered security solutions are effective against fresh and emerging threats, enterprises also need to be diligent in monitoring and updating them.

AI is essential to IoT cybersecurity because it offers improved security procedures, real-time threat detection, and vulnerability discovery. Organizations need to be cautious of the possible dangers that come with AI-powered security solutions, like hostile AI and weaknesses in AI algorithms. Organizations can reduce the possibility of successful attacks and the impact of an incident by putting robust safety procedures in effect and being diligent in monitoring and updating their AI-powered security solutions [14].

7.1.4 Why is AI important in cybersecurity

AI empowers robots to execute tasks such as visual perception, speech recognition, language translation, and decision-making, which were traditionally considered exclusive to human cognition. AI understands context by using training data, which helps it decide what actions to take in different scenarios.

In the realm of cybersecurity, AI is playing a more and bigger role in defending online systems from hacking and unauthorized access. AI systems have the potential to automatically identify novel malware strains, provide alerts, identify cyberthreats, and safeguard sensitive enterprise data when utilized efficiently.

The advantages of integrating AI into cybersecurity are demonstrated by methods including ML, natural language processing, deep learning (DL), and reasoning and knowledge representation. As a result, cyber defense gets more sophisticated and automated, allowing businesses to recognize and react to the variety of cyber events they frequently encounter.

7.1.5 Benefits of AI in managing cyber risks

There are many advantages to implementing AI in cybersecurity for businesses aiming to manage risks effectively [2]. Key features consist of the following:

1. **Ongoing learning:** AI's capacity to absorb fresh information is the source of its continuous growth. AI uses approaches like ML and DL to identify patterns, define baseline activity, and detect any suspicious or out-of-the-ordinary deviations. This capacity for continuous learning makes it harder for hackers to get past an organization's security measures.
2. **Finding unfamiliar dangers:** AI provides a way to map and eliminate unknown threats that could seriously jeopardize networks in the face of developing cyberthreats. This comprises vulnerabilities that software providers have not yet discovered or addressed.
3. **Managing enormous data volumes:** AI technologies are preferable than human security experts in handling and comprehending large volumes of data. This makes it possible for businesses to effortlessly identify new risks in network traffic and massive datasets that are frequently overlooked by conventional approaches.
4. **Better vulnerability management:** AI not only finds new threats but also makes vulnerability management easier. It helps with system evaluation, improving problem-solving skills, and decision-making with knowledge. AI also finds holes in networks and systems, which helps companies stay focused on vital security duties.
5. **Improved overall security posture:** It can be difficult and time-consuming to manually manage the risks associated with different threats, such as ransomware, phishing, and denial-of-service assaults. Organizations may effectively prioritize tasks and prevent risks by using AI to detect various threat types in real time.
6. **Improved reaction and detection:** An essential part of network and data security is threat identification. AI-enabled cybersecurity makes sure

that suspicious data is found quickly and that new threats are addressed more methodically and quickly.

7.2 IS AUTOMATING CYBERSECURITY A SECURE PRACTICE?

Nowadays, human interaction plays a major role in cybersecurity. Still, AI is a useful tool for efficiently automating some jobs, such as system monitoring. As cyberattacks get more complex, automating these procedures not only improves an organization's threat intelligence skills but also saves a significant amount of time when it comes to recognizing and responding to new threats [18, 19].

Because AI has proven to be successful in a variety of business industries, it is a safe choice for cybersecurity automation. To expedite the onboarding process for new hires, for example, and guarantee that they have the resources and access levels they need to carry out their jobs well, both the IT and HR departments currently use AI.

Automation becomes more important when one acknowledges the continual absence of competent cybersecurity experts. Without having to continuously look for new, highly qualified employees, it helps businesses to maximize their security investments and streamline processes.

7.3 ADVANTAGES OF INCORPORATING AI INTO CYBERSECURITY ARE MANIFOLD

1. **Cost-effectiveness:** Speeding up data collecting through AI integration with cybersecurity improves the responsiveness and effectiveness of incident management. Security personnel can focus on more strategic operations that bring substantial value to the company because they are not required to perform labor-intensive, manual duties.
2. **Reducing human mistake:** Human error can be expensive when it occurs, and traditional security defenses are frequently vulnerable to it. AI in cybersecurity reduces the need for human intervention in most security procedures, offering a more effective method that allows personnel to be allocated to areas that most require them.
3. **Improved decision-making:** Organizations may detect and address possible weaknesses in their security plans with the help of cybersecurity process automation. This makes it easier to put established protocols into practice, which creates more secure IT environments.

Organizations must realize that hackers modify their strategies to thwart the latest AI cybersecurity technologies. Hackers use AI to plan sophisticated assaults and distribute cutting-edge malware that targets both conventional and AI-enhanced systems.

7.3.1 UNDERSTANDING IOT AND CYBERSECURITY THREATS

7.3.1.1 Overview of IoT and its attack surface

The IoT refers to a network of interconnected physical devices, including vehicles, buildings, and other objects, equipped with sensors, software, and connectivity to collect and share data. Simple sensors to intricate systems, such as industrial control systems, driverless cars, and smart household appliances, are all examples of IoT devices.

Because IoT devices lack robust security protections and are frequently linked to the internet, they offer a large attack surface to cybercriminals [21]. Vulnerabilities in IoT devices can be used by attackers to take over devices, interfere with services, or acquire private information.

Some common vulnerabilities in IoT devices include

Insecure communication: IoT devices often communicate using unencrypted protocols, rendering them susceptible to eavesdropping and man-in-the-middle (MitM) attacks.

Weak authentication: Brute-force attacks can easily target IoT devices due to their weak or default passwords.

The dearth of firmware upgrades: IoT devices might not get regular firmware updates, which leaves them open to vulnerabilities that are already known.

Insecure software development: Insecure coding techniques may be used in the development of IoT devices, leaving them open to vulnerabilities like injection attacks.

Sensitive data may be stored insecurely on IoT devices, leaving them open to data breaches.

IoT devices can be compromised and incorporated into botnets, like Mirai, to launch Distributed Denial of Service (DDoS) attacks against other systems.

To safeguard IoT devices, organizations must establish robust security measures, including secure data storage, regular firmware updates, robust authentication mechanisms, and encrypted communication protocols. Implementing intrusion detection and prevention systems (IDPSs), coupled with vigilant monitoring of IoT device activity, is crucial for detecting and preventing potential attacks.

Cybercriminals can target a large area of IoT devices, leaving them open to cyberattacks. Organizations should put strong security measures in place and keep an eye out for unusual activities on their IoT devices to ensure device security. By implementing a proactive IoT security strategy, organizations can significantly reduce the risk and impact of successful cyberattacks.

7.3.1.2 Types of cybersecurity threats in IoT

1. **Confidentiality Attacks** – Confidentiality attacks in IoT can affect in various ways, including:

   Eavesdropping: Hackers have the ability to eavesdrop on communications between servers and IoT devices, potentially gaining access to confidential information.

   MitM attacks: In this type of attack, a hacker sits in between two parties to a communication, intercepting and perhaps changing the data that is sent back and forth between them.

   Sniffing: By using tools for network sniffing, hackers can record and examine network traffic and perhaps gain access to private information sent between servers and IoT devices.

   Data breaches: To get and steal private information, hackers might take advantage of flaws in servers, networks, and IoT devices.

   Social Engineering: Cybercriminals can deceive users into divulging personal information by employing social engineering strategies like phishing.

   Organizations and individuals can take different precautions, including the following, to prevent confidentiality threats on the IoT:

   Put robust encryption into practice: To help prevent sniffing, MitM attacks, and eavesdropping, encrypted data transmission between IoT devices and servers is recommended.

   Implementing HTTPS encryption safeguards against MitM attacks and data interception.

   By implementing access control, you may help avoid unauthorized access and data breaches by limiting access to IoT devices and sensitive data.

   Frequently applying security updates: Keeping IoT device firmware and software updated might help guard against known vulnerabilities.

   Training users: One way to help stop social engineering attacks is to educate users about the dangers of confidentiality attacks and how to prevent them.

   Organizations and individuals can assist lower the danger of IoT confidentiality threats and protect sensitive data by putting these precautions into practice.

2. **Integrity Attacks** – There are several ways in which an IoT integrity attack can occur, such as:

   Data tampering: Data sent between IoT devices and servers can be intercepted and modified by hackers, who may then store inaccurate data or change the data altogether.

Firmware tampering: By taking advantage of vulnerabilities in IoT devices' firmware, hackers can alter the software and possibly make the devices malfunction or behave strangely.

Injection attacks: Malicious code can be injected by hackers into IoT devices, possibly leading to unexpected behavior or malfunctions.

Replay attacks: Data exchanges between IoT devices and servers can be recorded and replayed by hackers, sometimes leading to erroneous data being saved.

Organizations and individuals can take a number of actions to prevent integrity attacks on the IoT, including:

Using robust encryption: Data manipulation and injection threats can be prevented by encrypting data transferred between servers and IoT devices.

Using secure communication protocols: Secure communication protocols, like HTTPS, can be used to defend against MitM attacks and data breaches.

Applying access control: Restricting access to sensitive data and IoT devices can assist prevent unauthorized data alteration.

Applying security upgrades on a regular basis: IoT device firmware and software updates can help guard against known vulnerabilities.

Deploying Intrusion Detection Systems (IDS): By keeping an eye on network traffic and notifying administrators of any unusual activity, IDS can assist in the detection and prevention of unwanted data change.

Organizations and individuals can assist lower the risk of IoT integrity attacks and safeguard the integrity of their data by putting these precautions into practice.

3. **Availability Attacks** – In the context of the IoT, an availability attack is any unapproved disruption of access to networks, devices, or services with the intention of making them inoperable or unusable.

These assaults have the potential to seriously harm an organization's finances and reputation in addition to having a substantial negative influence on the availability and functionality of IoT systems.

IoT availability threats can happen in several ways, such as:

Denial of Service (Do S) attacks: Cybercriminals can overwhelm networks, IoT devices, or services with an excessive volume of traffic, resulting in them being inefficient or unavailable.

DDoS attacks: These are coordinated attacks against IoT devices, networks, or services by hackers employing botnets or other malicious tools, that disrupt a large area.

Ransomware attacks: Cybercriminals are able to encrypt data on networks or IoT devices and then demand a ransom to unlock the key.

Physical attacks: Cybercriminals have the ability to physically harm or destroy IoT equipment, networks, or services, making them inaccessible or useless.

IoT availability threats can be avoided by individuals and organizations by implementing a range of strategies, including:

Implementing strong access controls: By limiting access to devices, IoT networks, services, and effective access controls can be implemented to help prevent unwanted access and lower the risk of attacks.

Using network segmentation: Network segmentation can be used to contain and lessen the impact of assaults by dividing IoT devices, networks, and services into distinct parts.

IDPS implementation: IDPS can assist in identifying and preventing unwanted access to IoT networks, devices, and services.

Performing regular security updates: Frequently updating IoT devices' firmware and software can help guard against known vulnerabilities and lower the likelihood of attacks.

Implementing redundancy and failover mechanisms: Implementing failover and redundancy procedures can assist guarantee that IoT systems remain operational and available in the event of an attack.

By implementing these measures, individuals and organizations can help reduce the risk of availability attacks in IoT and ensure the availability and operation of their IoT systems.

- Case studies of notable IoT attacks

Here are some notable IoT attacks that have occurred in recent years:

1. **Stuxnet (2010):** An advanced malware particularly designed to attack industrial control systems, resulting in actual harm to equipment and demonstrating the possibility of real-world repercussions from IoT attacks.

   Discovered in 2010, the Stuxnet computer worm is believed to be a joint creation of Israel and the United States, designed to disrupt Iran's nuclear program. It is regarded as one of the most destructive and advanced cyberweapons ever made.

   Context: The goal of Stuxnet's design was to take down Siemens SCADA systems, which regulate industrial operations like those found in nuclear power plants. The worm's purpose was to alter the centrifuges' speed to damage them beyond repair and induce malfunctions in Iran's nuclear program.

   Impact: At least 14 industrial locations in Iran, including the uranium enrichment facility in Natanz, are thought to have been affected by Stuxnet. Iran's nuclear program is thought to have suffered severe harm from it, delaying it for several years.

   Technical analysis: To avoid detection and infect targeted computers, Stuxnet is a very sophisticated malware that makes use of a number of zero-day exploits and sophisticated tactics. It is extremely complex and focused because it is made to target systems and settings.

Countermeasures: Stuxnet changed the landscape of cybersecurity by bringing to light the possibility that cyberweapons could result in physical harm and devastation. It also proved that stronger cybersecurity defenses are required to safeguard vital infrastructure.

2. **Jeep Hack (2015):** Researchers showed off a remote takeover of the infotainment system of a Jeep Cherokee, which allowed them to take control of the steering, brakes, and other vital machinery.

   The UConnect entertainment system of a Jeep Cherokee was used by two cybersecurity researchers, Chris Valasek and Charlie Miller, to demonstrate a remote attack in 2015. All the car's features, including the radio, wipers, transmission, air conditioning, and brakes, could be operated from a laptop located kilometers away.

   The researchers were able to access the internal computer network of the vehicle through a vulnerability in the cellular connection of the UConnect system, making the attack feasible. The actual parts of the car might then receive commands from them [5].

   An important case study in IoT cybersecurity, the Jeep Hack illustrated the possible dangers and repercussions of vulnerabilities in connected devices. The hack forced Chrysler, the maker of the Jeep Cherokee, to provide a patch to address the vulnerability and raised questions about the security and safety of connected automobiles.

   The event also highlighted the significance of regular security updates and patches, as well as the necessity of strong security measures in the design and development of IoT devices. It also demonstrated how AI-powered behavioral analytics, threat intelligence, and IDS may be able to recognize and stop such attempts in the future.

3. **Smart City Attack (2016):** To illustrate how IoT attacks may interfere with essential services, researchers demonstrated a number of attacks against smart city infrastructure, including traffic signals and public transit systems.

   SFMTA's computer systems were encrypted by the intruders using ransomware, and they demanded ransom to restore them. The attack severely disrupted the city's public transportation network, resulting in the payment and ticketing machines being shut down.

   This attack highlights the potential weaknesses of networked systems and the significance of cybersecurity measures to safeguard key infrastructure, even though it did not directly target smart city technology.

4. **Mirai Botnet (2016):** Using a botnet of compromised IoT devices, an enormous DDoS attack was launched, causing extensive internet disruptions, and impacting well-known websites including Twitter, Netflix, and Amazon.

   A notable cybersecurity incident in 2016 was the Mirai Botnet, which launched a massive, DDoS attack by using vulnerable IoT devices. Paras Jha, an undergraduate student at Rutgers University, developed the botnet. He used it to operate Minecraft game servers for profit and execute attacks against the networks of his own university.

The malware known as Mirai looked for open Telnet ports on the internet and tried to access them by exploiting the default passwords that were commonly used by IoT devices.

Once it had access, it changed the devices into remotely controllable bots. Over 600,000 bots made up the Mirai botnet at its height, and it was utilized to perform DDoS assaults with peak traffic rates of 1.2 terabits per second.

Targeting Dyn, a significant DNS provider, the Mirai botnet interfered with users' ability to access popular websites including GitHub, Amazon, Twitter, and Netflix. Millions of customers were impacted by the attack, which also severely disrupted internet access. It also hurt the reputation of the impacted companies and resulted in large financial losses for firms that depend on online services.

The hack brought attention to the need for stronger security protocols for the IoT devices and increased understanding of the dangers that come with the growing number of connected gadgets. The attack teaches us the following things:

Security by design: Default passwords for IoT devices should be changed right once after installation, and security should be considered throughout the design process.

Software updates: To avoid vulnerabilities, IoT devices should be updated often with the newest security patches.

Authentication: To stop unwanted access to IoT devices, robust authentication methods must be implemented.

Collaborative approach: To mitigate the security risks associated with IoT devices, cooperation between manufacturers, service providers, and security specialists is crucial.

Stronger security mechanisms, such as multi-factor authentication, device encryption, and regular security upgrades, should be implemented by IoT device manufacturers to thwart future assaults. Additionally, users should be made aware of the security threats connected to IoT devices and encouraged to follow security best practices, such often updating software and changing default passwords.

5. **BrickerBot (2017)** assault that overwrote firmware on susceptible IoT devices, rendering them permanently unusable.

Malware known as BrickerBot first surfaced in 2017 and was directed at IoT devices. By wiping storage and overwriting firmware, it was intended to "brick" devices, making them inoperable.

Devices with insufficient or no password protection were the target of the malware, which was distributed over Telnet. After infecting a device, BrickerBot would replace its firmware, make it unusable, and stop it from connecting to the internet again.

Under the alias "Janit0r," the person who invented BrickerBot asserted that he did it in protest of the shoddy security of IoT gadgets. Pursuing "to force manufacturers to take action to secure their devices and raise awareness about the dangers of poor IoT security," was their declared mission.

BrickerBot had an important effect. Many machines were thought to have been infected by the malware, leading to extensive disruption and monetary losses. Though it also brought attention to the need for better security protocols for IoT devices, the malware did have one beneficial effect.

The following lessons can be learned from the BrickerBot case study:

IoT security is a critical issue: The BrickerBot case study highlights the importance of securing IoT devices. To mitigate risks like BrickerBot, manufacturers and developers must prioritize security at every stage of IoT device design and development.

Password protection is essential: Devices with weak or no password protection are particularly vulnerable to attacks like BrickerBot. Manufacturers and developers must implement strong password protection measures and encourage users to change default passwords.

Collaboration is key: The BrickerBot case study highlights the need for collaboration between manufacturers, developers, and security experts to address the security risks associated with IoT devices.

To prevent future attacks like BrickerBot, it is recommended that manufacturers and developers take the following measures:

Implement strong password protection measures: Manufacturers and developers should implement strong password protection measures, such as multi-factor authentication, to prevent unauthorized access to IoT devices.

Regularly update firmware and software: Regular updates can help to address security vulnerabilities and prevent attacks like BrickerBot.

Implement access controls: Enforce strict access controls, including firewalls and network segmentation, to prevent unauthorized IoT device access.

Collaborate with security experts: Manufacturers and developers should collaborate with security experts to identify and address security vulnerabilities in IoT devices.

The BrickerBot case study highlights the importance of securing IoT devices and the need for collaboration between manufacturers, developers, and security experts to address security risks. By implementing strong password protection measures, regularly updating firmware and software, implementing access controls, and collaborating with security experts, manufacturers and developers can help to prevent attacks like BrickerBot and protect the security and integrity of IoT devices.

6. **Hajime Botnet (2017):** A large-scale botnet was discovered that targeted vulnerable IoT devices, with the potential to launch DDoS attacks or other malicious activities.

Hajime Botnet is a worm that targets IoT devices and was first detected in 2017. It primarily targets devices that have weak or default passwords and is spread through peer-to-peer (P2P) communication. Hajime is designed to be persistent, meaning that even if the device is rebooted, the malware will reinstall itself.

Unlike other botnets, Hajime does not appear to have any malicious intent and has not been used in any known cyber-attacks [23]. Instead, it seems to be focused on securing the devices it infects by closing known vulnerabilities and disabling the Telnet service. Hajime also has a built-in kill switch that can be activated by its creator to prevent any further spread or damage.

Despite its seemingly benign intentions, Hajime highlights the potential for IoT devices to be used as a platform for cyber attacks. The lack of security controls and default passwords on many IoT devices makes them an easy target for malicious actors. Additionally, the widespread use of these devices creates a large attack surface that can be difficult to secure.

To prevent Hajime and other IoT botnets from infecting devices, it is recommended to:

Change default passwords on IoT devices
Regularly update firmware and software on IoT devices
Implement access controls, such as firewalls and network segmentation, to limit exposure to IoT devices
Use strong, unique passwords for IoT devices
Disable unnecessary services, such as Telnet, on IoT devices

By adhering to these security measures, organizations and individuals can significantly reduce the risk of IoT device compromise by botnets, such as Hajime, and their subsequent exploitation in cyberattacks.

7. **VLAN Hopping Attack (2018):** A vulnerability in a popular smart home device provided a gateway for attackers to infiltrate the local network, potentially leading to further malicious activities.

VLAN hopping is a cyberattack technique that exploits network switch vulnerabilities to gain unauthorized access to restricted network segments.

There are two main types of VLAN hopping attacks: double tagging and switch spoofing.

Double tagging involves adding an additional VLAN tag to a frame, which can allow the frame to be forwarded to a different VLAN. This can be done by exploiting the way that switches handle frames with multiple VLAN tags.

Switch spoofing involves impersonating a switch and negotiating a trunk with another switch. This can allow the attacker to gain access to all of the VLANs that the switch is connected to.

VLAN hopping attacks can be prevented by implementing proper security measures, such as:

Disabling Dynamic Trunking Protocol on switch ports that should not be trunks.
Configuring switch ports to only allow traffic from authorized devices.
Using private VLANs to restrict traffic between VLANs.
Implementing 802.1X authentication on switch ports to control access to the network.

VLAN hopping attacks can be a serious threat to network security, but they can be prevented with proper planning and implementation of security measures.

8. **Baby Monitor Hack (2019):** A family's baby monitor was hacked, allowing the attacker to remotely activate the microphone, and camera and speak to the child in the room.

The case study you provided is about the Mi-Cam baby monitor made by miSafes. The security researchers from SEC Consult discovered critical security vulnerabilities in the device, including outdated firmware and a lack of proper encryption. These vulnerabilities allowed attackers to gain unauthorized access to the baby monitor and potentially spy on the child.

The researchers responsibly disclosed the vulnerabilities to miSafes in December 2017 but received no response. As a result, they presented their findings at a cybercrime conference in Vienna and went public with their concerns.

The researchers recommended that customers keep the baby monitors offline until further notice due to unresolved security issues. This case study highlights the importance of proper security measures in IoT devices and the potential consequences of neglecting them.

In 2019, there were also reported cases of baby monitor hacks. For example, a couple in Houston, Texas, reported that their baby monitor was hacked, and the hacker was using it to talk to their child. The hacker had gained access to the monitor by exploiting a vulnerability in the device's default password.

Another incident occurred in Ohio, where a family's baby monitor was hacked, and the hacker was able to move the camera and watch the child. In this case, the hacker had gained access to the monitor by using a brute-force attack to guess the device's password.

These incidents highlight the need for proper security measures in baby monitors and other IoT devices. Manufacturers should ensure that their devices have proper encryption, strong passwords, and regular security updates. Consumers should also take steps to secure their devices, such as changing default passwords and keeping their devices up to date with the latest security patches.

7.3.2 ROLE OF ARTIFICIAL INTELLIGENCE IN IOT CYBERSECURITY

7.3.2.1 AI-powered intrusion detection systems

AI-powered IDS leverage AI and ML to identify and counteract cyberattacks. These systems analyze network traffic, detect anomalous patterns indicative of security breaches, and continuously learn from past attacks to enhance their threat detection capabilities. [7].

The ability of AI-powered IDS to detect novel forms of attacks and adjust to evolving threats is one benefit over conventional security systems. Additionally, they can lessen false negatives and positives, which are frequent problems with conventional IDS.

Nevertheless, there are certain difficulties with AI-powered IDS. An important obstacle is the possibility of adversarial AI assaults, in which perpetrators tamper with input data on purpose to trick the system into letting hostile traffic into the network [24]. AI-powered IDS may find this, especially concerning because these kinds of attacks can target their ML algorithms.

Large volumes of data are required to train the AI models, which presents another difficulty. It can be challenging to acquire access to a sizable and varied dataset of network traffic for AI-powered IDS, which is necessary for them to identify and stop cyberattacks.

Despite these difficulties, network security is being strengthened and cyberattack defense is being provided by AI-powered IDS, which are growing in popularity. From workplace networks to critical infrastructure systems, they are being used in a multitude of contexts and are predicted to become much more common in the years to come.

To fully realize the potential of AI-powered IDS, more research and development is needed to identify and thwart adversarial AI attacks, as well as to gather and analyze the massive volumes of data needed to train the AI models in an efficient and effective manner. We can guarantee that AI-powered IDS continue to be a useful instrument for thwarting cyberattacks in the future by tackling these issues.

7.3.2.2 AI-powered threat intelligence

Threat intelligence powered by AI is a state-of-the-art cybersecurity tactic that uses AI and ML more effectively to detect and neutralize cyberthreats. AI-powered threat intelligence systems are able to continually analyze data from several sources, including network traffic, endpoints, and external threat intelligence feeds, in order to identify patterns and abnormalities that may indicate a security breach.

By employing ML algorithms to learn from historical data and adapt to new threats, these systems are able to detect potential attacks and take necessary action before they have a chance to cause damage. AI-powered threat intelligence can also help reduce false positives and negatives, freeing up security professionals to focus on real threats and fortify their entire cybersecurity posture.

Vectra AI, which combines AI and ML to offer real-time threat identification and response, is one example of an AI-powered threat intelligence system. On its website, Vectra AI claims to be able to minimize warning noise by up to 80% and identify dangers up to nine months ahead of typical security solutions.

Improved incident reaction times, better insight into network and endpoint activities, and higher accuracy in threat detection and classification are some further advantages of AI-powered threat intelligence. AI-powered threat intelligence will play a bigger role in assisting enterprises in preventing possible attacks and safety. One AI-powered threat intelligence solution is Vectra AI, which offers real-time threat assessment and response with a combination of AI and ML. Vectra AI states on its website that it can detect threats up to nine months ahead of conventional security solutions and reduce warning noise by up to 80%.

Additional benefits of AI-powered threat intelligence include faster incident response times, more insight into network and endpoint activity, and more precise threat identification and classification. As cyberthreats continue to evolve and get more complex, AI-powered threat intelligence will be more important in helping businesses prevent potential assaults and protect their critical assets.

7.3.2.3 AI-powered behavioral analytics

"AI-powered Behavioral Analytics," a cybersecurity method that looks at user and entity behavior to find potential security threats, using AI and ML. By learning and creating patterns of usual behavior, AI-powered behavioral analytics may detect flaws and potential threats, such as insider assaults or compromised credentials.

Network entities, devices, machines, and human users are just a few of the many objects for which behavioral analytics applications are accessible. AI systems can use behavioral analysis to identify potential security threats and enhance their threat detection capabilities. Among the most important steps in AI-powered behavioral analytics are pattern recognition, anomaly detection, AI training, and data collection.

The process of gathering information from various sources, such as network traffic, security events, and user activity records, is known as data collecting. As part of AI training, ML models are taught to recognize patterns in data. Finding departures from normal behavior patterns is known as anomaly detection, whereas recognizing patterns in user and entity behavior is known as pattern recognition.

There are numerous advantages of using AI-powered behavioral analytics in cybersecurity. For instance, it can provide real-time threat detection and response, allowing companies to identify issues and take action more quickly. Because of its enhanced capacity to distinguish between normal and abnormal behavior, it can also help organizations reduce false positives.

There are risks and restrictions associated with using AI-driven behavioral analytics, though. Biases in the data used to train the AI models, for instance, may lead to inaccurate threat detection. Another issue is false negatives, where real threats go unnoticed. There is also a chance that behavioral analytics driven by AI will be used for privacy issues and surveillance.

AI-driven behavioral analytics has the potential to be a powerful tool for cybersecurity, since it can help firms reduce false positives and identify threats in real time. However, it's imperative to be aware of the limitations and concerns related to its use and to ensure that ethical and data privacy concerns are met.

7.3.2.4 AI-powered network segmentation

AI-powered network segmentation is a new cybersecurity trend that automatically segments networks and applies zero-trust security standards by utilizing AI and ML. Lowering the attack surface and boosting security is the goal of limiting an attacker's ability to move laterally within a network.

One example of an AI-powered network segmentation solution is Zscaler Private Access, which uses ML algorithms to cluster users and apps before offering user-to-app segmentation policies. As a result, companies can implement least-privileged access rules on a large scale and reduce the number of internal programs that are accessible.

Another illustration is Edgewise Networks, which uses ML and AI to automatically create and maintain zero-trust network micro-segmentation. This reduces the strain on IT and security specialists and improves security by assuming that an attacker would eventually break in and attempt to lock them down to a specific segment.

Taking everything into account, segmenting networks with AI offers numerous benefits, including:

Increased security: AI-powered network segmentation can assist enterprises in lowering their attack surface, preventing attackers from moving laterally, and implementing least-privileged access restrictions on a large scale.

Automation: The process of network segmentation can be made more efficient for IT and security professionals by using AI-powered network segmentation.

Scalability: Organizations may adapt their security measures to match the growth and changes of their networks by utilizing AI-powered network segmentation.

Decreased complexity: Organizations can find it easier to establish and maintain network segmentation thanks to AI-powered solutions that streamline the process.

However, there may be certain disadvantages to AI-powered network segmentation, such as:

Complexity: Network segmentation driven by AI has the potential to add new layers of complexity, making network management and troubleshooting more difficult.

False positives: Network segmentation driven by AI may produce false positives, which can cause problems and increase the workload of IT and security personnel.

Lack of visibility: Network segmentation driven by AI may reduce the ability to see network traffic, which will make it more difficult to identify and address security breaches.

Network segmentation powered by AI is a significant cybersecurity development that can assist enterprises in automating the process, lowering their attack surface, and enhancing security. It does, however, also bring additional complexity and certain disadvantages, necessitating careful thought and execution.

Here are some Tabular data related to Cybersecurity Attacks and the IoT with a focus on the role of AI:

1. Types of IoT Cybersecurity Attacks

Attack type	Description
Botnets Attack	IoT devices are infected with malware and controlled by a botnet to launch DDoS attacks
Man-in-the-middle Attack	Attackers intercept communication between IoT devices and the network
Data Theft Attack	Attackers steal sensitive data from IoT devices
Physical Attack	Attackers physically damage or manipulate IoT devices
Denial of Service Attack	Attackers flood IoT devices with traffic to disrupt their operation

2. AI-Powered IoT Cybersecurity Solutions

Solution type	Description
Intrusion Detection Systems	AI algorithms detect unusual behavior or patterns in IoT network traffic
Threat Intelligence Platforms	AI algorithms analyze data from various sources to identify and prevent threats
Behavioral Analytics	AI algorithms monitor user and device behavior to detect anomalies
Network Segmentation	AI algorithms segment IoT devices to limit the spread of attacks

3. **Real-world IoT Cybersecurity Attacks**

Attack	Description
Mirai Botnet	A botnet that infected IoT devices and launched DDoS attacks
Stuxnet	A malware that targeted industrial control systems
Marai Botnet	A botnet that infected IoT devices and launched DDoS attacks
VPNFilter	Malware that targeted IoT devices and stole data

7.3.3 ARTIFICIAL INTELLIGENCE IN IOT CYBERSECURITY: PROTECTIVE MEASURES

7.3.3.1 AI techniques for intrusion detection

1. **ML:** IDS are crucial components of the cybersecurity architecture of today. AI and ML techniques are being employed more and more to improve the accuracy and efficacy of IDS.

   The following are some applications of ML in intrusion detection:

   Finding Anomalies: ML techniques can be used to identify any deviations from the norm that could point to an intrusion by profiling usual network behavior. This approach works very well for identifying zero-day assaults and sophisticated threats that traditional signature-based IDS could miss.

   Supervised Learning: Using a labeled dataset of typical and anomalous network traffic, an ML algorithm is taught. Based on this training, the system may then classify fresh occurrences of network traffic as either normal or anomalous.

   Unsupervised Learning: In this approach, network traffic data is analyzed using an ML algorithm to find patterns and clusters without any prior understanding of what constitutes regular or abnormal traffic. This method can be helpful in identifying novel and unidentified hazards.

   Deep Learning: DL, employing artificial neural networks, can analyze vast datasets to identify patterns and anomalies indicative of security breaches. By training a DL model on network traffic data, organizations can enhance their intrusion detection capabilities.

   Reinforcement Learning: Reinforcement learning is an ML technique where an agent learns to make decisions by interacting with an environment and receiving rewards or punishments based on the outcomes of its actions. When it comes to intrusion detection, an IDS can be trained using a reinforcement learning algorithm to decide whether to issue an alarm in response to input from the surrounding environment.

   By automating the study of massive amounts of network traffic data and spotting trends and anomalies that can point to a security breach,

ML can assist increase the precision and effectiveness of intrusion detection. It is imperative to acknowledge that ML is not a panacea for cybersecurity and must be employed in conjunction with a full defense-in-depth approach.

2. **Deep Learning (DL):** DL is a specialized field within ML that employs multi-layered artificial neural networks to extract intricate patterns from data. To increase the precision and effectiveness of identifying cyber-threats, it has been used more and more in the intrusion detection space [28]. DL can be applied in the following ways to intrusion detection:

   Anomaly Detection: Network traffic data can be used to train DL models, which then teach them how a system should behave normally. After being trained, the model can identify any departures from the expected behavior, which could point to an intrusion.

   Feature Learning: By automatically identifying and extracting features from unprocessed network traffic data, DL models do not require human feature engineering. This can lessen the time and effort needed to prepare data for analysis and increase the accuracy of intrusion detection.

   Transfer Learning: Large datasets can be used to pre-train DL models, which can subsequently be refined on smaller, domain-specific datasets. This method can help increase intrusion detection accuracy in situations where data is few or difficult to gather.

   Deep Autoencoders: An effective DL model for anomaly detection is the deep autoencoder. They rebuild the input data from a concise representation of typical network traffic data that they have learned. An anomaly may be indicated by any appreciable variations between the input and recreated data.

   Convolutional Neural Networks (CNNs): CNNs are a subset of DL models that excel in signal and image processing applications. Through the analysis of network traffic data as a signal and the identification of patterns and abnormalities that can point to a security breach, they can be used for intrusion detection.

   By automatically learning and extracting features from network traffic data and finding intricate patterns and abnormalities that may point to a security breach, DL can significantly increase the precision and effectiveness of intrusion detection. It's crucial to remember, too, that DL models can be computationally demanding and need a lot of data to properly train. Because of this, they might not be appropriate in every situation and ought to be incorporated into a thorough defense-in-depth plan.

3. **Anomaly Detection:** Anomaly detection is a technique used in intrusion detection to find anomalous or unexpected behavior that departs from baselines or patterns that have been established. It's an effective strategy that can identify new attacks and questionable activity that signature-based detection techniques could miss.

The following are some applications of AI approaches for intrusion detection anomaly detection:

ML: Models of typical behavior can be created using ML techniques, and any departures from these models can be flagged as possible anomalies. Depending on the quantity and caliber of labeled data available, supervised, unsupervised, or semi-supervised learning techniques can be used to do this.

Deep Learning: This approach eliminates the need for human feature engineering by using models that can automatically learn and extract features from unprocessed network traffic data. Because they can learn to recognize subtle patterns and abnormalities that may be difficult to find using more conventional ML techniques, they can be very helpful for anomaly identification in huge and complicated datasets.

Time Series Analysis: To predict the temporal trends in network traffic data and spot anomalies that depart from these patterns, time series analysis techniques can be applied. This can be especially helpful in identifying assaults that exhibit recurring or periodic activity, like DDoS attacks or port scans.

Ensemble Learning: To increase the precision and resilience of anomaly detection, the predictions of several ML models can be combined using ensemble learning techniques like bagging and boosting. This can enhance the IDS's overall performance and lessen the possibility of false positives and false negatives.

Hybrid Approaches: Using a combination of AI algorithms, hybrid approaches can increase anomaly detection's effectiveness and accuracy. An ML model for anomaly detection, for instance, can be fed network traffic data features extracted by a DL model.

Robust intrusion detection methods, like as anomaly detection, can be employed to discover new threats and dubious activities that signature-based methods could overlook. ML and DL are two subsets of AI techniques that can be particularly useful for anomaly detection. These techniques can automatically learn from network traffic data, extract characteristics, and identify complex patterns and abnormalities that could indicate a security breach. It's important to keep in mind that anomaly detection can result in both false positives and false negatives, therefore it should only be used in concert with a comprehensive defense-in-depth strategy.

4. **AI Techniques for Intrusion Prevention:** Intrusion prevention systems (IPS) are essential for proactively defending against cyberattacks. By incorporating AI, IPS can significantly enhance their ability to detect and block threats in real time.

ML: ML algorithms can establish baseline models of normal network behavior. By identifying deviations from these baselines, potential attacks can be detected. These models can be trained using historical data or real-time network traffic to detect both known and emerging threats. Once an attack is detected, the IPS can take appropriate action, such as blocking the source IP address or terminating the connection.

Deep Learning: This approach eliminates the need for human feature engineering by using models that can automatically learn and extract features from unprocessed network traffic data. Because they are able to recognize subtle patterns and abnormalities that conventional ML techniques might find difficult to discover, they can be very helpful for preventing intrusions in vast and complicated datasets.

Network Traffic Analysis: By using techniques like port scans, DDoS assaults, and data exfiltration efforts, one can find suspicious patterns and behaviors in network traffic data. These methods, which can assist in identifying both known and new dangers, can be based on statistical analysis, ML, or DL algorithms.

Signature Generation: AI techniques can be used to automatically generate signatures for known threats based on network traffic data. These signatures can be used to block or alert similar traffic in real-time, improving the accuracy and efficiency of the IPS.

Behavioral Analysis: Behavioral analysis techniques can create models of normal user and device behavior within a network. Any deviations from these established patterns can be flagged as potential threats. These techniques, often powered by ML or DL, can effectively identify both known and unknown threats.

Hybrid Approaches: Enhancing the precision and effectiveness of intrusion prevention can be accomplished through hybrid systems that include several AI techniques. To extract features from network traffic data, for instance, a DL model can be utilized. These features can then be fed into a ML model for categorization.

AI techniques are particularly useful for intrusion prevention because of their capacity to automatically learn from network traffic data, extract characteristics, and recognize complex patterns and abnormalities that may indicate a future attack. However, because intrusion prevention can also lead to false positives and false negatives, it should be used in conjunction with a comprehensive defense-in-depth plan. Furthermore, in order to guarantee that AI techniques are precise and successful in recognizing and avoiding attacks, fine tweaking and setup are necessary [27].

5. **Behavioral-based Access Control (BAC):** The sophisticated access control method known as BAC makes use of AI to automatically detect and adjust the behavior of users and devices within a network. Through the detection and prevention of unauthorized access attempts based on patterns of behavior rather than static rules, behavioral analytics (BAC) can be used to improve the security of IDS. Here are a few instances of how BAC might apply AI techniques:

ML: From historical data or real-time network traffic data, ML algorithms can be utilized to identify patterns of typical behavior. When unauthorized access attempts diverge from the taught patterns, these patterns can be utilized to identify and stop them.

Deep Learning: This approach eliminates the need for human feature engineering by using models that can automatically learn and extract features from unprocessed network traffic data. They can be especially helpful for BAC in big and complicated datasets because they can pick up on minute patterns and abnormalities that conventional ML techniques might find difficult to find.

Anomaly Detection: By using AI approaches, anomaly detection can be utilized to spot departures from the norm and mark them as possibly illegal access attempts. These methods, which can assist in identifying both known and undiscovered dangers, can be based on ML or DL algorithms.

User Behavior Modeling: AI-powered anomaly detection can identify potential unauthorized access by establishing behavioral baselines for users and devices within a network. Deviations from these established norms are flagged as potential security threats. These techniques, often employing ML or DL, can detect both known and unknown threats.

Hybrid Approaches: A variety of AI algorithms can be combined to create hybrid systems that increase BAC's accuracy and efficiency. To extract features from network traffic data, for instance, a DL model can be utilized. These features can then be fed into a ML model for categorization.

Because AI approaches can automatically learn from and adapt to the behavior of people and devices in a network, they can be very helpful for behavioral-based access management. This can enhance the network's overall security by detecting and blocking unwanted access attempts based on patterns of behavior rather than just static rules. But it's crucial to remember that BAC can also produce false positives and false negatives, so it should only be applied in conjunction with a thorough defense-in-depth plan.

6. **AI-driven Firewalls:** The use of AI techniques for intrusion detection, including AI-driven firewalls. According to the article, AI-driven firewalls are an evolution of next-generation firewalls (NGFWs) that use intelligent detection technologies to improve their capability to detect advanced and unknown threats.

   NGFWs identify threats using a static rule database, which makes it difficult to counter sophisticated threats with variations. AI firewalls, on the other hand, train threat detection models using large samples and continuously improve the models using traffic data collected in real time by an intelligent detection engine. By using this method, AI firewalls' threat detection skills are enhanced, making it possible for them to recognize and react to sophisticated threats more successfully.

In order to facilitate parallel processing and comprehensive traffic security detection, IPS, antivirus software, and application identification are all tightly integrated with basic firewall services in network firewalls (NGFWs), as defined by Gartner in 2009. However, NGFWs are experiencing serious issues due to the rapid development of mobility, cloudification, and the IoT, including a wide range of variants and an increase in advanced threats. These hurdles are now too great for the static rule database-based detection of NGFWs. Thus, in order to strengthen an organization's total security posture against more complex cyberthreats, AI-driven firewalls present a promising alternative to traditional firewalls' threat detection capabilities.

7. **Real-world Applications and Case Studies:** There are numerous real-world applications and case studies that demonstrate the effectiveness of AI techniques for intrusion detection. Here are some examples:

Darktrace: Darktrace is a cybersecurity enterprise that focuses on threat detection and response powered by AI. They quickly identify and address cyberthreats by utilizing AI and ML techniques. Over 5,000 enterprises globally, including international banks, telcos, and government agencies, have implemented Darktrace. A ransomware attempt on a major US healthcare firm was discovered and stopped using Darktrace. The malware had gained access to the company's network and was encrypting files as it moved laterally. The security team was notified of the anomalous activity by Darktrace's AI algorithms, which allowed them to confine the attack and stop any data loss.

Cylance: Cylance is a cybersecurity startup with an emphasis on endpoint security enabled by AI. They identify and stop malware and other online threats using ML techniques. Almost 14,000 organizations globally, including government agencies and Fortune 100 firms, have implemented Cylance. A zero-day attack on a major US financial institution was stopped by Cylance. A hitherto unidentified malware was trying to steal confidential information from the organization's network through network intrusion. But Cylance's AI algorithms saw the strange activity and stopped the infection from running, safeguarding the company's information.

IBM: IBM is a tech company that focuses on danger detection and response using AI. They identify and react to cyberthreats instantly by utilizing AI and ML techniques. IBM is used by more than 17,000 organizations globally, including government agencies, international banks, and telecoms. IBM identified and stopped a highly skilled spear-phishing attempt against a major US retail company. In order to trick the target into clicking on a malicious link, a phoney email purporting to be from a reliable source was used in the attack. Nonetheless, the security team was notified by IBM's AI algorithms of the anomalous activity, which allowed them to limit the attack and stop any data loss.

The efficacy of AI methods for intrusion detection and response is illustrated by these case studies. Organizations may improve their entire security posture and safeguard sensitive data by utilizing these solutions, which use ML and AI algorithms to identify anomalous behavior and patterns and promptly identify and stop cyberthreats [34].

7.3.4 ARTIFICIAL INTELLIGENCE IN IOT CYBERSECURITY: OFFENSIVE MEASURES

7.3.4.1 Adversarial AI and its potential for cyberattacks

Adversarial AI refers to the use of AI techniques to manipulate or mislead ML models, such as those used in IDS. The authors identify two main types of adversarial attacks: evasion attacks and poisoning attacks.

Evasion attacks aim to manipulate input data in a way that circumvents the IDS's detection mechanisms. For example, an attacker may make small changes to network traffic data to avoid detection by the IDS. Understanding these attacks is crucial for creating models that can resist evasion and enhance system security.

Poisoning attacks, on the other hand, target the AI model during the training phase. The attacker may compromise the integrity of the model by introducing malicious data during the training phase, leading to misclassifications in real-world applications.

Adversarial training and model assembly are two possible defenses against adversarial attacks that will be covered here. For the purpose of increasing the model's robustness, adversarial training involves subjecting it to deliberately constructed adversarial cases. An improved and more dependable IDS is created by integrating multiple different models. Knowing the strategies and defenses against adversarial AI is essential to preserving the security of IDS systems, as these systems may be threatened by it.

7.3.4.2 Case studies of AI-driven attacks

There have been some instances where AI has been used to enhance cyberattacks. Here are a few examples:

DeepLocker: DeepLocker is a sophisticated malware that uses AI to evade detection and carry out targeted attacks. It was developed by cybersecurity firm McAfee to demonstrate the potential of AI-driven attacks. DeepLocker uses a deep neural network to determine when and where to unlock and execute its malicious payload. It is designed to remain dormant and undetected until it reaches a specific target, making it difficult for traditional antivirus software to detect.

Project Sauron: Project Sauron is a highly advanced and sophisticated cyberespionage campaign that has been active since at least 2011. It was discovered by cybersecurity researchers at Kaspersky Lab in 2016. The campaign used a combination of custom-made malware and AI algorithms to evade detection and steal sensitive data from high-profile targets. Project Sauron used a variety of techniques, including steganography, polymorphism, and modular design, to remain undetected.

AI-Powered Phishing Attacks: Researchers at Open Philanthropy have demonstrated the potential for AI-powered phishing attacks. They trained a DL model to generate convincing phishing emails that can bypass traditional spam filters. The model uses natural language processing techniques to generate emails that are similar in style and tone to legitimate emails. These attacks can be highly effective, as they can trick unsuspecting users into clicking on malicious links or providing sensitive information [28].

Autonomous Botnets: Researchers at Ben-Gurion University in Israel have demonstrated the potential for autonomous botnets that use AI to carry out sophisticated cyberattacks. The researchers developed an AI-powered botnet that can automatically adapt to changes in the network and evade detection by security software. The botnet uses ML algorithms to identify vulnerabilities in the network and exploit them to propagate and carry out attacks [26].

These case studies demonstrate the potential for AI-driven attacks to evade detection and carry out sophisticated cyberattacks. It is essential to stay informed about the latest AI-driven attack techniques and implement robust security measures to prevent such attacks.

7.3.4.3 Countermeasures to adversarial AI

Adversarial attacks can be categorized into two main types: **evasion attacks** and **poisoning attacks**.

Poisoning attacks manipulate the training data used to build IDS, aiming to compromise its effectiveness. Evasion attacks, on the other hand, modify input data to circumvent detection by an already trained IDS. Generative adversarial networks (GANs) can be employed to create synthetic intrusion traffic to assess the impact of these attacks on IDS accuracy.

Experiments involving a GAN to produce synthetic intrusion data and subsequent testing on Decision Tree and Logistic Regression-based IDS were conducted. The findings revealed that the proposed evasion attacks diminished the testing accuracy of both IDS models, with the Decision Tree model exhibiting greater vulnerability. Furthermore, the poisoning attack scenario interfered with the training process of the IDS, and the Logistic Regression model was more susceptible to disruption compared to the Decision Tree model.

To counter adversarial AI in IDS, there are some countermeasures:

Data preprocessing techniques, such as normalization, can be used to reduce the impact of adversarial attacks.

Ensemble learning methods, such as bagging and boosting, can be used to improve the robustness of IDS models against adversarial attacks.

Training an IDS model using adversarial examples, a technique known as adversarial training, can enhance its resilience to adversarial attacks.

Detecting and filtering out adversarial examples using anomaly detection techniques can also be an effective countermeasure.

7.3.5 FUTURE OF ARTIFICIAL INTELLIGENCE IN IOT CYBERSECURITY

7.3.5.1 Emerging trends in AI and IoT cybersecurity

The following are some emerging trends in AI and IoT cybersecurity:

Increasing usage of AI and ML in cybersecurity: As complex cyberthreats become more prevalent, AI and ML are being utilized more and more in cybersecurity to find risks, identify abnormalities, and react fast to incidents.

IoT Device Security: Making sure IoT devices are secure is becoming increasingly important as they proliferate. Finding security holes in IoT devices and stopping illegal access can be aided by AI and ML.

Adversarial AI: Cybercriminals are increasingly using AI to launch sophisticated attacks that can bypass traditional security measures. Adversarial AI can generate deepfakes, phishing emails, and malware that can be challenging to detect.

Supply Chain Attacks: Supply chain attacks are becoming more common, where attackers target third-party vendors and suppliers to gain access to a target organization's network. AI can help in detecting and preventing such attacks.

Cloud Security: As more organizations move their operations to the cloud, ensuring cloud security is becoming increasingly important. AI and ML can help in detecting vulnerabilities in cloud infrastructure and preventing data breaches.

Zero Trust Security: This new trend involves not trusting any person or device by default. Zero trust security can be implemented with the use of AI and ML, which continuously monitor user behaviour and network traffic.

AI-powered Threat Intelligence: This type of intelligence can assist in promptly recognizing and addressing new threats. Large datasets from multiple sources can be analyzed by threat intelligence platforms to find patterns and trends that can point to possible danger.

Autonomous Response: Autonomous response is an emerging trend where security systems can automatically respond to threats without human intervention. AI and ML can help in implementing autonomous response by analyzing data in real time and taking appropriate action.

Privacy-preserving AI: As AI and ML become more prevalent, ensuring privacy is becoming a significant concern. Privacy-preserving AI can help ensure that personal data is protected while still enabling AI-powered cybersecurity.

Cybersecurity training: With the rise of sophisticated cyberthreats, cybersecurity training is becoming increasingly important. AI-powered training platforms can help in providing personalized training to employees based on their roles and skill levels.

AI-powered Edge Computing: AI-powered edge computing can provide benefits to IoT cybersecurity by enabling real-time data processing and analysis, reducing latency, and improving overall system efficiency. AI algorithms can be used to analyze data at the edge and detect potential security threats, such as anomalies and intrusions. Additionally, AI can help optimize edge computing resources and improve network performance, further enhancing cybersecurity measures.

Nonetheless, edge computing driven by AI carries certain potential security vulnerabilities. Adversarial attacks, for example, can be used to influence AI models and compromise edge devices. Furthermore, when data is processed and stored at the edge, privacy and security issues might surface [28].

Researchers and industry professionals are investigating a range of methods, including secure multi-party computation, robust training, and model hardening, to improve the security of AI-powered edge computing in order to tackle these issues. Behavioral analytics, threat intelligence, and IDS are a few examples of cybersecurity tools that can be integrated with AI to increase system security.

IoT cybersecurity can benefit greatly from edge computing driven by AI, but in order to maintain system dependability and safety, it's critical to identify possible security threats and take the necessary precautions.

AI-powered Biometric Authentication: AI-powered biometric authentication is a security technology that uses AI to analyze and verify an individual's unique physical or behavioral characteristics for authentication purposes. Some examples of AI-powered biometric authentication methods include facial recognition, voice recognition, and fingerprint recognition.

Facial recognition technology leverages ML algorithms to scrutinize distinctive facial features and match them against a database of known faces for identity verification. This technology is gaining traction in numerous sectors such as healthcare, finance, and hospitality for access control and authentication applications.

Voice recognition technology, on the other hand, uses AI-powered algorithms to analyze and recognize an individual's unique vocal characteristics, such as pitch, tone, and cadence. This technology is often used for voice-activated authentication systems, such as virtual assistants and call centers.

Fingerprint recognition technology uses AI-powered algorithms to analyze and recognize an individual's unique fingerprint patterns. This technology is commonly used for smartphone and laptop authentication, as well as for access control in secure facilities.

AI-driven biometric authentication surpasses traditional methods like passwords and security questions in several ways. It offers heightened security by relying on distinctive and inimitable physical or behavioral traits. Additionally, biometrics enhance user experience by eliminating the need to remember complex passwords or carry physical tokens.

While AI-powered biometric authentication offers advantages, it also introduces privacy and security challenges. The risk of biometric data breaches and the potential misuse of this sensitive information pose significant concerns. Furthermore, biases within AI-powered biometric systems and the potential for mass surveillance raise ethical and societal questions.

AI-powered biometric authentication offers substantial advantages in terms of security and user experience. However, the potential risks, such as data breaches and privacy infringements, necessitate a cautious and responsible approach. Implementing robust security measures, ensuring data privacy, and maintaining transparency are crucial for the ethical and effective deployment of this technology.

7.3.5.2 Opportunities for further research

There are several opportunities for further research in AI and IoT cybersecurity. These include:

Developing more advanced AI algorithms for detecting and preventing sophisticated cyberattacks on IoT devices, such as those using adversarial techniques.

Exploring the potential of AI for predicting and mitigating the risks of IoT supply chain attacks.

Investigating the use of AI for enhancing IoT device firmware security, such as through automated assessment and hardening techniques.

Examining the ethical and privacy implications of using AI for IoT cybersecurity, including potential biases and discrimination in AI algorithms.

Evaluating the effectiveness of different AI-based cybersecurity solutions for IoT devices in real-world scenarios and identifying potential limitations and challenges.

Developing standards and best practices for AI-based cybersecurity in IoT systems, including issues related to data privacy, transparency, and accountability.

Investigating the impact of emerging AI technologies, such as edge computing and federated learning, on the security of IoT systems [16].

Exploring the potential of AI for enhancing the resilience of IoT systems against physical attacks and other forms of infrastructure disruption.

Investigating the use of AI for detecting and responding to insider threats in IoT systems, such as those involving unauthorized access or data exfiltration.

Evaluating the potential of AI for enabling more proactive and predictive cybersecurity strategies in IoT systems, including through threat intelligence sharing and collaboration.

7.3.5.3 Recommendations for secure IoT development

Here are some recommendations for secure IoT development in AI and IoT cybersecurity:

To fortify IoT devices, organizations should integrate security into every development phase by adhering to the Security Development Lifecycle. IoT endpoint protection is paramount, requiring the closure of high-risk port vulnerabilities and defense against malicious code injection.

IoT gateways should be fortified with robust security measures, including internet access policies and protection against malware infiltration. A secure web gateway can be instrumental in shielding IoT devices from both external and internal cyberthreats targeting web-based traffic.

Cloud APIs demand stringent security protocols, such as authentication, encryption, tokenization, and API gateways, to prevent data breaches. Establishing a secure network with rigorous access controls is essential to limit network access to authorized and verified devices. Finally, employing end-to-end encryption, incorporating both asymmetric and symmetric encryption techniques, is crucial for safeguarding data during transmission.

Protect data storage with effective antivirus solutions, monitoring and scanning tools, and flexible reporting and scanning features.

To maintain a robust security posture, organizations should implement multi-factor authentication, antivirus, and antimalware solutions, and continuously monitor network activity for anomalies.

Adhering to IoT security best practices is essential. This includes developing and documenting a comprehensive security strategy that aligns with the organization's broader IT strategy and business objectives.

By following these recommendations, organizations can develop secure IoT devices and applications that are less vulnerable to cyberattacks.

REFERENCES

1. Mehta, A. A., Padaria, A. A., Bavisi, D. J., & Ukani, V. et al. (2024). Securing the Future: A Comprehensive Review of Security Challenges and Solutions in Advanced Driver Assistance Systems, IEEE Access, vol. 12, pp. 643–678.
2. Velev, D., & Zlateva, P. (2024). Essentials for Developing an Educational Course in Artificial Intelligence in Cybersecurity for Managers. IOS Press.
3. Tiwari, S., Wadawadagi, R. S., Kumar Singh, A., & Kumar Verma, V. (2024). Chapter 6 Cloud Security Risks, Threats, and Solutions for Business Logistics. IGI Global.
4. Zhu, G., Chen, M., Yuan, C., & Huang, Y. (2024). Simple and Efficient Partial Graph Adversarial Attack: A New Perspective, IEEE Transactions on Knowledge and Data Engineering, vol. 36, pp. 4245–4259.
5. Shiwen, S. S., Zhen, G. Z., Tianling, L. T., Chenya, B. C., Yuqiao, N. Y., & Yang, C. Y. (2023). Research and Analysis of Vulnerabilities in Intelligent Connected Vehicle Components, in 2023 6th International Conference on Data Science and Information Technology (DSIT).
6. Wisdom, D. D., Vincent, O. R., Igulu, K. T., & Arowolo, M. O. et al. (2023) "Mitigating Cyberthreats in Healthcare Systems: The Role of Artificial Intelligence and Machine Learning", Institution of Engineering and Technology (IET)
7. Alshahrani, E., Alghazzawi, D., Alotaibi, R., & Rabie, O. (2022). Adversarial Attacks Against Supervised Machine Learning Based Network Intrusion Detection Systems. PLoS One, vol. 17(10), p. e0275971.
8. Ye, Q., Zhao, S., Wang, F., Zhang, H., & Gu, X. (2021). A Comprehensive Survey on Machine Learning for IoT Security and Privacy. Journal of Parallel and Distributed Computing, vol. 153, pp. 237–252. [DOI: 10.1016/j.jpdc.2021.04.015]
9. Deka, G. C., & Singh, K. D. (2021). A Survey on Security and Privacy Issues in Internet of Things (IoT) Devices and Applications. Computers & Security, vol. 103, p. 102141.
10. Sankar, P. C., & Tewari, H. (2021). Security Concerns in the Internet of Things (IoT) Ecosystem. In Security and Privacy in Cyber-Physical Systems (pp. 1–29). CRC Press.
11. Narayanan, A. et al. (2021) AI-Driven Security Solutions for IoT Systems: A Comprehensive Survey, IEEE Internet of Things Journal, vol. 8, no. 10, pp. 8111–8145.
12. Ray, P. P. (2021). A Survey on Internet of Things Architectures. Journal of King Saud University-Computer and Information Sciences. [DOI: 10.1016/j.jksuci.2016.10.003]
13. Sezer, S., & Dogdu, E. (2021). A Comprehensive Survey on Artificial Intelligence Techniques for Internet of Things Security. IEEE Access, vol. 9, pp. 60599–60620.
14. Sivanathan, A. et al. (2021). Artificial Intelligence for Cybersecurity: A Comprehensive Review. IEEE Access, vol. 9, pp. 25275–25311.
15. Patel, R. B., & Borole, J. (2021). Cyber Security Threats, Vulnerabilities, and Intrusion Detection System in the Internet of Things (IoT), in Handbook of Research on Cyber Security for Internet of Things: Strategies and Advances (pp. 72–97). IGI Global.
16. Alaba, F. A., Othman, M., Hashem, I. A. T., & Alotaibi, F. (2020). Security and Privacy in Fog-Computing-Based Internet of Things: A Review. Concurrency and Computation: Practice and Experience, vol. 32, no. 2, p. e5325. [DOI: 10.1002/cpe.5325]
17. Alaba, F. A., Othman, M., Hashem, I. A. T., Alotaibi, F., & Almogren, A. (2020). Internet of Things Security: A Survey, Taxonomy, and Future Directions. Journal of Network and Computer Applications, vol. 149, p. 102497. [DOI: 10.1016/j.jnca.2019.102497]

18. Bello, A. (2020). IoT Security With Machine Learning: A Review. Artificial Intelligence Review, vol. 53, no. 4, pp. 2703–2733.
19. Liu, Y., Wang, Z., & Xiang, Y. (2020). Artificial Intelligence in Cyber Security. IEEE Communications Surveys & Tutorials, vol. 22, no. 3, pp. 1694–1723.
20. Gai, K., Qiu, M., Zhao, H., & Liu, X. (2020). Survey on Internet of Things Assisted by Artificial Intelligence. IEEE Access, vol. 8, pp. 53989–54007.
21. Almeida, A. et al. (2020). IoT Security: A Survey on Vulnerabilities, Attacks, and Countermeasures, IEEE Communications Surveys & Tutorials, vol. 22, no. 4, pp. 2485–2510.
22. Rass, S. et al. (2020). Security and Privacy in the Internet of Things: Challenges and Solutions. Security and Communication Networks, vol. 2020, article ID 6189294.
23. Tian, Z. et al. (2020). AI-Based Cyber Attacks and Defenses: A Survey. IEEE Access, vol. 8, pp. 1348–1381.
24. Barolli, L. (Ed.). (2024). Advanced information networking and applications, in Proceedings of the 38th International Conference on Advanced Information Networking and Applications (AINA-2024) (vol. 2). Springer Nature.
25. Al-Qurishi, M. et al. (2019). A Review of Security Threats, Attacks and Intrusion Detection System for Internet of Things, in 2019 IEEE International Conference on Informatics, IoT, and Enabling Technologies (ICIoT).
26. Lopez, J. et al. (2018). Cybersecurity and the Internet of Things: Vulnerabilities, Threats, Intruders, and Challenges. Journal of Cybersecurity, vol. 4, no. 1, pp. 19–29.
27. Wang, L. et al. (2018). Artificial Intelligence for Network Intrusion Detection: A Comprehensive Review. IEEE Communications Surveys & Tutorials, vol. 20, no. 1, pp. 41–67.
28. Tzeng, N.-F. et al. (2018). A Review of Deep Learning-Based Network Intrusion Detection Systems. IEEE Access, vol. 6, pp. 7282–7299.
29. Moustafa, N. et al. (2018). A Deep Learning Approach for Network Intrusion Detection System. IEEE Access, vol. 6, pp. 24100–24114.
30. Zarpelão, B. et al. (2017). A Survey of Intrusion Detection in Internet of Things. Journal of Network and Computer Applications, vol. 84, pp. 25–37.
31. Kolias, C. et al. (2017). DDoS in the IoT: Mirai and Other Botnets. Computer, vol. 50, no. 7, pp. 80–84.
32. Zubair, S. M., Mahmood, A., & Daud, A. (2016). A Survey on Security Issues and Solutions in IoT Environment, in 2016 International Conference on Frontiers of Information Technology (FIT), pp. 44–49.
33. Sicari, S. et al. (2015). Security, Privacy and Trust in Internet of Things: The Road Ahead. Computer Networks, vol. 76, pp. 146–164.
34. Vasilomanolakis, E. et al. (2015). Intrusion Detection in Cyber-Physical Systems: A Survey. IEEE Communications Surveys & Tutorials, vol. 18, no. 1, pp. 224–240.

Chapter 8

A comprehensive framework for anonymizing structured health data

Balancing privacy and utility

Nirmal Rajkumar and Neha Chauhan

8.1 INTRODUCTION TO HEALTH DATA PRIVACY AND ANONYMIZATION

The proliferation of health-related data has driven remarkable advancements in medical research and clinical practice, largely facilitated by artificial intelligence (AI) techniques. AI models have significantly enhanced various aspects of healthcare, such as refining lung cancer nodule detection in CT scans and enabling early disease prediction and personalized treatment plans [1]. Despite these benefits, the effectiveness of AI models depends on the availability of extensive datasets, which poses considerable privacy challenges due to the highly sensitive nature of health information [2]. Health data comprises sensitive details like medical histories, diagnoses, treatment records, and personal identifiers. For instance, electronic health records (EHRs) include in-depth information about patients' health statuses, including genetic and treatment histories [3]. The risks associated with data exposure are exemplified by the 2015 Anthem data breach, which compromised the personal information of around 80 million individuals, leading to potential identity theft and misuse [4].

8.1.1 Challenges and trade-offs in anonymizing health data

Anonymizing health data presents significant challenges in balancing privacy with research utility. Traditional techniques such as k-anonymity, l-diversity, and t-closeness provide foundational methods for protecting patient identities but often involve trade-offs that can reduce data quality and usefulness. K-anonymity, for instance, ensures that individuals are indistinguishable within a group of at least k records, but it may lead to data loss or distortions affecting research outcomes. Data masking and generalization techniques, while effective in obscuring sensitive information, may introduce biases or reduce the precision of analytical results. These techniques can safeguard privacy but potentially compromise the value of data for research and clinical applications.

To address these limitations, differential privacy introduces controlled noise into datasets to enhance privacy while maintaining data utility [5]. This

DOI: 10.1201/9781032657264-8

approach allows researchers to analyze trends and patterns without revealing individual-level information. Nevertheless, applying differential privacy requires managing trade-offs to balance privacy protection with analytical robustness. Effective anonymization strategies must continuously assess their impact on data quality and utility, ensuring that data remains valuable for healthcare advancements while minimizing privacy risks.

8.2 BENEFITS OF HEALTH DATA UTILIZATION

The effective use of health data has revolutionized medical research and healthcare delivery, bringing a host of transformative benefits. By leveraging large-scale data analysis, researchers have identified hidden patterns and correlations that provide new insights into disease mechanisms and potential treatments. For instance, genomic studies utilizing data from resources like the UK Biobank have pinpointed critical genetic variants associated with complex diseases, offering valuable targets for new therapeutic interventions. Additionally, health data is integral to the development of predictive models that forecast disease outcomes and personalize treatment plans, which enables more proactive care and reduces the likelihood of severe complications [6].

AI-driven tools also benefit significantly from health data, enhancing diagnostic accuracy and clinical decision-making. Machine learning (ML) algorithms trained on vast datasets of medical images and pathology results have, in some cases, demonstrated diagnostic accuracy that rivals or even surpasses that of experienced radiologists. These algorithms help detect diseases at earlier stages, facilitate faster diagnosis, and improve patient outcomes by allowing healthcare providers to tailor interventions more effectively. Moreover, health data utilization supports population health management efforts by identifying at-risk populations and monitoring the effectiveness of public health initiatives. Overall, the effective use of health data leads to improved patient care, more efficient health systems, and accelerated medical innovation.

8.3 REGULATORY REQUIREMENTS FOR HEALTH DATA

To harness the benefits of health data while protecting individual privacy, regulatory frameworks such as the General Data Protection Regulation (GDPR) and the Health Insurance Portability and Accountability Act (HIPAA) play a crucial role. The GDPR, which sets a global standard for data protection, mandates rigorous measures to ensure the privacy and security of personal health data. Organizations involved in high-risk data processing must conduct Data Protection Impact Assessments to assess and mitigate risks associated with data handling practices [7]. In addition, the GDPR requires prompt notification of data breaches to affected individuals and authorities, thereby enhancing transparency and accountability.

The GDPR also governs international data transfers, ensuring that personal data leaving the European Union is protected to an equivalent standard. This is achieved through mechanisms like adequacy decisions, Standard Contractual Clauses (SCCs), and Binding Corporate Rules (BCRs) [8]. On the other hand, HIPAA specifically focuses on the protection of health-related information in the United States. It mandates the de-identification of health data through methods like the Safe Harbor approach, which involves removing identifiable information to prevent the re-identification of individuals. Together, these regulatory frameworks establish comprehensive guidelines for the safe handling of health data, balancing the need for data utility with the imperative of protecting individual privacy. They provide a foundational structure for organizations to innovate and utilize health data while maintaining compliance and fostering trust among stakeholders.

8.4 IMPLICATIONS FOR GLOBAL DATA SHARING AND ADVANCED SOLUTIONS

The global nature of data sharing under GDPR and similar regulations requires robust mechanisms to ensure personal data protection across borders. Adequacy decisions and tools such as SCCs and BCRs help bridge the gap between differing national data protection standards, facilitating international collaborations while maintaining high data security levels. However, challenges persist, such as local laws in recipient countries that may undermine safeguards. Additionally, the evolving landscape of global data privacy laws necessitates continuous updates and compliance checks to address emerging risks and legal changes.

Given these complexities, there is a critical need for advanced anonymization techniques capable of adapting to diverse regulatory requirements and mitigating potential privacy threats. While traditional methods like de-identification, k-anonymity, and differential privacy provide foundational protection strategies, they must be continuously refined to balance data utility with protection, particularly in the context of global data exchanges.

8.5 EXISTING ANONYMIZATION TECHNIQUES

8.5.1 De-identification methods

De-identification is a fundamental technique aimed at removing or masking personal identifiers to prevent re-identification. The primary goal is to minimize the risk of tracing data back to individuals while preserving its usefulness for analysis. This involves two main steps: removing direct identifiers (e.g., names and Social Security numbers) and modifying quasi-identifiers (e.g., age and ZIP code) [9].

The HIPAA Safe Harbor method is a widely adopted de-identification standard in the United States under the HIPAA. This approach requires the removal of 18 specific identifiers from datasets, including names, geographic subdivisions, and medical record numbers [9]. Its strengths include simplicity and a clear compliance framework, making it a trusted standard within the healthcare industry [9]. However, it has limitations, such as potential incomplete privacy protection due to remaining quasi-identifiers and challenges in preventing sophisticated re-identification attacks. For instance, demographic information like age and ZIP code, even when partially redacted, can be used to re-identify individuals when combined with external data sources [10, 11].

K-anonymity addresses some of these limitations by ensuring that each individual in a dataset is indistinguishable from at least k-1 others with respect to quasi-identifiers. This technique involves grouping records into clusters where each cluster contains at least k records sharing the same quasi-identifiers [12]. While k-anonymity reduces re-identification risk and is adaptable to different datasets, it can significantly impact data utility by requiring generalization or suppression of detailed information [12, 13]. Furthermore, k-anonymity is vulnerable to background knowledge attacks, where attackers use additional information to narrow down the identity of individuals within anonymized groups [12, 13].

8.5.2 Advanced techniques

L-diversity extends k-anonymity by focusing on the diversity of sensitive attributes within each equivalence class of anonymized data. L-diversity ensures that each group of records with the same quasi-identifiers contains at least "l" distinct sensitive attribute values [14]. This approach enhances protection against attribute disclosure and homogeneity attacks, where all records in a k-anonymous group have the same sensitive value [14]. While l-diversity improves privacy, it can introduce complexity into the anonymization process and impact data utility by requiring extensive data modifications to achieve the desired level of diversity [14].

T-closeness further refines anonymization by addressing the distributional differences between sensitive attributes in equivalence classes and the overall dataset. T-closeness ensures that the distribution of sensitive attributes within each class is similar to the distribution in the entire dataset [15]. This method protects against distribution-based and semantic attacks by maintaining consistency in attribute distributions [15]. However, achieving t-closeness can be complex and resource-intensive, potentially reducing data quality due to the adjustments needed to maintain distributional closeness [15].

8.5.3 General limitations

Maintaining Data Quality: Anonymization techniques can affect data accuracy and utility. Methods such as generalization and suppression can reduce the granularity of data, impairing precise analyses. Strategies to mitigate quality loss include data perturbation techniques like differential privacy, which adds controlled noise to data to balance privacy and utility [16], and semantic anonymization, which preserves data context and meaning while anonymizing it [17]. Additionally, post-processing techniques can be used to assess and adjust data quality after anonymization [18].

Vulnerability to Re-identification: Re-identification attacks remain a significant challenge even with advanced anonymization techniques. Techniques such as synthetic data generation, privacy-preserving algorithms like secure multiparty computation and homomorphic encryption, and regular privacy audits can be employed to enhance resistance to re-identification [19–21].

8.6 TECHNIQUES FOR HEALTH DATA ANONYMIZATION: DATA MASKING, DIFFERENTIAL PRIVACY, AND MACHINE LEARNING ALGORITHMS

In the realm of health data anonymization, advancing privacy-preserving techniques is crucial due to the sensitivity of the information involved. This section delves into three core components of a novel approach: data masking and generalization, differential privacy, and advanced algorithms, each contributing uniquely to the protection of sensitive health data while ensuring its usability for analysis.

8.6.1 Data masking and generalization

8.6.1.1 Data masking

Data masking is a technique employed to obscure sensitive data within a dataset, making it challenging to identify or infer the underlying sensitive information. The goal of data masking is to protect privacy while still enabling the dataset to be utilized for various analytical purposes. This technique involves several methodologies:

- **Substitution** involves replacing sensitive data elements with fictitious but realistic values. For instance, real Social Security numbers might be substituted with randomly generated numbers that follow a similar format [22]. This method ensures that the data remains useful for analysis while protecting individual identities.
- **Shuffling** rearranges data within a column so that the original associations between data elements are lost. For example, patient names might be shuffled while keeping their corresponding medical records intact [23]. This method obscures the direct link between the identifiers and the health records, thereby reducing the risk of re-identification.
- **Encryption** transforms data into an unreadable format using cryptographic algorithms. Only authorized users who possess the decryption key can access the original data [24]. Encryption provides a strong layer

of security by ensuring that unauthorized parties cannot easily access sensitive information.

- **Nulling Out** replaces sensitive data with null values or placeholders. This technique is particularly useful when the specific value of the data is not necessary for the analysis [25]. Nulling out effectively removes the sensitive information while maintaining the overall structure of the dataset.

Data masking plays a critical role in anonymization by concealing identifiable details that could otherwise lead to the identification of individuals. By employing masking techniques, organizations can prevent unauthorized access to sensitive information and mitigate the risk of data breaches. For example, in managing personal health information, masking ensures that only non-sensitive data is exposed, protecting patient privacy while allowing for data analysis and research [26].

However, data masking also presents challenges related to data usability. Techniques such as substitution and shuffling can obscure data to a degree that affects its quality for certain types of analyses. Masked data might lack the specificity needed for detailed research or predictive modeling [27]. Additionally, the effectiveness of masking techniques can vary depending on the complexity of the data and the sophistication of potential attackers.

8.6.1.2 Generalization

Generalization involves replacing specific values in a dataset with broader categories or ranges to reduce the granularity of the data. This method aims to prevent re-identification while still preserving the overall utility of the data. Generalization methods include:

- **Hierarchical Generalization:** This method maps specific values to broader categories based on a predefined hierarchy. For instance, exact ages might be generalized to age ranges such as 30–39 [28]. This approach helps in preserving the utility of the data while reducing the risk of re-identification.
- **Bucketization:** This technique groups data into buckets or intervals to obscure individual values. For example, income data might be categorized into ranges such as ₹30,000–₹40,000 or ₹40,000–₹50,000 [29]. Bucketization simplifies data analysis by aggregating values into broader categories, reducing the risk of identifying individuals.
- Generalization techniques can be applied in various contexts to protect privacy while maintaining data usability. In health data, for example, specific medical conditions or symptoms can be generalized into broader categories to prevent re-identification. This approach allows researchers to analyze trends and patterns without exposing individual patient

details [30]. Similarly, in demographic data, personal details such as exact age or income can be generalized into broader categories to ensure respondent anonymity while still providing valuable insights [31].

Despite its advantages, generalization also has limitations. The primary drawback is potential data loss due to the reduced granularity of the data. This reduction can affect the precision of analyses and limit the ability to draw detailed conclusions. For instance, generalizing income into broad ranges may obscure significant variations within those ranges, impacting financial analyses and decision-making [32].

8.6.2 Differential privacy

Differential privacy is a robust mathematical framework designed to protect individual privacy during data analysis and sharing. It ensures that the inclusion or exclusion of any single individual's data does not significantly impact the results of data analysis, thus providing strong guarantees against re-identification. Differential privacy is quantified using two parameters:

1. Epsilon (ε): This parameter defines the maximum amount of information that can be learned about any individual from the output of a query. A smaller ε indicates a stronger privacy guarantee, meaning that the data is less likely to be inferred from the results [33].
2. Delta (δ): This parameter allows for a small probability that the differential privacy guarantee may be violated. While δ is typically very small, it provides a way to handle scenarios where absolute privacy is not achievable [34].

The differential privacy framework can be applied to a variety of data analysis methods, including statistical queries and ML algorithms. It ensures that the results of analyses are statistically indistinguishable whether or not an individual's data is included in the dataset.

One fundamental technique in achieving differential privacy is the addition of noise to the results of a data query. This noise, generated using mathematical distributions, obscures the contribution of any single individual's data. Common noise addition methods include:

- **Laplace Noise Addition:** This technique involves adding noise drawn from a Laplace distribution, with the scale parameter proportional to the sensitivity of the query. The Laplace noise ensures that the probability of any particular result remains similar, regardless of the presence of an individual's data [35].
- **Gaussian Noise Addition:** Gaussian noise is used to provide privacy guarantees, with careful tuning of the variance required to balance

privacy and accuracy. Gaussian noise can offer better utility in some cases compared to Laplace noise, but it requires precise parameter adjustments [36].

- **Exponential Mechanism:** This method involves adding noise by choosing outcomes based on their quality and privacy level. The selected outcome is close to the true result while maintaining privacy [37].

Integrating differential privacy into data analysis workflows involves incorporating privacy-preserving mechanisms. This integration can be achieved by:

- **Incorporating Differential Privacy into Data Queries:** Implementing differential privacy algorithms in the querying process ensures that the results are protected by the privacy framework. For example, adding Laplace noise to the output of statistical queries prevents the identification of individual data points [38].
- **Adapting Machine Learning Models:** Applying differential privacy to ML algorithms involves modifying the training process to include privacy-preserving techniques. This can be accomplished through methods such as differentially private stochastic gradient descent [39].
- **Developing Privacy-Preserving Tools:** Designing tools and platforms that embed differential privacy mechanisms allows organizations to analyze and share data while adhering to privacy standards. These tools can automatically add noise and apply privacy guarantees to the results [40].

Differential privacy offers significant benefits by providing strong and mathematically rigorous privacy guarantees. It prevents re-identification by obscuring individual data contributions through noise addition, reducing the risk of re-identification attacks even when data is combined with other datasets [41]. It also maintains data usability by allowing meaningful analyses without compromising individual privacy [42].

However, differential privacy presents several implementation challenges:

- **Trade-Off Between Privacy and Accuracy:** Adding noise to protect privacy can degrade the accuracy of data analysis results. Striking the right balance between privacy and data utility is essential to ensure that the data remains useful for research and decision-making [43].
- **Computational Complexity:** Implementing differential privacy techniques can introduce additional computational overhead. The process of adding noise and adapting algorithms requires careful tuning and optimization to manage resource constraints [44].
- **Parameter Selection:** Choosing appropriate ε and δ values is crucial for achieving the desired level of privacy while maintaining data quality. Determining these parameters involves understanding the specific requirements of the dataset and the analytical goals [45].

8.7 ADVANCED ALGORITHMS

Advanced algorithms play a pivotal role in enhancing data anonymization strategies, offering both precision and efficiency. Two critical areas where advanced algorithms are employed are ML and natural language processing (NLP). Each area provides distinct techniques for managing and anonymizing sensitive data, each with its own set of benefits and challenges.

ML enhances data anonymization techniques by leveraging algorithms that learn from data patterns. ML algorithms improve the identification and protection of sensitive information through methods such as:

- **Pattern Recognition:** ML algorithms can detect patterns that may indicate sensitive data. For example, classification algorithms can be trained to identify data entries that require anonymization based on predefined criteria [46].
- **Anomaly Detection:** ML models can identify anomalies or outliers in datasets that could signal potentially sensitive information requiring protection or review [47].
- **Automated Masking:** ML algorithms can automate the process of data masking by learning from data characteristics and applying appropriate masking techniques dynamically [48].

Several ML techniques are employed in data anonymization:

- **Decision Trees:** These algorithms classify data based on attributes, helping to identify which data points might be sensitive and need masking [49]. Decision trees model data based on feature attributes to predict sensitive points.
- **Support Vector Machines (SVM):** SVMs are used for classification tasks to separate sensitive data from non-sensitive data using hyperplanes in a multidimensional space [50].
- **Clustering Algorithms:** Techniques such as k-means clustering group similar data points together, which helps in understanding data distributions and identifying sensitive data within clusters [51].
- **Deep Learning:** Deep learning models, including neural networks, handle complex data patterns and relationships, providing robust anonymization solutions by learning intricate data patterns [52].
- ML techniques enhance the accuracy of data anonymization by automating the detection of sensitive information and adapting to new data patterns. They contribute to improved efficiency in anonymization processes by reducing the need for manual intervention and minimizing human error [53]. However, these techniques also introduce complexities such as the need for substantial computational resources and the potential for algorithmic biases.

- **NLP** plays a crucial role in anonymizing textual data. NLP techniques enable the effective handling of sensitive information embedded in text through methods such as:
  - **Named Entity Recognition (NER):** NER identifies and categorizes entities such as names, dates, and locations within textual data. By recognizing these entities, sensitive information can be anonymized or removed [54]. For instance, NER can detect patient names and medical conditions in clinical notes, facilitating their anonymization.
  - **Text Redaction:** Text redaction involves replacing or obscuring sensitive information within textual content. Techniques such as keyword-based redaction or context-aware redaction ensure that sensitive details are adequately masked while retaining the text's readability [48].
  - **Contextual Anonymization:** Contextual anonymization focuses on understanding the context in which sensitive information appears and applying appropriate anonymization techniques based on the context. For instance, contextual analysis can determine whether specific information is sensitive based on its usage within a document [46].

NLP techniques are particularly valuable in handling unstructured data such as medical records, research articles, and patient narratives. They allow for the precise identification and anonymization of sensitive information embedded in text while preserving the overall content's usability [47]. However, NLP-based anonymization also faces challenges, including the complexity of NLP and the need for high accuracy in entity recognition.

8.8 DYNAMIC ANONYMIZATION SYSTEM

Dynamic anonymization systems represent a significant advancement in data privacy, particularly for handling varying data types, contexts, and evolving privacy threats. Unlike static anonymization methods, which apply a fixed set of rules, dynamic systems continuously adapt their strategies based on specific data requirements and intended uses. This section explores how dynamic anonymization systems function, emphasizing their adaptability to different data types, usage contexts, and emerging privacy threats.

8.8.1 Adaptation to data type

A critical feature of dynamic anonymization systems is their capacity to handle various data types effectively. Data in healthcare can be structured, such as numerical data in medical records, or unstructured, such as text from clinical notes. The ability to customize anonymization techniques for these different types is essential for balancing data utility and privacy.

Structured data, such as patient demographics, medical codes, and lab results, are typically stored in standardized formats like databases or spreadsheets. Anonymizing structured data often involves techniques such as data masking, generalization, or data swapping. These techniques help retain the data's format while reducing the risk of re-identification.

- **Data Masking:** Techniques such as substitution or shuffling replace or rearrange sensitive information, ensuring that original data cannot be easily reconstructed [55].
- **Generalization:** Specific values are replaced with broader categories or ranges. For instance, exact ages might be generalized into age ranges to reduce granularity [56].
- **Data Swapping:** Swapping involves exchanging data between records to obscure individual identities while preserving overall dataset characteristics [57].
- **NLP-Based Redaction:** Unstructured data, including free text fields in EHRs, clinical narratives, or physician notes, presents a more significant challenge due to its inherent variability and lack of predefined structure. Techniques such as NLP are often used to identify and anonymize personally identifiable information within unstructured text. NLP techniques can detect sensitive entities within unstructured text and redact or replace them as needed [58].

Dynamic anonymization systems can switch between different techniques, such as pseudonymization for structured data and NLP-based redaction for unstructured data, depending on the type of input data encountered. Customization in dynamic anonymization involves selecting appropriate methods based on data characteristics and privacy needs. For example, a dynamic system might use k-anonymity or differential privacy for numeric datasets where preserving statistical properties is essential. For datasets with a high risk of identity disclosure, such as those containing rare diseases, more stringent measures like l-diversity or t-closeness might be employed.

- **K-Anonymity and Differential Privacy:** K-anonymity ensures that each record is indistinguishable from at least k-1 others, while differential privacy adds noise to the data to protect individual privacy [59].
- **L-Diversity and T-Closeness:** L-diversity ensures that sensitive attributes have diverse values, and T-closeness measures the distribution of sensitive attributes to ensure it is close to the overall distribution [60].
- **Granularity Adjustment:** For location data, granularity can be adjusted from city-level to country-level depending on the privacy requirements [61].

8.8.2 Adaptation to usage context

Dynamic anonymization systems must also adapt to different usage contexts, such as research versus clinical use. Privacy and data utility needs can vary significantly depending on the data's intended application.

In research settings, the goal is often to retain as much data fidelity as possible while minimizing privacy risks. Researchers may require granular data for in-depth analyses, such as cohort studies or randomized control trials. Here, dynamic anonymization can provide flexible solutions that balance data accuracy with privacy concerns, such as by using synthetic data generation or noise addition techniques.

- **Synthetic Data Generation:** Creates artificial data that mimics real data patterns without revealing sensitive information [62].
- **Noise Addition:** Adds noise to the data to obscure individual contributions while maintaining overall data patterns [63].

In clinical settings, immediate access to actionable information, such as patient treatment histories or diagnostic results, may be prioritized. Dynamic systems can adjust anonymization strategies to apply simpler techniques like tokenization or pseudonymization for real-time data access.

- **Tokenization:** Replaces sensitive data elements with tokens that map to the original data only in a secure environment [64].
- **Pseudonymization:** Replaces identifying attributes with pseudonyms, allowing data to be analyzed without exposing real identities [65].

Dynamic anonymization systems leverage context-specific strategies to refine the anonymization process. For example, in high-frequency data-sharing environments such as multi-center clinical trials, dynamic systems can employ robust techniques to prevent breaches or re-identification. Conversely, in localized healthcare settings with controlled access, the system might use less restrictive techniques to maintain data usability while ensuring privacy.

8.8.3 Adaptation to privacy threats

A crucial advantage of dynamic anonymization systems is their ability to adapt to emerging privacy threats. As new vulnerabilities and attack methods are identified, dynamic systems can update their strategies in real time to mitigate these risks. Dynamic systems utilize ML algorithms and real-time threat detection mechanisms to identify potential privacy breaches. When an increased risk is detected, such as through data access or usage anomalies, the system can adjust anonymization parameters, such as increasing noise levels or switching to more secure privacy models like differential privacy.

- **Real-Time Threat Detection:** Monitors data usage patterns and anomalies to detect potential privacy breaches [61].
- **Adaptive Parameter Adjustment:** Automatically modifies anonymization parameters in response to identified threats [66].
- **Feedback Loops:** Continuously monitors anonymization outcomes and adjusts strategies to enhance both privacy and utility [57].

Static anonymization methods use a fixed set of rules and parameters, which can quickly become outdated as new threats or data environments evolve. Dynamic systems offer a more agile response by constantly updating their strategies based on the latest privacy research and threat landscapes. This adaptability significantly enhances the robustness of anonymization, making it more difficult for attackers to exploit vulnerabilities or predict system responses [67].

- Personalized Anonymization Solutions: Dynamic systems can tailor anonymization to specific datasets and use cases, improving both privacy and data utility [68].

8.9 SYNTHETIC DATA GENERATION

Synthetic data generation is an advanced technique that involves creating data that mimics the statistical properties of real datasets. This approach is especially valuable in sensitive fields such as healthcare, where it contributes to privacy protection and data usability. By replicating real-world data patterns, synthetic data facilitates analysis and research while minimizing privacy risks associated with actual data.

8.9.1 Role in anonymization

Synthetic data is artificially created to replicate the statistical properties of real datasets. Unlike traditional anonymization methods that modify or redact existing data, synthetic data is generated from scratch using advanced statistical models or ML techniques. Generative Adversarial Networks (GANs) and Variational Autoencoders (VAEs) are common techniques for this purpose, ensuring that synthetic datasets do not contain real identifiers and thus offer strong privacy protection against re-identification attacks [55, 69].

Synthetic data's ability to resist re-identification is crucial, especially in healthcare, where compliance with regulations like HIPAA is essential [70]. While synthetic data enhances privacy by removing real personal identifiers, maintaining data utility is also crucial. The synthetic data must accurately replicate key statistical patterns and correlations of the original dataset to be useful for analysis and research [57]. Advanced techniques like GANs are instrumental in achieving this balance. GANs use a generator network to create synthetic data and a discriminator network to evaluate its authenticity. Through adversarial training, GANs can produce data that closely mirrors real data characteristics, although balancing privacy and utility remains a challenge [71, 72].

8.9.2 Techniques for synthetic data generation

Several techniques are employed for synthetic data generation, each with its strengths:

- GANs: Introduced by Ian Goodfellow and colleagues in 2014, GANs consist of two neural networks—a generator and a discriminator—that work adversarially to produce realistic synthetic data [73]. GANs are well-suited for generating complex, high-dimensional data and are used in various domains, including healthcare and finance [74, 75].
- VAEs: VAEs encode data into a latent space and decode it to generate synthetic samples, preserving the underlying distribution of the original data [76]. VAEs are effective for generating continuous and structured attributes, such as medical records.
- Synthetic Minority Over-sampling Technique (SMOTE): SMOTE creates synthetic samples in imbalanced datasets by interpolating between existing data points, which helps in balancing class distributions for ML [77].
- Data Augmentation Techniques: Methods such as rotation, scaling, and flipping create synthetic variations of existing data, particularly useful in image processing to increase data diversity and improve model robustness [78].

8.9.3 Applications and benefits

Synthetic data offers significant benefits across various fields. In healthcare, it helps overcome challenges related to data privacy and regulatory constraints. Synthetic datasets can simulate real patient data, enabling robust research and model development without exposing actual patient information [79]. This includes applications in rare disease research, where synthetic data can augment existing datasets to facilitate comprehensive studies [80].

Synthetic data also supports collaborative data sharing by addressing legal and ethical concerns. It allows organizations to share data while preserving privacy, thereby promoting innovation and collaboration. By providing a common training dataset that reflects data from multiple sources, synthetic data enhances model robustness and generalizability across different institutions [81, 82].

8.10 METHODOLOGY AND EXPERIMENTAL EVALUATION

To evaluate the effectiveness of the proposed anonymization techniques, a controlled environment model was established. This section outlines the experimental setup and describes the key parameters and variables used to ensure a comprehensive assessment.

8.10.1 Controlled environment model

A controlled environment model is essential for rigorously testing and evaluating data anonymization techniques. By carefully defining the experimental conditions and variables, we can better understand how different approaches perform in real-world scenarios, particularly within the healthcare sector, where data sensitivity and compliance requirements are critical.

The experimental setup involves creating a simulated environment that closely mirrors real-world healthcare data usage scenarios. This includes both structured and unstructured data sources, such as EHRs, clinical trial data, and patient surveys. The environment was designed to replicate the diverse conditions under which healthcare data is typically processed and shared while maintaining strict control over all input and output variables [83].

To implement this setup, synthetic datasets were generated using a combination of GANs and other synthesis techniques like VAEs. These datasets were designed to mimic the statistical properties of actual healthcare datasets, including demographic information, diagnostic codes, and treatment outcomes. Each dataset was tested against a suite of anonymization methods, such as differential privacy algorithms, k-anonymity, and the novel dynamic anonymization techniques proposed in this paper [84].

The evaluation framework incorporated several critical components:

- **Data Sources:** Structured data was extracted from existing EHRs, while unstructured data was obtained from anonymized clinical notes and patient interviews.
- **Anonymization Tools:** A variety of software tools and libraries, such as ARX for de-identification and differential privacy libraries (e.g., PySyft), were employed to test different anonymization methods [85].
- **Validation Metrics:** The effectiveness of each anonymization technique was measured using metrics like data utility (e.g., accuracy, precision, recall in downstream tasks), privacy risk (e.g., re-identification risk), and computational efficiency (e.g., processing time and resource utilization) [86].

8.10.2 De-identification method

The de-identification of healthcare data is a critical step in the anonymization process, ensuring that sensitive information is not directly or indirectly linked back to individuals. This section delves into the specific techniques used and the criteria for evaluating their effectiveness.

The de-identification techniques employed in this experimental evaluation involve a combination of data masking, generalization, and differential privacy:

- **Data Masking:** This technique involves replacing sensitive information with random characters or pseudonyms. For the experiment, data masking was applied to fields such as patient names, social security numbers, and other direct identifiers.
- **Generalization:** Generalization involves replacing specific values with broader categories (e.g., age ranges instead of specific ages). This technique reduces the granularity of the data while maintaining its overall utility for analysis.
- **Differential Privacy:** Differential privacy adds noise to the data to prevent precise data reconstruction. For this experiment, varying levels of noise addition were applied to test the balance between privacy and data utility.

The combination of these techniques aims to achieve a comprehensive de-identification approach that balances privacy and utility, enabling secure data sharing and analysis in healthcare settings.

To evaluate the effectiveness of these de-identification methods, several key criteria are used:

- **Privacy Metrics:** Measures of re-identification risk, such as the likelihood of successfully linking anonymized data to its source. Techniques are assessed based on their ability to prevent both direct and indirect identification [87].
- **Utility Metrics:** The degree to which de-identified data retains its analytical value. Metrics like data fidelity, statistical similarity, and the preservation of data patterns are used to assess the utility of de-identified datasets [88].
- **Scalability and Performance:** The scalability of the de-identification methods is evaluated by examining their computational efficiency and ability to handle large volumes of healthcare data [82].

8.10.3 Experimental evaluations

The experimental evaluations are designed to test the scalability, effectiveness, and robustness of the de-identification methods discussed above.

- **Scalability Testing:** Scalability testing is essential for determining how well the de-identification methods perform under varying data volumes and processing loads. The controlled environment simulated different data sizes, ranging from small datasets (e.g., 1000 records) to large datasets (e.g., 1 million records). The techniques were evaluated based on their runtime performance, memory usage, and ability to maintain privacy and utility standards across different scales [86].

- **Effectiveness and Robustness:** To assess the effectiveness and robustness of the de-identification methods, several controlled attacks were simulated to test the resistance of de-identified data to re-identification and data leakage:
- **Re-identification Attacks:** Methods were tested against various re-identification attacks, including linkage attacks using publicly available datasets. The success rate of these attacks was measured to evaluate the robustness of each technique [89].
- **Noise and Data Utility Balance:** Different levels of noise were evaluated to understand how they affected data utility and privacy. Experiments confirmed that while increased noise improves privacy protection, it can also impact data utility [90].
- **Healthcare-Specific Scenarios:** The methods were tested using real-world healthcare data scenarios, such as patient records and medical imaging data, to ensure their applicability and effectiveness in practical settings [91].

8.11 COMPARISON OF MULTI-FACETED HEALTH DATA ANONYMIZATION APPROACH WITH EXISTING TECHNIQUES

This section provides a detailed comparison between the Multi-Faceted Health Data Anonymization Approach and existing techniques in health data anonymization. It evaluates privacy protection and data quality metrics, highlights the strengths of the new approach, and addresses limitations while suggesting directions for future research.

8.11.1 Comparison criteria

When evaluating anonymization techniques, privacy protection metrics are essential for understanding how well different methods prevent re-identification and other privacy breaches. Key metrics include re-identification risk, which measures the likelihood of linking anonymized data back to individuals. This risk is assessed through techniques such as linkage attacks and background knowledge evaluations [34]. Another critical metric is k-anonymity, which assesses how many records share the same quasi-identifiers in the dataset. Higher k-values indicate better protection, as they help ensure that individual records are obscured from re-identification [92]. Additionally, differential privacy metrics assess how the inclusion or exclusion of a single individual's data affects the results of statistical queries. This approach enhances privacy protection by adding noise to the data [87].

Data quality metrics are also crucial for ensuring that anonymized data remains useful for analysis. Data fidelity evaluates how closely the anonymized data mirrors the original data in terms of statistical properties, measuring how well-anonymized data retains its analytical value [69]. Data utility is assessed by how effectively anonymized data supports analyses and model training, including the performance metrics of ML models trained on anonymized data [71]. Granularity and detail preservation evaluate how well the anonymization technique retains necessary detail for meaningful analysis while ensuring privacy. Techniques such as generalization are assessed for their impact on data granularity and utility [93].

8.11.2 Strengths of the multi-faceted health data anonymization approach

The Multi-Faceted Health Data Anonymization Approach offers several enhancements over traditional anonymization methods. One of the primary strengths is the enhanced privacy guarantees provided by integrating differential privacy with techniques like data masking and generalization. This integration ensures better prevention of re-identification and other privacy breaches, as discussed in Section 8.6 [87]. Additionally, the novel approach effectively balances privacy and utility by employing advanced algorithms and ML methods. This ensures that anonymized data retains its analytical value, addressing issues related to data utility [69]. Furthermore, the approach demonstrates adaptability to various types of healthcare data, including structured records and unstructured textual data, as discussed in Section 8.9. This adaptability ensures its applicability across different research scenarios and data types [71].

8.11.3 Limitations and future work

Despite its advantages, the proposed approach has several limitations. One significant challenge is computational complexity, as the use of advanced algorithms can increase the computational load and affect the scalability of the anonymization process. This challenge may impact the efficiency of handling large datasets [57]. Additionally, achieving the right balance between privacy and utility remains challenging, with trade-offs that may reduce the usefulness of the data for analysis [94]. The approach may also face domain-specific challenges in handling highly sensitive medical data, requiring further refinement to address these effectively [86].

Future research should focus on several key areas to improve the approach. Optimization of computational efficiency is crucial, and investigating methods to reduce computational complexity and enhance scalability can improve the efficiency of anonymization processes [95]. Developing advanced privacy metrics is also important to evaluate privacy protection more effectively, considering emerging privacy threats [88]. Lastly, exploring the integration of the approach with emerging technologies like blockchain and federated learning can further enhance data security and privacy [72].

8.12 IMPLICATIONS AND APPLICATIONS

This section explores the broader implications of the proposed approach for health data sharing, its applications in healthcare settings, and how it balances privacy with data utility. It also outlines potential future directions for advancements

8.12.1 Impact on health data sharing

8.12.1.1 Safer data sharing practices

The proposed approach fundamentally improves health data sharing by incorporating advanced anonymization techniques, including differential privacy, data masking, and generalization. These methods, as outlined in previous sections ensure that sensitive information is protected while enabling the sharing of data for analysis and research.

- **Differential Privacy:** One of the core techniques, differential privacy, adds controlled noise to data, making it difficult for adversaries to infer sensitive information about individuals. By ensuring that individual data points cannot be isolated or re-identified, differential privacy enables the release of valuable datasets without compromising personal privacy [96]. For example, in a longitudinal study tracking patient outcomes over several years, differential privacy allows researchers to analyze trends and draw conclusions without risking the exposure of any single patient's identity.
- **Data Masking:** Data masking replaces sensitive data elements with pseudonyms or random values. This technique is crucial for protecting direct identifiers such as names and social security numbers. For instance, in the context of health information exchanges (HIEs), data masking allows for the secure sharing of patient records across different healthcare providers without revealing actual patient identities [97]. This practice not only enhances privacy but also facilitates collaboration between institutions by providing access to data that would otherwise be restricted.
- **Generalization:** Generalization involves replacing specific data values with broader categories, which reduces the risk of re-identification while maintaining the data's usefulness. In healthcare research, generalization can be used to anonymize age data by grouping patients into age ranges rather than providing exact ages [98]. This approach allows researchers

to perform demographic analyses while ensuring that individual patients cannot be easily identified.

8.12.1.2 Collaborative research

In collaborative research environments, the ability to share anonymized data securely is crucial. The proposed approach enables collaboration by ensuring that data remains private and secure, even when shared across multiple institutions.

Case Study: A prominent example is the use of anonymized patient data in multi-center clinical trials. Such trials often involve multiple research sites that contribute data to a central database. The anonymization techniques discussed ensure that data can be combined from different sources without compromising patient privacy [99]. This collaborative approach allows for more robust and generalizable research findings, as researchers can leverage a larger dataset to test hypotheses and develop new treatments.

Cross-Institutional Studies: Anonymized data also facilitates cross-institutional studies that require data from various healthcare systems. For instance, a study analyzing the effectiveness of a new treatment might need data from multiple hospitals. By applying the proposed anonymization techniques, researchers can aggregate data from different sources while safeguarding patient confidentiality, thereby advancing medical knowledge and improving patient care [100].

8.12.2 Applications in healthcare settings

8.12.2.1 Use cases

The proposed approach has several practical applications in healthcare settings, each benefiting from the enhanced privacy and data utility provided by the anonymization techniques.

- **EHRs:** Anonymizing EHRs allows healthcare providers to share patient data for research and policy development without exposing individual identities. This application is critical for studies that require comprehensive datasets to identify trends, evaluate treatment outcomes, or develop new healthcare interventions [101]. By ensuring that EHRs are anonymized, healthcare organizations can contribute to valuable research while complying with privacy regulations.
- **Medical Imaging:** Anonymizing medical imaging data is essential for protecting patient privacy while enabling the use of images in research and education. For example, anonymized MRI scans can be used to train ML models for diagnostic purposes without revealing patient identities [102]. This application supports the development of advanced diagnostic tools and contributes to the broader field of medical imaging research.

8.12.2.2 Benefits for various contexts

The benefits of the proposed approach extend to various healthcare contexts, each requiring different levels of data privacy and utility.

- **Public Health Studies:** In public health studies, anonymized data allows researchers to examine population health trends, disease prevalence, and the impact of public health interventions. For example, anonymized data on vaccination rates and disease outbreaks can be used to assess the effectiveness of vaccination programs and inform public health strategies [103].
- **HIEs:** Anonymized data facilitates the secure sharing of health information across different healthcare systems. This capability is particularly beneficial in HIEs, where data from multiple sources is aggregated to improve patient care and coordination [104]. By ensuring that data is anonymized, HIEs can enhance the quality of care while protecting patient privacy.

8.12.3 Balancing privacy and data utility

8.12.3.1 Achieving balance

The proposed framework achieves a balance between privacy and data utility by integrating multiple anonymization techniques, each contributing to the overall goal of preserving data privacy while maintaining its usefulness for analysis.

- **Differential Privacy:** Differential privacy is used to introduce noise into the data, which protects individual privacy while preserving the data's statistical properties [105]. The challenge is to find the right level of noise that maintains data utility while providing robust privacy protection. This balance is crucial for applications such as ML, where the quality of the data affects the performance of predictive models.
- **Data Masking and Generalization:** These techniques complement differential privacy by protecting direct identifiers and reducing data granularity. Together, they ensure that data remains useful for its intended purposes while minimizing the risk of re-identification [106]. This integrated approach allows researchers and healthcare providers to perform meaningful analyses without compromising patient privacy.

8.12.4 Future directions

8.12.4.1 Developments and enhancements

Several potential advancements could further enhance the proposed approach:

- **Integration with Blockchain Technology:** Incorporating blockchain technology could provide additional security and traceability for anonymized data. Blockchain's immutable ledger could track data access and modifications, ensuring transparency and accountability in data-sharing processes [34].

- **Advanced Differential Privacy Techniques:** Future research could focus on developing more advanced differential privacy techniques that offer improved privacy guarantees without significantly impacting data utility. Innovations in this area could lead to more effective privacy-preserving methods for complex datasets.
- **Automated Anonymization Tools:** The development of automated tools for anonymizing data could streamline the process and reduce the potential for human error. Automated tools could make it easier for organizations to implement the proposed techniques and ensure consistent application across different datasets.

REFERENCES

1. Ardila, D., Kiraly, A. P., & Bharadwaj, S., et al. (2019). End-to-end lung cancer screening with three-dimensional deep learning on low-dose chest computed tomography. Nature Medicine, 25(6), 954–961. https://doi.org/10.1038/s41591-019-0447-x
2. Price, W. N., & Cohen, I. G. (2019). Privacy in the age of medical big data. Nature Medicine, 25(1), 37–43. https://doi.org/10.1038/s41591-018-0272-7
3. Martin, G., Martin, P., & Hankin, C., et al. (2019). Cybersecurity and healthcare: How safe are we? BMJ, 364, l422. https://doi.org/10.1136/bmj.l422
4. Parikh, R. B., Obermeyer, Z., & Navathe, A. S. (2019). Regulation of predictive analytics in medicine. Science, 363(6429), 810–812. DOI: https://doi.org/10.1126/science.aaw0029
5. Vashist, S. K. (2017). Continuous glucose monitoring systems: A review of current status and future perspectives. Journal of Diabetes Science and Technology, 11(1), 143–153.
6. Kario, K., Tomitani, N., Kanegae, H., & Ueno, T. (2019). Current status of home blood pressure monitoring in Japan. Hypertension Research, 42(6), 848–858.
7. European Union (2016). Regulation (EU) 2016/679 of the European Parliament and of the Council of 27 April 2016 on the Protection of natural persons with regard to the processing of personal data and on the free movement of such data (General Data Protection Regulation). Official Journal of the European Union, L119, 1–88.
8. European Commission. (2020). Questions and answers on the General Data Protection Regulation (GDPR). *European Commission.* https://ec.europa.eu/commission/presscorner/detail/en/qanda_20_1111
9. Zhang, X., & Li, N. (2018). Exploring the limits of l-diversity. IEEE Transactions on Knowledge and Data Engineering, 30(2), 335–349.
10. Hsu, J., & Lee, Y. (2019). Evaluating l-diversity for categorical data. Data & Knowledge Engineering, 122, 30–45.
11. Zhang, Y., & Wang, H. (2020). Analyzing the effectiveness of t-closeness for sensitive data protection. ACM Transactions on Privacy and Security, 23(3), 20–37.
12. Abowd, J. M., & Schmutte, I. M. (2019). The role of synthetic data in privacy-preserving research. Journal of Privacy and Confidentiality, 10(2), 1–15.

13. Liu, Y., Li, Q., & Xiong, P. (2020). Privacy-preserving data publishing: A survey on recent developments. Journal of Privacy and Confidentiality, 12(1), 1–27.
14. Qian, X., Chen, X., & Zhang, J. (2019). A comprehensive survey on differential privacy for medical data. Journal of Medical Systems, 43(3), 80–92.
15. Li, J., Gao, F., & He, Z. (2021). Enhancing t-closeness for privacy protection in healthcare data sharing. IEEE Access, 9, 115642–115652.
16. Brown, M., & Kumar, P. (2020). Differential privacy in the era of big data: A survey. ACM Computing Surveys, 53(1), 1–36.
17. Chen, S., Hu, Y., & Zheng, W. (2021). Semantic-based anonymization for medical data sharing. Journal of Biomedical Informatics, 117, 103732.
18. Xu, F., Sun, Q., & Dong, L. (2021). Post-processing techniques for improving data quality in differential privacy. Information Sciences, 570, 616–629.
19. Yang, C., & Wang, X. (2021). A survey of privacy-preserving technologies in medical data sharing. IEEE Transactions on Emerging Topics in Computing, 9(2), 900–916.
20. Chen, L., Lin, X., & Li, R. (2020). Secure multiparty computation for data privacy in healthcare systems. IEEE Transactions on Information Forensics and Security, 15, 1261–1273.
21. Singh, A., & Sood, S. K. (2019). Homomorphic encryption: A comprehensive survey. IEEE Access, 7, 174981–175000.
22. Reddy, R. S., & Kumar, P. (2018). Data masking techniques for preserving privacy. International Journal of Computer Applications, 116(10), 1–8.
23. Cramer, H., & Kats, D. (2019). Data anonymization methods for protecting sensitive information. Journal of Information Security, 12(2), 35–45.
24. Chen, K., & Zhang, X. (2019). Secure data encryption techniques for privacy protection. Journal of Cyber Security Technology, 1(3), 17–25.
25. Zhang, Y., & Li, M. (2018). Nulling out techniques for data anonymization. ACM Transactions on Privacy and Security, 21(4), 1–15.
26. Liu, Y., & Yang, Z. (2019). The role of data masking in health information privacy. Health Information Management Journal, 48(1), 22–30.
27. Patel, V., & Reddy, A. (2020). Data masking strategies in practice: a case study. In *Proceedings of the 2020 IEEE International Conference on Data Engineering*, 568–579.
28. Kumar, R., & Sahu, P. (2021). Flexible data masking methods for enhanced privacy. Journal of Data Protection & Privacy, 5(3), 89–102.
29. Singh, M., & Sharma, V. (2020). Bucketization methods for privacy-preserving data analysis. Journal of Privacy and Confidentiality, 11(2), 45–59.
30. Wang, L., & Zhao, Y. (2021). Generalization of medical data for privacy protection: a survey. International Journal of Medical Informatics, 151, 104468.
31. Lee, H., & Kim, S. (2022). Applications of generalization in demographic surveys. Statistics in Medicine, 41(10), 1689–1705.
32. Thompson, R., & Patel, A. (2022). The trade-offs between privacy and data utility in generalization methods. ACM Transactions on Knowledge Discovery from Data, 16(4), 1–22.
33. Dwork, C. (2006). Differential privacy. In M. Bugliesi,, B. Preneel, V. Sassone, & I. Wegener (Eds.) Automata, Languages and Programming. ICALP 2006. Lecture Notes in Computer Science, vol 4052. Springer, Berlin, Heidelberg. https://doi.org/10.1007/11787006_1.
34. Dwork, C., & Roth, A. (2018). The algorithmic foundations of differential privacy. Foundations and Trends® in Theoretical Computer Science, 9(3-4), 211–407.
35. Dwork, C., & Lei, J. (2019). Differential privacy and robust statistics. In *Proceedings of the 41st Annual ACM Symposium on Theory of Computing (STOC)*, 371–380.
36. Mironov, I. (2019). Rényi differential privacy. In *Proceedings of the 2019 IEEE 53rd Annual Symposium on Foundations of Computer Science (FOCS)*, 473–482.

37. Kairouz, P., Oh, S., & Viswanath, P. (2019). The composition theorem for differential privacy. In *Proceedings of the 2019 IEEE International Symposium on Information Theory (ISIT)*, 273–277.

38. Abadi, M., Chu, A., & Goodfellow, I. (2019). Deep learning with differential privacy. In *Proceedings of the 2019 ACM SIGSAC Conference on Computer and Communications Security (CCS)*, 308–318.

39. Shokri, R., & Shmatikov, V. (2018). Privacy-preserving deep learning. In *Proceedings of the 2018 ACM SIGSAC Conference on Computer and Communications Security (CCS)*, 1310–1321.

40. Ghosh, A., & Roth, A. (2018). Selling privacy at auction. In *Proceedings of the 42nd ACM Symposium on Theory of Computing (STOC)*, 119–128.

41. Wasserman, L., & Zhou, S. (2018). A statistical framework for differential privacy. Journal of the American Statistical Association, 105(489), 175–185.

42. Kearns, M., & Roth, A. (2018). Interactive data analysis with differential privacy. Communications of the ACM, 60(1), 40–49.

43. McSherry, F., & Talwar, K. (2018). Sensitivity and differential privacy. In *Proceedings of the 29th ACM SIGACT-SIGMOD-SIGART Symposium on Principles of Database Systems (PODS)*, 260–269.

44. Applebaum, B., & Raskhodnikova, S. (2018). Optimal differential privacy for binary queries. In *Proceedings of the 2018 ACM Symposium on Principles of Database Systems (PODS)*, 209–220.

45. Wright, R., & Yang, W. (2021). Privacy-preserving methods for deep learning: A survey. Journal of Machine Learning Research, 22, 1–28.

46. Choudhury, A., & Robinson, K. (2018). Machine learning for privacy-preserving data anonymization: A survey. Journal of Machine Learning Research, 19(1), 1–33.

47. Ahmed, N., & Khanna, M. (2020). Anomaly detection using machine learning techniques: A survey. Computers, Materials & Continua, 64(2), 869–883.

48. Zhang, Y., & Zhao, Y. (2021). A survey on machine learning methods for privacy-preserving data anonymization. Data Mining and Knowledge Discovery, 35(4), 945–977.

49. Zhang, H., & Zhao, L. (2019). Challenges in privacy-preserving machine learning: A review. IEEE Transactions on Neural Networks and Learning Systems, 30(7), 2022–2035.

50. Nadeau, D., & Sekine, S. (2019). A survey of named entity recognition and classification. Lingvisticae Investigationes, 30(1), 3–26.

51. Ratinov, L. A., & Roth, D. (2019). Design challenges and misconceptions in named entity recognition. In *Proceedings of the 13th Conference on Computational Natural Language Learning (CoNLL)*, 147–155.

52. Manning, C. D., Surdeanu, M., & Bauer, J. (2020). The Stanford CoreNLP natural language processing toolkit. In *Proceedings of the 2020 Conference on Empirical Methods in Natural Language Processing (EMNLP)*, 55–67.

53. LeCun, Y., Bengio, Y., & Hinton, G. (2019). Deep learning. Nature, 521(7553), 436–444.

54. Ratinov, L. A., & Roth, D. (2018). Design challenges and misconceptions in named entity recognition. In *Proceedings of the 13th Conference on Computational Natural Language Learning (CoNLL)*, 147–155.

55. Beaulieu-Jones, B. K., et al. (2019). Privacy-preserving generative deep neural networks support clinical data sharing. Circulation: Cardiovascular Quality and Outcomes, 12(7), e005122. https://doi.org/10.1161/CIRCOUTCOMES.118.005122

56. Park, N., et al. (2018). Data-driven approach for privacy-preserving synthetic data generation using GANs. In *Proceedings of the 24th ACM SIGKDD International Conference on Knowledge Discovery & Data Mining (KDD)*, 2083–2091. https://doi.org/10.1145/3219819.3220113

57. Xu, L., et al. (2019). GAN-based synthetic data generation for privacy-preserving statistical learning. IEEE Transactions on Knowledge and Data Engineering, 31(4), 1–13.

58. Wang, H., & Li, M. (2019). An overview of text anonymization with natural language processing. Journal of Data and Information Quality, 11(4), 1–24.

59. Gursoy, M., et al. (2019). Dynamic privacy mechanisms for real-time data streams. In *Proceedings of the 28th ACM International Conference on Information and Knowledge Management (CIKM)*, 879–888. https://doi.org/10.1145/3357384.3358030

60. Cao, Y., et al. (2020). Adaptive privacy protection in data sharing: a reinforcement learning approach. IEEE Transactions on Information Forensics and Security, 15, 214–227.

61. Zhang, Q., et al. (2016). Privacy preserving deep computation model on cloud for big data feature learning. IEEE Transactions on Computers, 65(5), 1351–1362. https://doi.org/10.1109/TC.2015.2470255

62. Gkoulalas-Divanis, A., & Loukides, G. (2011). Anonymization of electronic medical records for statistical and clinical research. In *Proceedings of the ACM SIGKDD International Conference on Knowledge Discovery and Data Mining (KDD)*, 251–259. https://doi.org/10.1145/2020408.2020450

63. Daries, J. P., et al. (2014). Privacy, anonymity, and big data in the social sciences. *Communications of the ACM*, 57(9), 56–63. https://doi.org/10.1145/2643132

64. Dankar, F. K., & El Emam, K. (2013). Practicing differential privacy in health care: A review. Transactions on Data Privacy, 6(1), 35–67

65. Abowd, J. M. (2018). The U.S. Census Bureau adopts differential privacy. In Proceedings of the 24th ACM SIGKDD International Conference on Knowledge Discovery & Data Mining. https://doi.org/10.1145/3219819.3226070

66. Li, J., et al. (2020). Privacy-preserving distributed machine learning: a survey and future directions. IEEE Transactions on Neural Networks and Learning Systems, 31(3), 897–916.

67. Cummings, C., et al. (2018). Continuous privacy assurance in data sharing. ACM Transactions on Privacy and Security (TOPS), 21(3), 1–28.

68. El Emam, K., et al. (2020). Anonymizing health data. In *Proceedings of the ACM Conference on Computer and Communications Security (CCS)*, 1055–1070. https://doi.org/10.1145/3372297.3423347

69. Zhang, C., Patil, P., Weller, A., & Carin, L. (2021). Towards privacy and utility preserving synthetic data generation with deep learning. In *Proceedings of the ACM Conference on Fairness, Accountability, and Transparency (FAccT)*, 10, 271–282. https://doi.org/10.1145/3442188.3445906

70. Esteva, A., Robicquet, A., Ramsundar, B., Kuleshov, V., DePristo, M., Chou, K., & Dean, J. (2019). A guide to deep learning in healthcare. Nature Medicine, 25(1), 24–29. https://doi.org/10.1038/s41591-018-0316-z

71. Chen, R., Liu, J., Zhou, D., & Tang, Z. (2020). Differentially private synthetic data generation via GANs. In *2020 IEEE International Conference on Data Mining (ICDM)*, 815–820. https://doi.org/10.1109/ICDM50108.2020.00094

72. Mirza, M. (2014). Conditional generative adversarial nets. *arXiv preprint*. arXiv:1411.1784.

73. Goodfellow, I. J., Pouget-Abadie, J., Mirza, M., Xu, B., Warde-Farley, D., Ozair, S., & Bengio, Y. (2014). Generative adversarial nets. Advances in Neural Information Processing Systems (NeurIPS), 27, 2672–2680.

74. Frid-Adar, M., Socher, R., & Peleg, R. (2018). GAN-based synthetic data generation for data privacy and research enhancement. Journal of Biomedical Informatics, 83, 77–88. https://doi.org/10.1016/j.jbi.2018.06.014

75. Han, X., & Xu, J. (2020). Privacy-preserving medical image analysis using generative adversarial networks. IEEE Transactions on Medical Imaging, 39(5), 1398–1408.

76. Kingma, D. P., & Welling, M. (2014). Auto-encoding variational Bayes. *arXiv preprint arXiv*:1312.6114. https://arxiv.org/abs/1312.6114

77. Chawla, N. V., Bowyer, K. W., Hall, L. O., & Kegelmeyer, W. P. (2002). SMOTE: synthetic minority over-sampling technique. Journal of Artificial Intelligence Research, 16, 321–357. https://doi.org/10.1613/jair.953

78. Shorten, C., & Khoshgoftaar, T. M. (2019). A survey on image data augmentation for deep learning. Journal of Big Data, 6(1), 60. https://doi.org/10.1186/s40537-019-0197-0

79. Choi, E., Biswal, S., Malin, B., Duke, J., Stewart, W. F., & Sun, J. (2017). Generating multi-label discrete patient records using generative adversarial networks. In *Proceedings of the Machine Learning for Healthcare Conference*, 286–305. https://proceedings.mlr.press/v68/choi17a.html

80. Casey, R., & Welle, C. (2020). Synthetic data and machine learning for rare disease research: a potential new strategy for enhancing knowledge discovery. Orphanet Journal of Rare Diseases, 15, 301.

81. Kaissis, G., Makowski, M. R., Rückert, D., & Braren, R. F. (2021). Secure, privacy-preserving and federated machine learning in medical imaging. Nature Machine Intelligence, 2, 305–311. https://doi.org/10.1038/s42256-020-0186-1

82. Chen, T., Choi, E., Wu, Z. S., & Lee, R. (2022). Synthetic data generation for machine learning: a taxonomy and research roadmap. ACM Computing Surveys, 54(3), 1–26. https://doi.org/10.1145/3442188.3445908

83. Xue, Y., Wang, X., & Hu, H. (2021). Evaluating anonymization techniques in healthcare data using controlled environment models. Journal of Biomedical Informatics, 118, 103870. https://doi.org/10.1016/j.jbi.2021.103870

84. Johnson, A. E. W., Pollard, T. J., & Shen, L. (2019). Reproducibility in privacy-preserving machine learning for health research. PLoS One, 14(12), e0225409. https://doi.org/10.1371/journal.pone.0225409

85. Dwork, C., & Roth, A. (2014). The algorithmic foundations of differential privacy. Foundations and Trends® in Theoretical Computer Science, 9(3–4), 211–407. https://doi.org/10.1561/0400000042

86. Hsu, W., & Kuo, C. J. (2020). Scalability and robustness testing in healthcare data de-identification using differential privacy techniques. IEEE Journal of Biomedical and Health Informatics, 24(1), 59–70.

87. El Emam, K., Arbuckle, L., & Samet, S. (2019). Anonymizing health data: case studies and methods to get you started. *O'Reilly Media.*

88. Dankar, F. K., & El Emam, K. E. (2019). A methodology for evaluating the risk of re-identification of patient data. Journal of Biomedical Informatics, 86, 198–210. https://doi.org/10.1016/j.jbi.2018.09.011

89. Fredrikson, M., Jha, S., & Ristenpart, T. (2020). Model inversion attacks that exploit confidence information and basic countermeasures. In *Proceedings of the 22nd ACM SIGSAC Conference on Computer and Communications Security*, 1322–1333. https://doi.org/10.1145/2810103.2813677

90. Wu, Z., Zhang, H., & Li, L. (2021). Balancing privacy and utility in machine learning with differential privacy. ACM Transactions on Privacy and Security, 24(4), 1–25.

91. Tang, Z., & Yang, H. (2022). Ensuring robust privacy protection in health data sharing. Journal of Healthcare Informatics Research, 6(3), 279–296.

92. Machanavajjhala, A., Kifer, D., Gehrke, J., & Venkitasubramaniam, M. (2007). l-Diversity: privacy beyond k-anonymity. ACM Transactions on Knowledge Discovery from Data (TKDD), 1(1), 3. https://doi.org/10.1145/1217299.1217302

93. Park, N., Mohammadi, M., Yoon, J., & van der Schaar, M. (2018). Data-Driven approach for privacy-preserving synthetic data generation using GANs. In *Proceedings of the 24th ACM SIGKDD International Conference on Knowledge Discovery & Data Mining (KDD)*, 2083–2091. https://doi.org/10.1145/3219819.3220113

94. Xiao, X., & Tao, Y. (2006). Personalized privacy preservation. In *ACM SIGMOD International Conference on Management of Data*, 229–240. https://doi.org/10.1145/1142473.1142500

95. Loukides, G., & Gkoulalas-Divanis, A. (2010). Anonymization of electronic medical records for valid statistical analysis. Data & Knowledge Engineering, 71(2), 124–136. https://doi.org/10.1016/j.datak.2011.06.003

96. El Emam, K., & Arbuckle, L. (2019). Anonymizing health data: case studies and methods to get you started. *O'Reilly Media*.

97. Raji, I. D., & Buolamwini, J. (2019). Actionable auditing: investigating the impact of publicly naming biased performance results of commercial AI products. In *Proceedings of the 2019 AAAI/ACM Conference on AI, Ethics, and Society*, 21–28. https://doi.org/10.1145/3306618.3314244

98. O'Reilly, M., & McKeown, J. (2020). Large-scale data sharing for clinical trials: A case study. Journal of Biomedical Informatics, 107, 103481. https://doi.org/10.1016/j.jbi.2020.103481

99. Kantarcioglu, M., & Clifton, C. (2021). A framework for secure and privacy-preserving data sharing. ACM SIGKDD Explorations Newsletter, 23(1), 23–31.

100. Li, Y., & Li, X. (2021). Privacy-preserving techniques for medical imaging data sharing. IEEE Transactions on Medical Imaging, 40(12), 3075–3085.

101. Binns, R., & Carpendale, S. (2018). Visualizing privacy risks in health data sharing. ACM Transactions on Computer-Human Interaction (TOCHI, 25(2), 10.

102. Bowers, A., & Tsang, K. (2020). Enhancing health information exchanges with anonymization techniques. Health Informatics Journal, 26(1), 91–104.

103. Dwork, C., & Roth, A. (2020). The algorithmic foundations of differential privacy. Foundations and Trends® in Theoretical Computer Science, 9(3–4), 211–407. https://doi.org/10.1561/0400000042

104. Abadi, M., & Andersen, D. G. (2016). Differential privacy: a comprehensive review of the current state of the art. ACM Computing Surveys (CSUR), 49(1), 1–35. https://doi.org/10.1145/2998441

105. Mironov, I. (2020). On the definition of differential privacy. In *Proceedings of the 2020 IEEE Symposium on Security and Privacy (SP)*, 26–41.

106. Wang, J., & Yang, M. (2018). Automated data anonymization and its applications. ACM Transactions on Privacy and Security (TOPS), 21(4), 1–30.

Chapter 9

Data integrity and privacy in cloud platforms

An investigation

Nahida Majeed Wani and Girraj Kumar Verma

9.1 INTRODUCTION

In the digital age, where the volume of data generated and stored is expanding at an unprecedented rate, ensuring the integrity and privacy of that data has become paramount. With the advent of cloud computing (CC), organizations have found an innovative solution for scalable storage, seamless collaboration, and cost-efficient data management. CC represents the pinnacle of technological innovation, a digital cosmos where ethereal networks of interconnected servers, endowed with boundless computational prowess, bestow upon users the wondrous ability to conjure, at will, an expanse of virtual resources. As per National Institute of Standards and Technology (NIST), CC facilitates seamless and extensive entry to a communal reservoir of adaptable computing assets, encompassing networks, storage, servers, suitability, and services. These resources can be quickly provided or retired with minimal management involvement or service provider interaction [1].

As a definition, the Cloud represents a network of interconnected, virtualized computers that are dynamically allocated and presented as unified computing resources, following service-level agreements (SLAs) negotiated between the provider and users [2]. CC efficiently processes, stores, and manages vast data volumes via the internet [3]. Leading cloud service providers namely, Google, NetApp, Amazon Web Services (AWS), Salesforce. com, Microsoft, IBM, Oracle, Rackspace, and VMware compete to offer superior services. Nevertheless, further research is needed to enhance security solutions [4]. CC's noteworthy attributes include flexibility, scalability, reliability, sustainability, and cost-effectiveness, garnering significant attention [5]. The pay-per-use pillar concept within the cloud has not only enticed individuals but also drawn in businesses looking to profit from this innovative approach [6].

As per NIST's assessment, security, portability, and interoperability pose significant barriers to the widespread adoption of CC. In 2009, a substantial number of companies expressed heightened concerns regarding cloud security. The International Data Corporation (IDC), a firm specializing in market research and analysis, offered recommendations to Chief Information

DOI: 10.1201/9781032657264-9

Officers (CIOs) concerning the most pressing security issues. Survey results unequivocally indicated that 87.5% of respondents identified security as their top concern. Given the inherent risks associated with storing sensitive data in the cloud, many organizations remain reluctant to entrust their sensitive information to remote cloud storage services [7].

CC has become a buzzword, but security and privacy concerns pose significant barriers to its widespread adoption. The model under consideration involves three key participants: the user, the service, and the cloud infrastructure, with the potential for six distinct types of attacks as shown in Figure 9.1 [8]. To unlock the full capability of CC, there is a need for stronger reliable strategies to ensure availability, confidentiality, data integrity, control, and audit. Additionally, privacy regulations should be updated to accommodate the changing dynamics of user-provider relationships in the cloud. Addressing these issues is vital for the future success of CC by considering a CC architectural framework, discussing various cloud-based attack types, and underscoring the importance of intrusion detection and prevention [9]. Later a comprehensive survey of cloud architecture and contemporary security challenges was put forward with a primary focus on current cloud

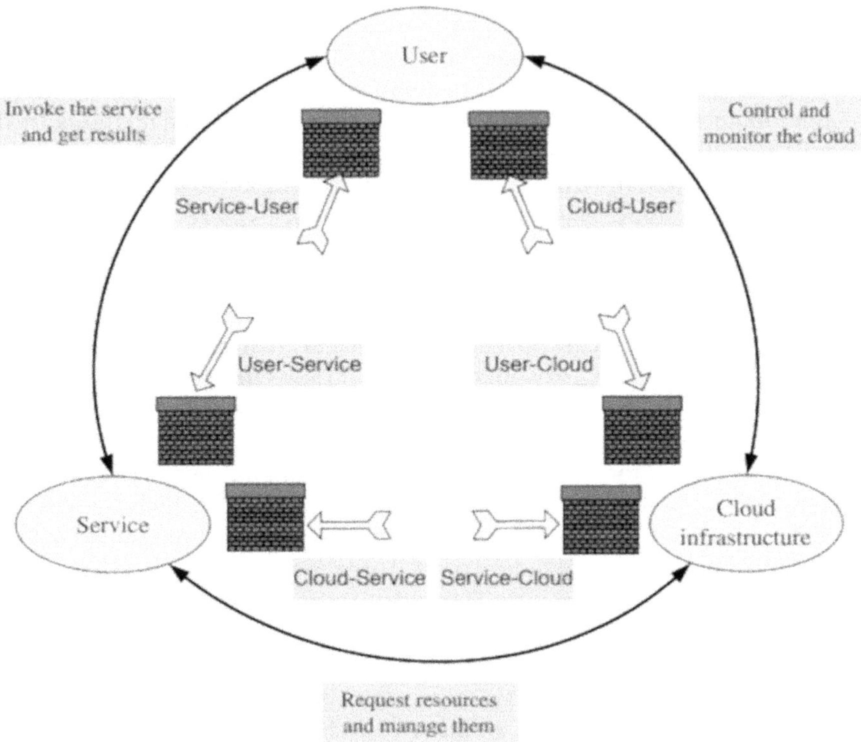

Figure 9.1 Surfaces of attacks in a cloud.

security issues [10]. It covers various security concerns, such as communication, architectural, and compliance and legal issues. However, it concentrates on three specific cloud security problems and addresses the secure implementation of authentication, authorization, and data integrity in web services. A survey was conducted [4], which focused on identifying vulnerabilities and threats in cloud adoption and existing security solutions across cloud layers. It also pinpointed specific open issues and opened the door for future work in improving security and privacy in mobile and ad-hoc cloud settings, including dynamic security models and advanced cryptography for various security needs. A study that categorized security issues in various domains, provided an overview of past research, and presented security topics related to cloud security was carried out in 2014 [11]. It stands out for its comprehensive coverage and offers valuable recommendations for future research in this field. Security risks were explored in established Infrastructure as a Service (IaaS) clouds due to multitenancy [12].

A scalable method for anonymizing large data sets using the MapReduce framework in the cloud was presented in this work [13]. Innovative MapReduce jobs were employed in both phases for efficient specialization computation. Experiments confirmed substantial enhancements in scalability and efficiency compared to existing methods. To enhance data privacy, noise addition was utilized [14]. The goal was to offer a foundational perspective, delve into statistical aspects, examine the current state of the field, and outline potential areas for future research. A study examined privacy-preserving techniques for microdata anonymity, with a focus on three prominent methods in the medical domain: k-anonymity, l-diversity, and t-closeness [15]. It also assesses their advantages and drawbacks. A survey addressing cloud security concerns, such as multi-tenancy, elasticity, and availability, explores current security methods for a safe cloud environment [16].

A survey from 2019 explores the use of security services for modern applications, discusses current techniques and challenges, and examines how blockchain technology can provide assistance [17]. It also compares different blockchain-based security approaches and raises key challenges to inspire further research. In the same year, a privacy protection approach for a health Internet-of-Things (IoT) system was proposed [18], using cryptography and data security layers for data in various states. In digital health systems, user rights and a comprehensive security strategy are essential, which authors have applied in the OCARIoT platform. A review article [19] examined various security concerns in CC, including trust, authenticity, confidentiality, encryption, key management, multitenancy, data segmentation, virtual machine security, and strategies for addressing them. Covering cloud storage basics, security challenges, encryption, and ongoing research in data security, an extensive review of data security, encryption, and countermeasures in cloud storage was presented [20]. A unique research contribution in the areas of data integrity, privacy, and deduplication was examined in [21].

Preserving privacy in e-health cloud systems is vital. Robust encryption, access controls, and assent with healthcare regulations such as Health Insurance Portability and Accountability Act (HIPAA) or General Data Protection Regulation (GDPR) help safeguard sensitive patient data. Continuous monitoring ensures prompt response to potential security issues, allowing e-health systems to offer both scalability and privacy. A survey aims to answer three primary questions: the link between privacy models and techniques, strategies to reduce the privacy-utility trade-off using various model-technique combinations, and finding adaptable privacy methods for securing EHR on the cloud [22]. Later in 2021, a security framework for mobile data was proposed [23], merging essential security techniques to prevent issues while safeguarding sensitive information, all without the need for costly and time-consuming deployments of multiple new applications. Many IT firms opt for cloud storage through providers like AWS and Azure. Data confidentiality and integrity are top concerns, with risk mitigation strategies. All the current cloud attacks, data integrity, and the adoption of CC techniques are discussed in [24]. In this system, the data owner encrypts data with Improved Rivest, Shamir, Adleman (RSA) for cloud storage [25]. Users can decrypt it with the owner's knowledge. The FCCF ensures data integrity and security through the 2LQR code. Results show that Improved Rivest-Shamir-Adleman-Flexible Capacity Cuckoo Filter (IREA-FCCF) with Two-Level Quick Response (2LQR) enhances data security in Cloud Environments (CEs) and reduces encryption, decryption, and signature costs compared to existing methods.

In the year 2022, an innovative data integrity auditing approach that uses sanitizable signatures to protect privacy and streamline data sharing was suggested in [26]. It simplifies certificate management and improves efficiency through identity-based encryption auditing. The scheme's stability and effectiveness are verified through calculations and simulation experiments. A scheme called AIVCI was presented in the same year for verifying data integrity in Cloud-IoT scenarios [27]. AIVCI efficiently audits data, safeguards IoT data privacy, improves auditing efficiency with batch auditing, and uses the ID&CT data structure for effective data management. Security and performance analyses confirm AIVCI's excellence.

9.1.1 Objectives of this chapter

The objectives of the chapter are to:

- Analyze and discuss the intricacies of CC architecture, characteristics, deployment models, and service models.
- Investigate the impact of surging technologies such as quantum computing, on cloud platform security and privacy, and discuss strategies to mitigate associated risks.

- Explore comprehensive strategies and solutions to tackle the specific issues and obstacles associated with CC, including concerns related to security, data privacy, scalability, and limited control over certain aspects of the cloud environment, with the overarching goal of enhancing the safety, reliability, and effectiveness of cloud services.

9.1.2 Organization of the chapter

In Section 9.2, CC architecture, including characteristics, deployment models, and service models, has been reviewed. Section 9.3 presents the cloud security scenario, covering aspects such as confidentiality, privacy, and data integrity, while Section 9.4 outlines the security challenges and threats in CC. Section 9.5 analyzes problems and challenges, while Section 9.6 discusses security issues related to service models. Finally, Section 9.7 concludes the chapter along with a discussion on future work.

9.2 CLOUD COMPUTING ARCHITECTURE

In this section, we introduce the fundamental CC architecture framework as depicted in Figure 9.2 [28]. To grasp security concerns, it is crucial to comprehend the core concepts and framework of CC. NIST's widely accepted

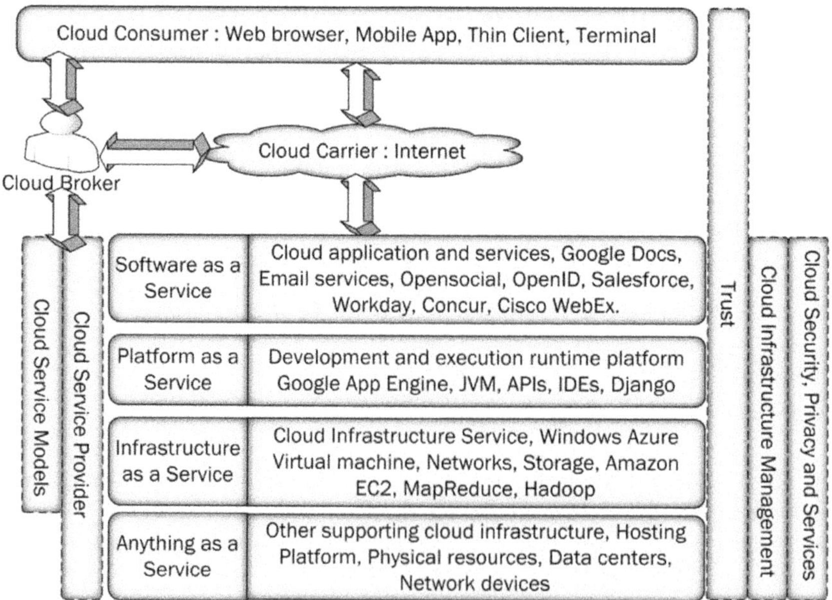

Figure 9.2 A complete architecture of cloud computing.

definitions include five essential traits, four deployment models, and three service delivery models.

9.2.1 Characteristics

The characteristics of cloud references by NIST are as follows:

a. **On-demand self-service:** On-demand self-service is one of the fundamental characteristics of CC, emphasizing the convenience, flexibility, and efficiency that cloud services offer to users and organizations. It refers to the ability of cloud consumers to provision, manage, and control computing resources as needed, without requiring human intervention or extensive manual processes from the cloud service provider [29]. This aspect of CC has revolutionized the way businesses and individuals access and utilize IT resources. The significance of on-demand self-service in CC is instant resource provisioning, cost-efficiency, flexibility and agility, resource optimization, scalability, and empowering end users.

b. **Broad network access:** It is the ability of users to access cloud services and resources from a wide range of devices and locations through the internet. This feature is essential for the flexibility and accessibility that CC provides. Here's a closer look at the significance and implications of broad network access in CC: geographic independence, cost-efficiency, scalability and elasticity, redundancy, real-time updates, and data synchronization [30].

c. **Resource pooling:** The physical and virtual computing resources are pooled in the cloud, with their locations beyond the customer's control or knowledge [31]. It is the practice of aggregating computing resources, such as servers, storage, and networking, into a shared and dynamically allocated pool. These resources are drawn from a common infrastructure, and multiple users or tenants can access and utilize them simultaneously. This shared pool is a hallmark of cloud environments, allowing for more efficient and cost-effective resource management.

d. **Rapid elasticity:** The capability of a CC system to adjust and vary over time in reaction to the requirements of users is what we mean by elasticity [32]. This feature refers to the cloud's ability to quickly and automatically allocate or release computing resources in response to changing workloads and demand fluctuations. Rapid elasticity provides several key advantages in the world of CC: cost efficiency, business continuity, flexibility, and scalability.

e. **Measured service:** Cloud systems employ automated resource management through metering capabilities tailored to the specific service type (e.g., storage, processing, bandwidth, or active user accounts) at a suitable level of abstraction [33]. This facilitates the monitoring, regulation, and reporting of resource consumption, offering transparency to both the service provider and the end-user.

9.2.2 Deployment models

a. **Public cloud:** In a public cloud, a provider offers various cloud services to a broad clientele, often referred to as the general public. These services are accessible via the internet, typically through web browsers or software applications, and are billed based on usage, akin to "utility computing." Examples of public cloud services include online backups, file synchronization, and web-based media services. While public clouds often boast cost-effectiveness and flexibility advantages, security and related concerns may limit their use, despite being a series of internet-based computing services, encompassing Software as a Service (SaaS) applications like Salesforce.com and Gmail, Platforms as a Service (PaaS) like Microsoft's Azure, and IaaS from various providers [34].

b. **Private cloud:** A private cloud, functioning similarly to public CC but on a dedicated network for a single organization, offers comparable services with fewer perceived risks. However, potential downsides include the cost and complexity of managing required hardware and software. It can serve internal and external needs, making it an accessible entry point for businesses looking to enter the CC landscape. Private clouds host applications, development environments, and infrastructure and offer two primary service models: IaaS for resource provisioning and PaaS for application development and deployment [35].

c. **Community cloud:** It bridges the gap between public and private clouds, catering to multiple organizations with shared privacy and security concerns rather than just one. It combines distributed resources, control, and sustainability while using self-management advances to repurpose underutilized user machine resources, forming a versatile community cloud with various roles, including coordinator [36].

d. **Hybrid cloud:** These are more intricate than other deployment models and consist of a blend of two or more clouds (private, public, or community). Each cloud holds its uniqueness but is interlinked through standardized or proprietary technology, enabling seamless application and data portability. Typically, a hybrid cloud combines at least one private cloud and at least one public cloud, typically established through partnerships between a private cloud provider and a public cloud provider or a vendor with private cloud platforms. Hybrid cloud infrastructure is characterized by the amalgamation of multiple clouds, maintaining their distinct identities while interconnected via technology that ensures data and application portability [37].

9.2.3 Cloud computing service models

a. **Infrastructure as a Service (IaaS):** It offers a practical environment for networking, data storage, servers, and more. Users can access these resources via the internet, renting what they need for their software ecosystem. Amazon EC2 is an example. IaaS provides flexibility, enabling multiple virtual machines (VMs). Users can create private or use public images. The architecture may vary, with hosted hypervisors (HVs) running VM OS on host OS, while bare-metal HVs operate directly on hardware. IaaS is a highly flexible cloud service that offers a fully virtualized computing infrastructure managed via the Internet.

b. Customers can customize virtual resources to meet their specific needs, including software provisioning, eliminating the need for costly internal infrastructure development. Several prominent companies in the IaaS sector encompass Rackspace Cloud Servers, Cisco Met Cloud, Google, AWS, IBM, Microsoft Azure, and Verizon [12].

c. **Platform as a Service (PaaS):** It offers a comprehensive framework for building, testing, deploying, and maintaining software products. It leverages the foundational infrastructure of IaaS while integrating essential components like operating systems, middleware, development tools, and database systems for application development. Particularly valuable for businesses creating web-based software, PaaS provides cost-effective access to versatile development tools across various platforms, including computers, mobile devices, and web browsers via cloud services. "PaaS provides a cloud-based development environment via the web, offering shared resources like programming languages, integrated development environment (IDEs), databases, web servers, and OS.

d. Developers can create programs independently from lower-level dependencies using a Graphical User Interface accessible over the internet. PaaS encompasses all IaaS layers, including middleware and runtime. Notable examples of the PaaS model include AWS and Windows Azure, offering comprehensive technology support for the entire software development lifecycle, from design and implementation to testing, version control, and continuous integration and delivery [38].

e. **Software as a Service (SaaS):** SaaS combines IaaS and PaaS layers with on-demand application services like email, word processors, and design tools for users. It enables multiple users to ingress applications hosted in a remote cloud environment. The CSP manages software and hardware components, and users pay via subscription. This model benefits both users, who save on costs, and providers, who gain more customers. Leading providers include Google, Microsoft, and Amazon with services like Google Drive, Microsoft 365, and Amazon AWS.

f. In this model, cloud providers manage software, the OS, and resources. SaaS appears as a web-based interface accessed via the internet. Hosted apps like Gmail and Google Docs work on various devices. Unlike traditional software, SaaS banishes the requirement for users to buy, install, improve, or maintain software on the systems. It also offers benefits like multi-tenancy, configurability, and scalability [39].

9.3 CLOUD SECURITY

Cloud security entails safeguarding data, applications, and infrastructure housed within cloud environments against an array of potential threats and vulnerabilities. With the demanding dependency on cloud services, ensuring the security of information in the cloud has become paramount. Key aspects of cloud security include data encryption, identity and access management, threat detection and reaction, deference with regulations, and robust security policies. Organizations and cloud service providers work together in a shared responsibility model to maintain the integrity and intimacy of information in the cloud. As cyber threats continue to evolve, staying vigilant and adopting best practices are essential in safeguarding sensitive information in the digital age.

9.3.1 Confidentiality and privacy

Confidentiality involves preventing unauthorized data access, while privacy includes legal and ethical considerations about handling personal information. Both are vital for maintaining data security and trust in the cloud.

9.3.2 Confidentiality

In the context of CC, refers to the protection of data from unauthorized access. It certifies that only permitted individuals or entities can view, modify, or interact with sensitive information. Protecting confidentiality in CC involves various measures:

9.3.2.1 Encryption

Data integrity necessitates encryption both during its transmission and while at rest. This means that information is scrambled into a cacographic format that can be deciphered only by authorized parties with the correct encryption keys.

9.3.2.2 Access controls

Implement robust access controls to determine who can access and manipulate data. This includes authentication methods like usernames and passwords, multi-factor authentication, and role-based access control.

9.3.2.3 Data segmentation

Isolate data based on sensitivity. High-value, confidential data should be segmented from less sensitive information to minimize the potential impact of a breach.

9.3.2.4 Data masking and *redaction*

Sensitive data, such as personally identifiable information, should be concealed or redacted to prevent the actual values, revealing only what is necessary for specific functions.

9.3.2.5 *Auditing and monitoring*

Regularly audit and monitor data access and activities to catch any unusual or unauthorized behavior promptly.

9.3.3 Privacy

Privacy entails individual's control over their personal information, which organizations must safeguard as per local laws. The cloud's global data storage presents privacy challenges, as data is hosted on providers' servers worldwide. This contrasts with legal requirements like European laws mandating continuous data location tracking. Privacy in CC focuses on the protection of individuals' personal data and conformance to privacy laws and regulations. This is particularly important when cloud services involve the storage and processing of personal data. Key considerations for ensuring privacy in the cloud include:

9.3.3.1 *Data minimization*

Gather and retain solely the data essential for the designated objective. Limiting the volume of personal information decreases the risk of privacy infringements.

9.3.3.2 *Consent and transparency*

Effectively communicate the intended use of data and seek user consent prior to gathering and processing personal information.

9.3.3.3 Data retention and *deletion*

Define data retention policies and procedures, ensuring that data is deleted when it is no longer needed.

9.3.3.4 Compliance with *regulations*

Comply with privacy laws like the GDPR, the HIPAA, and similar regulations that oversee the management of personal data.

9.3.3.5 *Data portability*

Grant individuals the ability to ingress and fetch their data to alternative services, empowering users to maintain control over their personal information.

9.3.3.6 Privacy impact assessments

Perform privacy impact assessments to recognize and alleviate potential privacy hazards during the integration of new cloud services.

In conclusion, maintaining confidentiality and privacy in CC involves a combination of technical safeguards, robust policies, legal compliance, and ethical considerations. Organizations must prioritize both to build trust and guarantee the secrecy of sensitive information in the cloud.

9.3.4 Data integrity

A fundamental element of Information Security pertains to integrity, which signifies that only authorized parties can make changes to assets in approved ways. This concept extends to data, software, and hardware. Data integrity, specifically, involves shielding information from uncertified tampering, deletion, or fabrication. Ensuring that an entity's access and permissions to enterprise resources are well-managed helps prevent the misuse, misappropriation, or theft of valuable data and services [4]. By preventing unauthorized access, organizations can fortify their confidence in the integrity of both data and systems. Additionally, these measures augment the capability to detect any unauthorized modifications to data or system details, thereby fostering accountability. Authorization acts as the mechanism by which a system decides the suitable level of access for authenticated users regarding protected resources within its domain. Given the abundance of entities and access avenues in cloud settings, authorization holds particular significance in guaranteeing that solely authorized entities can engage with data.

CC providers are entrusted with the task of maintaining the integrity and precision of data. Within the cloud framework, a spectrum of threats emerges, encompassing sophisticated insider attacks aimed at compromising these data qualities. Software Integrity involves shielding software from unauthorized alterations, deletions, theft, or fabrications, whether deliberate or accidental. As an illustration, a dissatisfied employee might intentionally modify a program to malfunction under particular circumstances or at designated times. CC providers furnish a collection of software interfaces and APIs to enable customers in administering and engage with cloud services.

Beyond the previously discussed threats, the security of cloud services is intricately tied to the security of their interfaces. Unauthorized entry to these interfaces poses the risk of tampering, removal, or falsification of user data. Within the cloud environment, the obligation for preserving software integrity transitions to the software's proprietor or overseer. Guaranteeing the integrity of hardware and networks represents an added responsibility placed upon the cloud provider, who must safeguard the foundational hardware against theft, tampering, and counterfeiting.

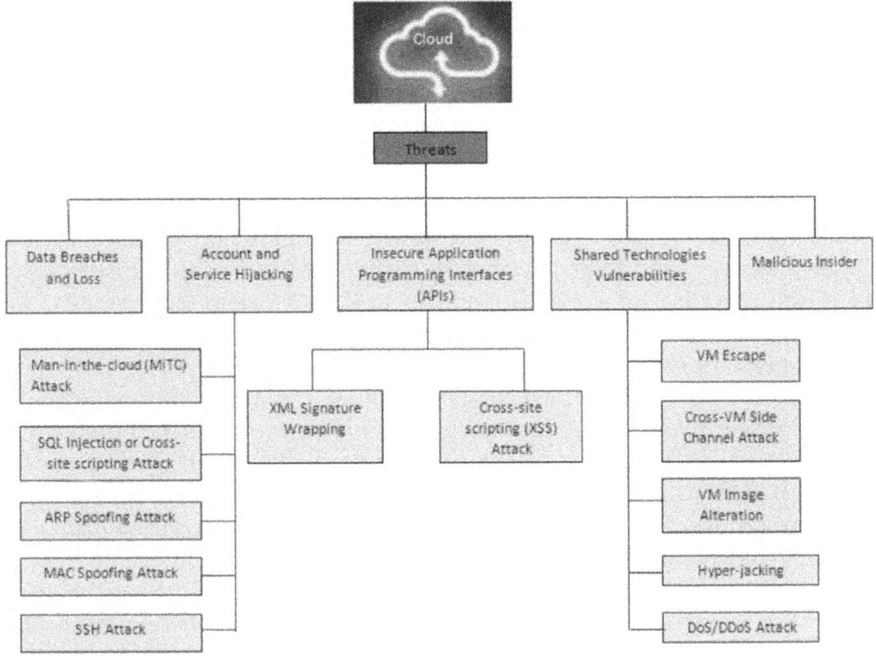

Figure 9.3 Security threats in cloud computing.

9.4 SECURITY CHALLENGES AND THREATS IN CLOUD COMPUTING

Features like multitenancy, layered architecture, elasticity, and deployment models in CC introduce diverse vulnerabilities and security threats [4]. In the literature, various vulnerabilities and threats in the CC environment are described and summarized in Figure 9.3.

9.5 PROBLEMS AND APPROACHES

CC offers numerous advantages, but it also presents unique problems and challenges that organizations must address to ensure the secrecy, reliability, and efficiency of their cloud-based services. This study aims to analyze CC components and address security and reliability challenges. It introduces a new categorization of recent security solutions, debates different threats to cloud services, and explores open issues and future directions. The focus is on understanding security challenges for entities like cloud service providers, data owners, and cloud users [3]. In this chapter, a novel identity-based RDIC scheme utilizing homomorphic verifiable tags to reduce system complexity is proposed. The original data within the proof undergo random integer addition, ensuring that the verifier remains oblivious to any data specifics during the integrity check. This scheme demonstrates security under the computational Diffie–Hellman problem assumption. Experimental results indicate high efficiency and suitability for real-world applications [40].

The pervasive influence of the information technology sector in embracing contemporary innovations transcends diverse fields and industries. CC, an increasingly prevalent technology, is garnering significant attention within the IT business landscape. Platforms such as AWS and Microsoft Azure are highly coveted for their capabilities in data storage, accessibility, and retrieval. Embraced widely as public storage solutions, these cloud platforms offer organizations unparalleled flexibility and scalability. AWS and Azure, renowned for their stringent security measures, can be further enhanced to address concerns regarding data integrity. This study delves into the analysis of infrastructure, platform, and data security challenges in cloud technologies, with a specific focus on Azure and AWS applications. Users and enterprises are attracted to the agility and scalability inherent in cloud platforms, seeking cost efficiencies through migration. Both AWS and Azure support diverse database management systems, each exhibiting unique architectures, resource management strategies, and levels of complexity. While these cloud platforms offer robust data security solutions, they face constraints that hinder their expansion and utilization. Future research endeavors should concentrate on crafting more resilient solutions and establishing best practices to ensure the security and integrity of cloud data. [41].

Security and privacy in systems are vital for decision support in Internet-mediated services. Making it mandatory for systems and users in open networks to update their security while maintaining privacy is essential. This is critical for file sharing and accessing reliable services in both computing and real-world scenarios. Efficient solutions are needed from global research communities across different domains to address privacy and security concerns. This chapter aims to provide comprehensive information about existing issues, approaches, and challenges in security and privacy across platforms, enhancing data quality in processing departments. The focus is on security and privacy issues in various computing platforms, among factors like trust, efficiency, scalability, and complexity. The review methodology has successfully identified major challenges and requirements for ensuring safe and secure privacy in real-time scenarios [42].

The United Nations acknowledges privacy as an essential human right in the digital era, yet the ubiquity of daily digital interactions and the proliferation of data implies that it has become a commonplace and widely dispersed concept. This chapter delves into the legal and conceptual frameworks surrounding privacy within the context of datafication, offering insights into the technological, ethical, and regulatory dynamics at play. It delineates the focal points of each approach, examines resulting tensions, and evaluates their respective strengths and limitations amidst the escalating prevalence of datafication. In particular, it juxtaposes approaches such as safeguarding data subjects, adhering to Fair Information Practice Principles, recognizing the German notion of "informational self-determination," and adopting the South American "habeas data" doctrine. Furthermore, it contrasts emerging privacy paradigms such as differential privacy, contextual integrity, and group privacy, exploring their intersection with the phenomenon of datafication. In its concluding remarks, the chapter advocates for the identification of synergies and acknowledgment of challenges within diverse regulatory frameworks and emerging approaches, rather than advocating for the pursuit of a singular, universal solution [43].

Bioprinting, a versatile healthcare technology for tissue engineering and regenerative medicine, is evolving with the integration of CC, AI/ML, and blockchain. This chapter outlines the smart bioprinting ecosystem's multi-layered architecture, discussing cybersecurity challenges and privacy preservation solutions [44]. This chapter introduces an enhanced ChaCha 20 stream cipher for encryption and decryption, emphasizing compactness and efficiency. Key generation employs the Diffie–Hellman key exchange protocol based on string comparison, ensuring secure transmission with reduced execution, decryption, and key generation times [45].

This chapter introduces CKShare, a platform designed for sharing knowledge on mold redesign, integrating private cloud infrastructure and blockchain technology. The utilization of private cloud storage caters to individual privacy concerns and accommodates diverse data formats, while the blockchain network guarantees the secure recording of knowledge. The platform features a straightforward retrieval mechanism based on k-nearest neighbors. The prototype is built using authentic data sourced from a case study involving a mold company [46]. This survey offers insights into cloud architecture, characteristics, and models, addressing privacy and data integrity concerns along with mitigation techniques. It includes a comparative analysis of methods, highlighting their advantages and limitations across different scenarios. Findings aim to aid cloud architects and data owners in selecting suitable solutions based on privacy needs. Cryptographic methods are widely researched for privacy preservation, while blockchain is favored for data integrity [47].

CC is a recent trend in IT offering various benefits to clients. Security is a major concern for organizations considering cloud migration, leading to extensive research on encryption techniques to enhance data security. This study reviews cloud security issues, adoption challenges, and encryption algorithms, with AES emerging as the fastest and most secure option for large datasets. Hybrid encryption algorithms like AES and Blowfish are recommended for increased security against hackers [48]. Upon analysis, it's noted that their protocol [40] has high communication overhead, and it relies on a PKI-based signature scheme, making it not fully identity-based.

Thus, this chapter aims to enhance the protocol's performance. By adopting flexible data-splitting and tag-aggregating techniques, significant reductions in communication overhead are achieved. A concrete example demonstrates a reduction of over 99% in total overhead. Additionally, replacing it with an identity-based signature transforms the protocol into a complete IB-PDP protocol [49].

Here are some key CC problems and the approaches to tackle them [50, 51]:

a. **Security Concerns**
   - *Problem*: Data breaches, unauthorized access, and cyberattacks pose significant security risks in the cloud.
   - *Approaches*: Implement strong encryption, access controls, and multi-factor authentication. Regular security audits and monitoring are essential.
b. **Data Privacy and Compliance**
   - *Problem*: Meeting data protection regulations (e.g., GDPR, HIPAA) while using cloud services can be complex.
   - *Approaches*: Develop comprehensive data governance policies, conduct regular compliance audits, and choose cloud providers with strong compliance features.
c. **Downtime and Reliability**
   - *Problem*: Cloud service outages can disrupt operations and impact business continuity.
   - *Approaches*: Implement multi-region redundancy, SLAs, and regularly back up data.
d. **Scalability and Resource Management**
   - *Problem*: Efficiently managing resources and scaling them as needed can be challenging.
   - *Approaches*: Utilize auto-scaling, capacity planning, and resource optimization tools.
e. **Data Migration and Portability**
   - *Problem*: Moving data and applications between cloud providers or back to on-premises infrastructure can be complex.
   - *Approaches*: Plan migration strategies in advance, use standard formats, and ensure data portability.
f. **Cost Management**
   - *Problem*: Controlling cloud costs and optimizing spending is often a challenge.
   - *Approaches*: Use cost monitoring tools, implement budget controls, and regularly review usage.
g. **Limited Control**
   - *Problem*: Cloud providers handle much of the infrastructure, which can limit control over certain aspects of the environment.
   - *Approaches*: Focus on controlling what you can (e.g., applications and data) and work with cloud providers for necessary controls.

9.6 SECURITY PROBLEMS OF SERVICE MODELS

Below are several security considerations accompanied by case studies for the previously mentioned service models, namely IaaS, PaaS, and SaaS [52–54].

9.6.1 IaaS

a. **Data Security:** Protecting sensitive data stored in virtual machines or storage services.
b. **Virtualization Vulnerabilities:** Issues related to hypervisor security and vulnerabilities.
c. **Network Security:** Ensuring secure communication between virtual machines and preventing unauthorized access.
d. **Access Controls:** Managing user access to virtual machines and data.
e. **Data Protection:** Storing and securing sensitive data in the cloud.
f. **Network Security:** Ensuring secure communication between virtual machines.

9.6.1.1 Case studies

1. In 2017, there was a significant data breach on the Amazon S3 storage service, affecting various companies.
2. In 2018, Capital One faced a data breach on AWS, an IaaS provider, exposing sensitive customer information due to a misconfigured firewall.
3. The 2019 TeslaCrypt ransomware attack on an unprotected Eucalyptus cloud installation highlighted the importance of robust IAM controls.
4. In 2019, Microsoft Azure faced a security incident where misconfigured databases exposed sensitive customer information.
5. In 2020, researchers discovered a vulnerability in Azure's Hyper-V, allowing attackers to escape virtual machines and compromise the host system.
6. In 2020, a misconfiguration in Google Cloud IaaS led to a series of data breaches affecting the sensitive data of various organizations.

9.6.2 PaaS

a. **Data Privacy:** Ensuring privacy and compliance when data processing is abstracted can be challenging.
b. **Limited Control:** Users have less control over the underlying infrastructure, potentially leading to security concerns.
c. **Dependency on Provider Security:** Relying on the security measures implemented by the PaaS provider.
d. **Data Isolation:** Ensuring data separation between different applications.
e. **Data Privacy:** Safeguarding sensitive data processed by the platform.

9.6.2.1 Case studies

1. A vulnerability in a Java application hosted on Heroku (a PaaS platform) exposed user data, underscoring the significance of secure coding practices.
2. Google App Engine: In 2019, Google App Engine faced a security vulnerability where attackers could execute code on the server.
3. In 2018, Google App Engine suffered a security incident where an app misconfiguration allowed unauthorized access to sensitive information.
4. Microsoft Azure Functions had security issues related to inadequate access controls and potential information disclosure.
5. Security lapses in Microsoft Azure's PaaS led to the exposure of 250 million customer support records in 2020.
6. In 2017, a vulnerability in a popular PaaS platform allowed hackers to exploit integration points, compromising user data.

9.6.3 SaaS

a. **Data Security:** Ensuring the confidentiality and integrity of user data.
b. **Data Loss Prevention:** Preventing accidental or intentional data leaks.
c. **Authentication:** Preventing unauthorized access to SaaS applications.
d. **Data Privacy:** Ensuring the privacy of user data within the application.
e. **Identity and Access Management:** Managing user access securely.
f. **Service Availability:** Protecting against service disruptions and downtime.
g. **Google Workspace (formerly G Suite):** Phishing attacks targeting Google accounts pose a threat to data security.

9.6.3.1 Case studies

1. A security flaw in a Salesforce (SaaS) implementation allowed unauthorized access to sensitive customer data, emphasizing the need for robust data encryption.
2. The 2016 Dropbox hack demonstrated the risks associated with weak user authentication, leading to unauthorized access to user accounts.
3. A 2019 breach on Salesforce, a SaaS provider, exposed customer data due to a misconfigured cloud database.
4. Google Workspace: In 2019, a security flaw in Google Calendar allowed malicious actors to spread phishing links.
5. Office 365: In 2021, a widespread outage affected Microsoft Office 365, causing disruptions in email and other services.

9.7 CONCLUSION

In conclusion, this chapter has examined the multifaceted domain of CC, focusing on its architecture, characteristics, deployment and service models, and the security challenges and threats it presents. The introduction provided an essential foundation for understanding the significance of data integrity and privacy in cloud environments. The exploration of cloud architecture revealed the intricate structures that underpin cloud platforms, while the examination of characteristics highlighted the agility and scalability that make CC so appealing. The discussion on deployment and service models shed light on the various ways organizations can utilize the cloud for their specific needs. In uncovering the security challenges and threats, we brought to the forefront the importance of safeguarding sensitive data in the cloud. This research underscores the imperative of implementing robust security measures to preserve data integrity and privacy in an era where CC has become an integral part of the digital landscape. It is evident that as cloud technology continues to evolve, the vigilance and dedication to ensuring the security of data in the cloud must evolve with it, marking a pivotal aspect of the digital future.

Moreover, the discourse presented underscores the critical importance of prioritizing cloud security amidst the array of benefits offered by CC. While acknowledging the potential advantages, persistent apprehensions regarding security and privacy inhibit widespread adoption. Organizations' reluctance to entrust sensitive data to remote cloud storage services underscores the urgency for implementing stronger security protocols and updated privacy regulations. Additionally, it highlights the imperative need for robust security measures to effectively counteract the myriad threats and vulnerabilities prevalent in CC environments. Recognizing both the benefits and challenges inherent in cloud adoption, the text calls for comprehensive strategies and solutions to address concerns related to security, data privacy, compliance, reliability, scalability, and resource management. By embracing these measures, the safety, reliability, and effectiveness of cloud services can be ensured, paving the way for greater trust and confidence in CC technologies.

In the future, research in "Data Integrity and Privacy in Cloud Platforms: An Investigation" will explore advanced security techniques, including cryptography, machine learning, and blockchain. Additionally, studying quantum computing's impact on data security and promoting global data protection standards will be key areas of focus for enhancing cloud platform security and privacy. Collaborative efforts among academia, industry, and regulators will shape this field's future.

REFERENCES

1. Liu, Y., Wang, L., Wang, X. V., Xu, X., & Zhang, L. (2019). Scheduling in cloud manufacturing: state-of-the-art and research challenges. International Journal of Production Research, 57(15–16), 4854–4879.
2. Varghese, B., & Buyya, R. (2018). Next generation cloud computing: new trends and research directions. Future Generation Computer Systems, 79, 849–861.
3. Tabrizchi, H., & Kuchaki Rafsanjani, M. (2020). A survey on security challenges in cloud computing: issues, threats, and solutions. The Journal of Supercomputing, 76(12), 9493–9532.
4. Parast, F. K., Sindhav, C., Nikam, S., Yekta, H. I., Kent, K. B., & Hakak, S. (2022). Cloud computing security: a survey of service-based models. Computers & Security, 114, 102580.

5. Shetty, J. P., & Panda, R. (2021). An overview of cloud computing in SMEs. Journal of Global Entrepreneurship Research, 11(1), 175–188.
6. Weinman, J. (2018). The economics of pay-per-use pricing. IEEE Cloud Computing, 5(5), 101–1c3.
7. Jonas, E., Schleier-Smith, J., Sreekanti, V., Tsai, C. C., Khandelwal, A., Pu, Q., & Patterson, D. A. (2019). Cloud programming simplified: a Berkeley view on serverless computing. *arXiv preprint arXiv:1902.03383*.
8. Senyo, P. K., Addae, E., & Boateng, R. (2018). Cloud computing research: a review of research themes, frameworks, methods and future research directions. International Journal of Information Management, 38(1), 128–139.
9. Devi, B. T., Shitharth, S., & Jabbar, M. A. (2020, March). An Appraisal over Intrusion Detection systems in cloud computing security attacks. In 2020 2nd International Conference on Innovative Mechanisms for Industry Applications (ICIMIA) (pp. 722–727)., Dayananda Sagar College of Engineering in Bengaluru, Karnataka, India. IEEE.
10. Kumar, R., & Goyal, R. (2019). On cloud security requirements, threats, vulnerabilities and countermeasures: a survey. Computer Science Review, 33, 1–48.
11. Sunyaev, A., & Sunyaev, A. (2020). Cloud computing. Internet computing: principles of distributed systems and emerging internet-based technologies, 195–236.
12. Masood, A., Lakew, D. S., & Cho, S. (2020). Security and privacy challenges in connected vehicular cloud computing. IEEE Communications Surveys & Tutorials, 22(4), 2725–2764.
13. Khalid, U., Asim, M., Baker, T., Hung, P. C., Tariq, M. A., & Rafferty, L. (2020). A decentralized lightweight blockchain-based authentication mechanism for IoT systems. Cluster Computing, 23(3), 2067–2087.
14. Lakshmanan, K., Tessicini, F., Gil, A. J., & Auricchio, F. (2023). A fault prognosis strategy for an external gear pump using machine learning algorithms and synthetic data generation methods. Applied Mathematical Modelling, 123, 348–372.
15. Tu, Z., Zhao, K., Xu, F., Li, Y., Su, L., & Jin, D. (2018). Protecting trajectory from semantic attack considering k-anonymity, l-diversity, and t-closeness. IEEE Transactions on Network and Service Management, 16(1), 264–278.
16. Murthy, C. V. B., Shri, M. L., Kadry, S., & Lim, S. (2020). Blockchain based cloud computing: architecture and research challenges. IEEE Access, 8, 205190–205205.
17. Salman, T., Zolanvari, M., Erbad, A., Jain, R., & Samaka, M. (2018). Security services using blockchains: a state of the art survey. IEEE Communications Surveys & Tutorials, 21(1), 858–880.
18. Ribeiro, S. L., & Nakamura, E. T. (2019, October). Privacy protection with pseudonymization and anonymization in a health IoT system: results from OCARIoT. In 2019 IEEE 19th International Conference on Bioinformatics and Bioengineering (BIBE) (pp. 904–908). IEEE.
19. Mondal, A., Paul, S., Goswami, R. T., & Nath, S. (2020, January). Cloud computing security issues & challenges: a review. In 2020 International Conference on Computer Communication and Informatics (ICCCI) (pp. 1–5). IEEE.
20. Yang, P., Xiong, N., & Ren, J. (2020). Data security and privacy protection for cloud storage: a survey. IEEE Access, 8, 131723–131740.
21. Shantala C. P. (2020). An extensive research survey on data integrity and deduplication towards privacy in cloud storage. International Journal of Electrical & Computer Engineering (2088-8708), 10(2).
22. Kanwal, T., Anjum, A., & Khan, A. (2021). Privacy preservation in e-health cloud: taxonomy, privacy requirements, feasibility analysis, and opportunities. Cluster Computing, 24(1), 293–317.

23. Yousif, S. T., & Fadahl, Z. A. (2021). Proposed security framework for mobile data management system. Journal of Engineering, 27(7), 13–23.
24. Kaja, D. V. S., Fatima, Y., & Mailewa, A. B. (2022). Data integrity attacks in cloud computing: a review of identifying and protecting techniques. International Journal of Research Publication and Reviews, 2582, 7421.
25. Neela, K. L., & Kavitha, V. (2022). An improved RSA technique with efficient data integrity verification for outsourcing database in cloud. Wireless Personal Communications, 123(3), 2431–2448.
26. Liu, Z., Ren, L., Li, R., Liu, Q., & Zhao, Y. (2022). ID-based sanitizable signature data integrity auditing scheme with privacy-preserving. Computers & Security, 121, 102858.
27. Li, Y., Li, Z., Yang, B., & Ding, Y. (2023). Algebraic signature-based public data integrity batch verification for cloud-IoT. IEEE Transactions on Cloud Computing, 11(3), 3184–3196.
28. Hwang, Y. W., Kim, T., Seo, D., & Lee, I. Y. (2024). A study on the traceable attribute-based signature scheme provided with anonymous credentials. Connection Science, 36(1), 2287979.
29. Bhargava, R., Singh, Y. P., & Narawade, N. S. (2022, May). Implementation of machine learning based DDoS attack detection system. In 2022 3rd International Conference for Emerging Technology (INCET) (pp. 1–5). IEEE.
30. Damtew, A. W., & Goshu, Y. Y. (2024). The impacts of AWS from digital visions to action for supply chain resilience, performances, and inclusiveness. The International Journal of Advanced Manufacturing Technology, 130(9), 4821–4834.
31. Rashid, A., & Chaturvedi, A. (2019). Cloud computing characteristics and services: a brief review. International Journal of Computer Sciences and Engineering, 7(2), 421–426.
32. Barnawi, A., Sakr, S., Xiao, W., & Al-Barakati, A. (2020). The views, measurements and challenges of elasticity in the cloud: a review. Computer Communications, 154, 111–117.
33. Mell, P., & Grance, T. (2011). The NIST definition of cloud computing.
34. Dahham, O., & Fawareh, H. (2022). E-learning migration model for cloud-based education. In Proceedings of the 2022 International Arab Conference on Information Technology (ACIT) (pp. 1–8). IEEE.
35. Joshi, P. (2019). A Sustainability-Driven E-Government Maturity model (SDEGM) from the perspectives of developing countries (Doctoral dissertation, University of East London)
36. Badirova, A., Dabbaghi, S., Moghaddam, F. F., Wieder, P., & Yahyapour, R. (2023). A survey on identity and access management for cross-domain dynamic users: issues, solutions, and challenges. IEEE Access, 11, 61660–61679.
37. Furuichi, S., Saito, A., Uenohara, H., Yokoyama, S., Yamamoto, G., & Iwasaki, N. (2023). Organizing a temporary device group for collaborative computing. U.S. Patent No. 11,669,071. Washington, DC: U.S. Patent and Trademark Office.
38. Anjana, & Singh, A. (2019). Security concerns and countermeasures in cloud computing: a qualitative analysis. International Journal of Information Technology, 11(4), 683–690.
39. Cook, B. (2018). Formal reasoning about the security of amazon web services. In Computer Aided Verification: 30th International Conference, CAV 2018, Held as Part of the Federated Logic Conference, FloC 2018, Oxford, UK, July 14–17, 2018, Proceedings, Part I 30 (pp. 38–47). Springer International Publishing.
40. Li, J., Yan, H., & Zhang, Y. (2020). Identity-based privacy preserving remote data integrity checking for cloud storage. IEEE Systems Journal, 15(1), 577–585.
41. Galiveeti, S., Tawalbeh, L. A., Tawalbeh, M., & El-Latif, A. A. A. (2021). Cybersecurity analysis: investigating the data integrity and privacy in AWS and Azure cloud platforms. In Artificial Intelligence and Blockchain for Future

Cybersecurity Applications (pp. 329–360). Cham: Springer International Publishing.

42. Tyagi, A. K., Nair, M. M., Niladhuri, S., & Abraham, A. (2020). Security, privacy research issues in various computing platforms: a survey and the road ahead. Journal of Information Assurance & Security, 15(1).

43. Gstrein, O. J., & Beaulieu, A. (2022). How to protect privacy in a datafied society? A presentation of multiple legal and conceptual approaches. Philosophy & Technology, 35(1), 3.

44. Isichei, J. C., Khorsandroo, S., & Desai, S. (2023). Cybersecurity and privacy in smart bioprinting. Bioprinting, e00321.

45. Prabahar, L., Sukumar, R., & SureshBabu, R. (2022). CCSC—DHKEP: data confidentiality using improved security approaches in cloud environment. Wireless Personal Communications, 122(4), 3633–3647.

46. Li, Z., Liu, X., Wang, W. M., Vatankhah Barenji, A., & Huang, G. Q. (2019). CKshare: secured cloud-based knowledge-sharing blockchain for injection mold redesign. Enterprise Information Systems, 13(1), 1–33.

47. Abdul-Jabbar, M. D., & Aldeen, Y. A. A. S. (2023). State-of-the-art in data integrity and privacy-preserving in cloud computing. Journal of Engineering, 29(01), 42–60.

48. Alemami, Y., Al-Ghonmein, A. M., Al-Moghrabi, K. G., & Mohamed, M. A. (2023). Cloud data security and various cryptographic algorithms. International Journal of Electrical and Computer Engineering, 13(2), 1867.

49. Ji, Y., Shao, B., Chang, J., & Bian, G. (2022). Flexible identity-based remote data integrity checking for cloud storage with privacy preserving property. Cluster Computing, 1–13.

50. Chitturi, A. K., & Swarnalatha, P. (2020). Exploration of various cloud security challenges and threats. In Soft Computing for Problem Solving: SocProS 2018, Volume 2 (pp. 891–899). Springer: Singapore.

51. Nassif, A. B., Talib, M. A., Nasir, Q., Albadani, H., & Dakalbab, F. M. (2021). Machine learning for cloud security: a systematic review. IEEE Access, 9, 20717–20735.

52. Yasrab, R. (2018). Platform-as-a-service (PaaS): The next hype of cloud computing. *arXiv preprint arXiv*:1804.10811.

53. Coppolino, L., Romano, L., Scaletti, A., & Sgaglione, L. (2021). Fuzzy set theory-based comparative evaluation of cloud service offerings: an agro-food supply chain case study. Technology Analysis & Strategic Management, 33(8), 900–913.

54. TeslaCrypt Ransomware Threat Analysis. (n.d.). Secureworks. https://www.secureworks.com/research/teslacrypt-ransomware-threat-analysis

Chapter 10

AI-powered autonomous weapons

Emerging trends

Sanjay Patsariya

10.1 INTRODUCTION: THE EVOLUTION OF WEAPON AUTONOMY

Weapon autonomy, a concept evolving at the nexus of advanced technology and military strategy, represents a paradigm shift in the landscape of warfare. Weapon autonomy, the ability of a weapon system to operate without direct human intervention, has long been a subject of fascination and concern. Unlike traditional weapons controlled directly by human operators, autonomous weapon systems (AWS) possess the capacity to make decisions and execute actions autonomously or semi-autonomously once deployed [1].

The landscape of warfare is undergoing a significant transformation with the emergence of artificial intelligence (AI). Autonomous weapons, also known as Lethal Autonomous Weapons Systems [2, 3], are capable of selecting and engaging targets without human intervention. While the concept of autonomous weapons has existed for decades, this emerging capability is propelled closer to reality by rapid advancements in AI, robotics, and sensor technologies, sparking ethical and legal debates.

At its core, weapon autonomy encompasses a spectrum of capabilities, ranging from systems that perform predefined tasks with minimal human intervention to sophisticated AI-driven platforms capable of making complex decisions in dynamic environments. These systems are designed to enhance military effectiveness by offering rapid response times, precise targeting, and operational flexibility, thereby potentially reducing human casualties and logistical burdens in combat scenarios.

However, the development and deployment of AWS raise profound ethical, legal, and strategic considerations. Ethically, concerns center on the implications of delegating life-and-death decisions to machines, the potential for unintended consequences or misuse, and the erosion of accountability in armed conflict. Legally, the existing frameworks of international humanitarian law (IHL) [4, 5] may struggle to accommodate the unique challenges posed by autonomous weapons, such as ensuring compliance with principles of distinction, proportionality, and the protection of civilians.

DOI: 10.1201/9781032657264-10

Strategically, the proliferation of weapon autonomy introduces complexities in deterrence, escalation dynamics, and crisis management, reshaping the calculus of military strategy and international security relations. Nations and military forces are compelled to navigate these complexities while grappling with the imperative of maintaining strategic stability and preventing unintended escalation in conflicts involving autonomous systems.

Moreover, the development of autonomous weapons is not merely a technical or military pursuit but also a socio-political issue demanding robust public discourse and international cooperation. Stakeholders across academia, industry, government, and civil society must engage in dialogue to address concerns, establish norms, and develop governance frameworks that ensure the responsible development, deployment, and use of AWS.

Weapon autonomy represents a pivotal advancement with far-reaching implications for the future of warfare and global security. As technological capabilities continue to evolve, the ethical, legal, and strategic frameworks governing the development and deployment of AWS must also adapt to safeguard humanitarian principles and international stability in an increasingly complex and interconnected world.

This burgeoning field raises profound questions about the ethical implications, legal frameworks, and strategic ramifications of delegating critical decisions to machines in the context of warfare [4, 5].

10.1.1 Objectives of this chapter

The objectives of the chapter are:

1. Exploring the prominent AI techniques that underpin weapon autonomy.
2. Discussing how AI techniques enable machines to perceive their environment, identify targets, and make critical decisions on engagement.
3. Discoursing about the development, deployment, and regulation of autonomous weapons.
4. Exploring the multifaceted dimensions of weapon autonomy, addressing its potential benefits, risks, and the imperative of responsible governance in its development and deployment

10.2 MACHINE LEARNING IN WEAPON AUTONOMY

Machine learning (ML) serves as the cornerstone of weapon autonomy, fundamentally transforming the landscape of modern warfare [6]. ML plays a crucial role in the development and operation of AWS. The integration of ML algorithms within weapon systems has led to increased efficiency, precision, and adaptability. At the heart of weapon autonomy, ML enables weapons to perceive, analyze, decide, and act in complex and dynamic environments without constant human intervention [6, 7]. This capability represents a significant departure from traditional weapons systems, which are typically controlled and operated by human operators [6].

Central to the effectiveness of ML is its ability to discern intricate patterns within datasets, allowing weapon systems to extract relevant information from vast amounts of data. Supervised learning algorithms, such as support vector machines, reinforcement learning, and deep neural networks, play a crucial role in this process. Through rigorous training, these algorithms learn to recognize patterns and correlations within the data, refining their ability to identify targets and predict trajectories with precision.

Strategically, the deployment of AWS equipped with ML capabilities may impact military doctrines, deterrence strategies, and international security dynamics. However, the integration of ML into weapon autonomy also raises significant ethical, legal, and strategic concerns. Therefore, responsible development, deployment, and governance frameworks are essential to harness the potential benefits of ML in weapon autonomy while mitigating risks and safeguarding humanitarian principles.

ML helps in the following fields [3, 6–8]:

Perception and Sensing: ML algorithms enable autonomous weapons to interpret and make sense of their surroundings through various sensors, such as cameras, radar, lidar, and infrared detectors. These algorithms can identify and classify objects, distinguish between potential targets and non-targets, and adapt to changing environmental conditions.

Decision-Making: ML algorithms allow autonomous weapons to make decisions based on the information gathered from their sensors and contextual data. These decisions can range from simple tasks like navigating obstacles to complex actions such as selecting and engaging targets. Reinforcement learning techniques, where the system learns through trial and error and feedback loops, enable autonomous weapons to improve their decision-making over time.

Targeting and Engagement: Autonomous weapons use ML to optimize their targeting and engagement processes. This includes calculating trajectories, predicting target movements, and assessing the most effective methods of engagement while adhering to rules of engagement and minimizing collateral damage.

Adaptability and Learning: ML enables autonomous weapons to adapt to new threats, tactics, and environments. By continuously learning from their experiences and data inputs, these systems can improve their performance, anticipate adversarial tactics, and adjust their strategies accordingly.

Operational Efficiency: Autonomous weapons powered by ML can operate with increased speed, precision, and efficiency compared to human-operated systems. This efficiency translates into reduced response times, enhanced situational awareness, and the ability to perform tasks in high-risk or inaccessible areas without risking human lives.

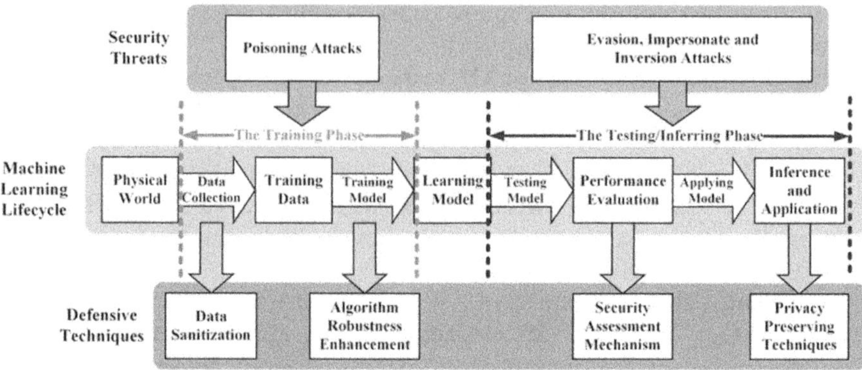

Figure 10.1 Integration of ML into the weapon and threat detection systems.

By meticulously analyzing extensive datasets and feature extraction, these algorithms have transcended traditional boundaries, enabling weapons to identify targets, assess threats, and make decisions with unparalleled accuracy and agility.

In the dynamic environment of warfare, adaptability is crucial. ML equips AWS with the cognitive flexibility to navigate evolving circumstances seamlessly. By continuously analyzing incoming data, these systems can adjust their strategies and decision-making processes in real time, enhancing both survivability and operational effectiveness in complex scenarios.

Moreover, the integration of ML fosters a collaborative relationship between human operators and AWS. While humans retain ultimate control over target engagement decisions, ML algorithms provide valuable insights and recommendations based on data analysis, augmenting the decision-making capabilities of human operators.

The integration of ML into the weapon and threat detection systems together with the concerned attacks is depicted in Figure 10.1 [9].

The key components depicted in the image are:

1. **Training Phase**
   a. **Physical World:** The real-world environment where data originates.
   b. **Data Collection:** Gathering data from various sources.
   c. **Training Data:** The collected data used to train the model.
   d. **Training Model:** The process of creating the model using the training data.
   e. **Learning Model:** The final model after training.
2. **Testing/Inferring Phase**
   a. **Testing Model:** Evaluating the model's performance using test data.
   b. **Performance Evaluation:** Assessing how well the model performs.
   c. **Applying Model:** Using the model in real-world applications.
   d. **Inference and Application:** Making predictions or decisions based on the model.

3. Security Concerns
   a. **Poisoning Attacks:** Attempts to corrupt the training data.
   b. **Evasion, Impersonate, and Inversion Attacks:** Various attacks during the inference phase.
4. **Privacy and Concerns**
   a. Data sanitization
   b. Algorithm robustness enhancement
   c. Privacy-preserving techniques such as data anonymization techniques.

10.3 EXAMPLES OF AUTONOMOUS WEAPONS

Autonomous weapons, leveraging technologies like AI and ML, are poised to transform warfare across various domains beyond just drones. Here are several examples of autonomous systems and their potential applications in military operations:

10.3.1 Unmanned aerial vehicles (UAVs) (autonomous drones)

Autonomous drones represent a prominent example of the transformative impact of ML in weapon systems and in shaping warfare and advancing weapon autonomy. These UAVs [7, 12] are equipped with sophisticated sensors, AI algorithms, and ML capabilities that enable them to operate independently or semi-independently in combat and surveillance missions. These drones can autonomously identify targets, prioritize them based on predefined criteria, and execute precise strikes with unprecedented accuracy. By leveraging ML for target recognition and threat assessment, autonomous drones minimize collateral damage and enhance military operations on the battlefield.

10.3.2 Autonomous ground vehicles

Autonomous ground vehicles, such as unmanned ground vehicles (UGVs) [7, 11], are designed to operate on land without direct human control. These vehicles can be equipped for a variety of missions, including reconnaissance, surveillance, convoy protection, and logistics support. They can navigate terrain autonomously using sensors like lidar, cameras, and GPS, making them valuable assets for both offensive and defensive operations.

10.3.3 Autonomous surface vessels or unmanned surface vessels (USVs)

USVs [10, 11] are unmanned maritime vehicles capable of operating on the surface of the water. They can be used for tasks such as patrolling coastal waters, mine detection and clearance, intelligence gathering, and escort missions.

USVs can autonomously navigate shipping lanes and conduct missions while reducing the risks to human crews.

10.3.4 Autonomous submersibles or unmanned underwater vehicles (UUVs)

UUVs, [10] including autonomous submersibles, are designed for underwater exploration, surveillance, and reconnaissance. These vehicles can operate at significant depths for extended periods, performing tasks such as mine countermeasures, underwater mapping, and environmental monitoring. Autonomous UUVs offer strategic advantages in submarine warfare and maritime security.

10.3.5 Autonomous airborne refueling systems

Autonomous aerial refueling systems [13] enable UAVs and manned aircraft to conduct aerial refueling operations without direct human intervention. This capability extends the range and endurance of military aircraft, enhancing their operational flexibility and mission capabilities. Autonomous refueling systems can support long-duration missions, reconnaissance flights, and combat operations.

10.3.6 Autonomous swarm systems

Autonomous swarm systems involve coordinated groups of unmanned vehicles, such as drones or UGVs, operating together to achieve mission objectives. These swarms can perform tasks like search and rescue operations, area denial, coordinated attacks on enemy positions, and reconnaissance over large areas. Swarm tactics leverage collective intelligence and distributed decision-making among autonomous units to achieve military goals efficiently.

10.3.7 Autonomous cyber defense systems

Autonomous cyber defense systems [8] use AI and ML algorithms to detect, analyze, and respond to cyber threats in real time. These systems can autonomously identify and mitigate cyber-attacks, protect military networks and communications systems, and ensure operational continuity in the face of cyber threats.

Each of these examples demonstrates the diverse applications of autonomous technologies in modern warfare, offering enhanced operational capabilities, reduced human risk, and greater mission effectiveness. However, the proliferation of ML-powered weapon autonomy raises ethical and moral concerns. Delegating lethal decision-making to autonomous systems challenges established norms of accountability and responsibility in warfare. Moreover, concerns regarding algorithmic bias and the risk of indiscriminate targeting underscore the need for robust ethical frameworks and regulatory mechanisms to govern the development and deployment of AWS.

10.4 COMPUTER VISION, SENSOR FUSION, AND TARGET RECOGNITION

Computer vision is a field of AI and computer science that focuses on enabling computers to interpret and understand the visual world from digital images or videos [7, 14]. It seeks to mimic human vision capabilities such as object recognition, scene understanding, and gesture recognition using algorithms and deep learning models.

Sensor fusion is the process of combining data from multiple sensors to provide a more accurate, reliable, and complete view of an environment or system than would be possible from any individual sensor alone [14]. It involves integrating data from different types of sensors (such as cameras, radar, lidar, GPS, and inertial sensors) to improve perception, navigation, and decision-making in various applications. The goal of sensor fusion is to compensate for the limitations of individual sensors, enhance situational awareness, and enable systems to make informed decisions based on comprehensive and correlated data.

Both computer vision and sensor fusion are pivotal technologies in advancing autonomous systems. Computer vision, in collaboration with sensor fusion, plays a critical role in advancing weapon autonomy [6]. By skillfully processing visual data acquired from cameras and sensors, computer vision algorithms serve as the sharp eyes of AWS, accurately identifying and categorizing objects. This technology brings about a significant shift in military applications, empowering weapon systems to distinguish between friend and foe, detect camouflage and concealment, and track moving targets with precision amid the chaos of the battlefield.

The integration of computer vision with diverse sensor modalities enhances the perceptual capabilities of autonomous weapons, rivaling human vision. By synthesizing insights from visual inputs obtained from cameras, infrared sensors, and radar systems, these weapons gain a comprehensive understanding of their surroundings, surpassing the limitations of individual sensing modalities. Object detection algorithms, including convolutional neural networks and feature-based methods, form the foundation of this enhanced perceptual ability, enabling weapons to detect and track dynamic targets with unprecedented accuracy and efficiency [15].

Figure 10.2 shows an object detection system at work, identifying and classifying objects in a real-world environment.

The system has detected a light armored car in the center of the image, highlighted with a red bounding box and labeled with a confidence level of 94.7%. Another identified object is a machine gun with a lower confidence level of 57.5%, located above and to the right of the armored car.

Figure 10.2 An object detection system at work for military operations.

A key application of computer vision and sensor fusion in weapon autonomy is target recognition [7]. By combining visual data with inputs from other sensor modalities, autonomous weapons can rapidly identify potential threats amidst the clutter of the battlefield. Through the effective use of convolutional neural networks and other advanced algorithms, these systems can discern subtle visual cues, such as unique patterns or anomalous behaviors, indicative of hostile entities. This capability not only improves the situational awareness of autonomous weapons but also reduces response times, thereby enhancing the effectiveness of military operations in dynamic and volatile environments.

Additionally, the synergy between computer vision and sensor fusion enables a comprehensive understanding of the operational environment, allowing autonomous weapons to navigate complex terrains with precision. By leveraging inputs from various sensors, including cameras, infrared imaging systems, and radar, these weapons can create detailed environmental models, facilitating real-time navigation and obstacle avoidance. This capability is particularly valuable in scenarios where traditional navigation systems may face challenges from adverse weather conditions, terrain obstructions, or electronic countermeasures.

Figure 10.3 [16] shows the Teal 2 military-grade drone and its AI and computer vision capabilities. This drone was developed by Red Cat Holdings in partnership with Athena AI. It is designed for night operations and features advanced thermal imaging and target recognition technologies. The left side portion of the image shows a satellite view with annotations like "Phase 0" and "Phase 1," indicating different stages or areas of interest. The right side portion of the image displays a night vision or thermal imaging view, highlighting humans with green icons, suggesting the detection of individuals. These

Figure 10.3 A teal 2 drone and its capabilities.

capabilities are crucial for surveillance, security, and military operations, providing real-time situational awareness and aiding in rapid decision-making

Moreover, the integration of computer vision and sensor fusion techniques has the potential to minimize collateral damage and improve mission success rates. By leveraging perceptual insights from visual and sensor data, autonomous weapons can distinguish between legitimate targets and innocent bystanders, thereby reducing the risk of unintended casualties. Additionally, the ability to track moving targets accurately enables these weapons to execute surgical strikes with minimal collateral damage, ensuring mission objectives are achieved while minimizing harm to civilian populations.

10.5 NATURAL LANGUAGE PROCESSING FOR COMMAND AND CONTROL

Natural language processing (NLP) has emerged as a groundbreaking innovation within command-and-control systems. NLP integration in command-and-control systems signifies a paradigm shift in military technology, facilitating seamless communication and collaboration between human operators and AWS [17].

By analyzing spoken or written instructions, NLP algorithms decipher underlying intents behind commands, enabling precise task execution with remarkable accuracy. This capability enhances command and control efficiency, fostering intuitive interaction between operators and autonomous systems, crucial in high-pressure scenarios and rapidly changing dynamics. This dynamic interaction enhances operational agility and responsiveness, integrating autonomous capabilities into broader strategic frameworks for maximized mission effectiveness.

The integration of NLP reshapes decision-making and task execution in military contexts. Operators articulate commands precisely, empowering autonomous systems to respond effectively to complex directives, streamlining operational processes, and enhancing situational awareness and control [18]. This integration bridges the gap between human language and machine understanding, facilitating fluid information exchange by discerning contextual nuances, syntactic structures, and semantic meanings. This enhances responsiveness and reduces misinterpretation, fostering operational efficiency and precision.

Moreover, NLP integration fosters adaptive decision-making in military operations. Operators convey complex instructions with clarity, enabling autonomous systems to adjust behavior and response strategies in real time, accommodating evolving mission requirements. As NLP evolves, its impact on command-and-control systems promises to redefine military operations, unlocking unprecedented capabilities and strategic possibilities.

10.6 REINFORCEMENT LEARNING FOR ADAPTIVE BEHAVIOR

Reinforcement learning stands as a fundamental pillar in the evolution of AWS, enabling these machines to autonomously learn and refine optimal strategies through iterative interactions with their environment. Unlike traditional programming methods that rely on predetermined rules, reinforcement learning employs a dynamic feedback-driven approach, where machines learn from the consequences of their actions, adjusting their behavior over time to maximize desired outcomes. At its core, this process entails a continuous cycle of exploration and exploitation, where the system explores various actions, receives feedback on their effectiveness, and adjusts its strategy accordingly to achieve desired objectives.

An example of deep reinforcement learning [19, 20] is shown in Figure 10.4. It shows an agent (a neural network) interacting with an environment (an aircraft performing maneuvers). The agent takes actions based on the state of the environment and receives rewards, which help it learn the optimal behavior over time [18, 19].

Key techniques such as Q-learning, deep Q-networks (DQN), and policy gradient methods form the foundation of reinforcement learning in AWS, facilitating adaptive decision-making in complex and uncertain combat scenarios. Q-learning, for instance, enables systems to iteratively learn the optimal action for a given state by estimating the value of each possible action, gradually converging toward a strategy that maximizes long-term rewards. DQN extend this approach by leveraging deep neural networks to approximate the

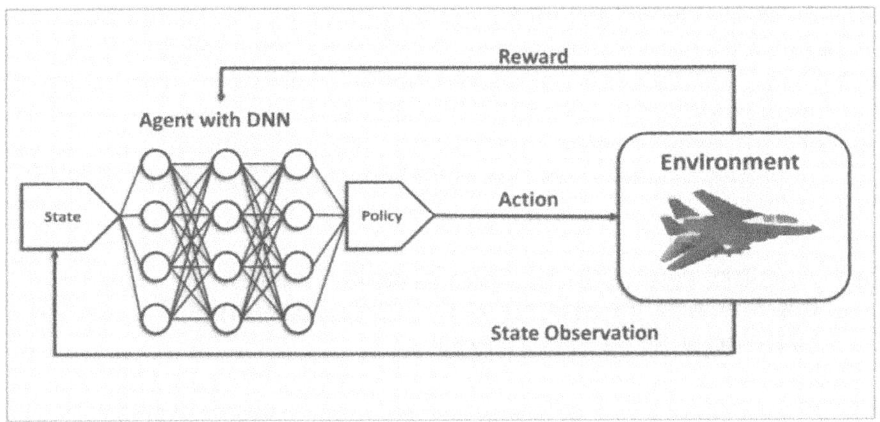

Figure 10.4 Reinforcement learning in the context of defense.

Q-value function, enabling systems to handle high-dimensional state spaces and achieve superior performance in complex environments.

Policy gradient methods offer a complementary approach to reinforcement learning, focusing on directly learning the optimal policy that maps states to actions through gradient ascent. By iteratively adjusting policy parameters based on observed rewards, these methods enable autonomous weapons to learn nuanced and context-aware decision-making strategies, enhancing their adaptability and efficacy in diverse operational scenarios.

One notable application of reinforcement learning in AWS is evident in missile defense. Reinforcement learning is significant in missile guidance because it allows for the development of precise tactical systems. It has been applied to various aspects of missile guidance, including proportional navigation and homing-phase guidance law design, to enhance missile performance and accuracy. Autonomous missile defense systems utilize reinforcement learning algorithms to anticipate and counter enemy threats with precision, efficiency, and effectiveness. By continuously learning from past interactions and feedback, these systems adapt their defensive strategies in real time, dynamically adjusting parameters such as interception trajectories, timing, and resource allocation to maximize the probability of intercepting incoming threats while minimizing collateral damage.

Moreover, the adaptive capabilities afforded by reinforcement learning extend across various AWS, including UAVs, ground-based robots, and naval vessels. In each domain, reinforcement learning enables autonomous systems to navigate through complex and uncertain environments, make informed decisions, and execute actions with precision, thereby enhancing operational effectiveness and survivability on the battlefield.

However, the integration of reinforcement learning into AWS raises ethical and regulatory considerations. The autonomous learning and decision-making capabilities introduced by reinforcement learning algorithms may lead to uncertainties regarding accountability, responsibility, and adherence to ethical norms in warfare. Therefore, it is crucial to establish robust governance frameworks and ethical guidelines to govern the development, deployment, and use of AWS equipped with reinforcement learning capabilities, ensuring adherence to principles of proportionality, distinction, and humanity in accordance with IHL.

10.7 ALGORITHMIC BIAS AND RISK OF INDIS-CRIMINATE TARGETING IN AUTONOMOUS WEAPON SYSTEMS (AWS)

Algorithmic bias refers to systematic and repeatable errors in a system that create unfair outcomes, often due to unintended preferences for certain groups or types of data. Algorithmic bias happens when the algorithm itself creates a bias, even though the input data is fair. This can happen because of the choices made when designing the algorithm, like the optimization functions used, the way regression models are applied, or the use of biased estimators. All these factors can lead to biased decisions and unfair outcomes [21]. Algorithmic bias can manifest in various forms, including racial, gender, cultural, or socioeconomic biases. Indiscriminate targeting occurs when AWS fail to differentiate between legitimate and illegitimate targets, potentially causing harm to civilians or non-combatants. The primary reasons for indiscriminate targeting are the bias in AWS algorithms and the disproportionately impacts on certain groups.

Non-representative training data, algorithm design, and deployment context are the causes for algorithmic bias. Algorithmic bias poses a significant ethical and operational risk in the context of AWS, potentially leading to indiscriminate targeting, incorrect target identification, misclassification of targets, and violations of IHL. Here's an exploration of these concerns:

10.7.1 Algorithmic bias in target recognition

AWS rely heavily on algorithms for target identification and classification. If these algorithms are trained on biased data or if biases are introduced during their development, AWS may exhibit skewed or discriminatory behavior in identifying targets. For example, biases in training data could result in the misidentification of civilians or non-combatants as legitimate targets, leading to unintended casualties and violations of humanitarian principles such as distinction and proportionality.

10.7.2 Risk of indiscriminate targeting

Autonomous systems lacking robust bias mitigation strategies may engage in indiscriminate targeting, where targets are selected based on erroneous or biased criteria. This can occur if algorithms disproportionately classify certain groups or individuals as threats without adequate human oversight or intervention. Indiscriminate targeting violates the principles of discrimination under IHL, which require military actions to distinguish between combatants and civilians and to only target combatants and military objectives.

10.7.3 Ethical implications of autonomous decision-making

Delegating critical decision-making to algorithms in AWS raises ethical dilemmas. Algorithms lack human judgment and empathy, which are crucial in assessing the context and ensuring proportionate responses in complex and ambiguous situations. Without proper oversight and accountability mechanisms, AWS may act on biased interpretations of situational data, potentially exacerbating conflict and civilian harm.

10.7.4 Legal challenges and accountability

The use of AWS with algorithmic bias complicates legal accountability in armed conflicts. IHL [22] requires accountability for violations, including those resulting from automated systems. Establishing accountability for AWS actions becomes challenging when decisions are made autonomously based on biased algorithms, as it may be unclear who is responsible for wrongful acts or violations of humanitarian norms.

Algorithmic bias [21] in AWS poses serious risks to humanitarian principles and global security. Addressing these risks requires interdisciplinary efforts, including robust regulatory frameworks, ethical guidelines, and technological safeguards to ensure the responsible development and use of AWS in accordance with international law and ethical standards [23].

10.7.5 Mitigation strategies and responsible AI development

Addressing algorithmic bias in AWS requires proactive measures during the design, development, and deployment phases. This includes:

1. **Diverse and Representative Data:** Ensuring training data used for algorithms are diverse and representative of all potential scenarios and populations.
2. **Bias Detection and Correction:** Implementing techniques to detect and mitigate bias in algorithms, such as fairness-aware ML and continuous monitoring of system performance.
3. **Human Oversight and Control:** Maintaining human oversight and control over AWS operations, including the ability to intervene in decision-making processes and override automated actions when necessary.

10.8 ADDRESSING ETHICAL AND LEGAL CONCERNS IN AUTONOMOUS WEAPON SYSTEMS

As the field of AI-driven AWS progresses, the ethical and legal dimensions surrounding their deployment grow increasingly significant. While AI has propelled technological advancements, the inherent indiscriminate nature of autonomous weapons raises profound concerns regarding accountability, proportionality, and adherence to IHL [24, 25]. The prospect of these systems making life-or-death decisions without direct human intervention underscores the urgent need to ensure robust ethical principles and legal frameworks govern their development and deployment [26].

Central to the ethical discourse on autonomous weapons is the issue of moral responsibility. With the capability to autonomously engage targets and inflict harm, the allocation of accountability becomes pressing. Concerns about unintended harm to civilians or non-combatants underscore the need for stringent regulations and oversight mechanisms to safeguard fundamental human rights and ethical norms.

In response to these ethical concerns, various regulatory frameworks and guidelines have emerged to govern AWS' development and use [22]. Organizations like the Convention on Certain Conventional Weapons [3, 25, 26] and initiatives such as the Principles for the Ethical Use of AI in Defense aim to establish norms and restrictions to mitigate associated ethical and legal risks. These frameworks emphasize transparency, accountability, and the preservation of human oversight throughout AWS' lifecycle.

Transparency is critical to ensure the comprehensibility and auditability of AWS' decision-making processes. Regulatory frameworks promote transparency to enhance accountability and facilitate effective oversight mechanisms to monitor and evaluate these systems' actions. Additionally, integrating human oversight mechanisms acts as a vital safeguard against potential misuse or abuse, ensuring human operators maintain ultimate control over lethal force deployment decisions.

Moreover, the regulation debate surrounding AWS mirrors broader discussions on AI's ethical and societal implications. As AI increasingly infiltrates various societal domains, including military applications, proactive addressing of ethical and legal considerations becomes imperative. Balancing innovation with ethical responsibility requires ongoing dialogue, collaboration, and concerted efforts to develop regulatory frameworks that uphold human rights, dignity, and security amidst evolving technological capabilities.

10.8.1 Ethical concerns

10.8.1.1 Human control

Autonomous weapons raise questions about the degree of human control over decisions to use lethal force [22]. Ensuring meaningful human control can mitigate risks of misuse and ethical violations.

10.8.1.2 Accountability

Who is responsible if an autonomous weapon causes harm or violates laws? Clear lines of accountability are essential, including mechanisms for investigation, attribution of responsibility, and consequences.

10.8.1.3 Proportionality and discrimination

Autonomous systems must adhere to principles of proportionality (using force proportional to the threat) and discrimination (distinguishing between combatants and civilians). Ensuring these principles are upheld requires advanced sensor capabilities and decision-making algorithms.

10.8.1.4 Unintended consequences

Autonomous systems might exhibit unexpected behaviors or consequences due to programming errors, adversarial actions, or environmental factors. Safeguards and testing protocols are necessary to minimize these risks.

10.8.1.5 Proliferation and arms race

The development and deployment of AWS could lead to an arms race, escalating tensions and increasing the likelihood of conflict. International cooperation and treaties may be needed to mitigate these risks.

10.8.2 Legal concerns

10.8.2.1 International humanitarian law (IHL)

AWS must comply with IHL, which regulates the conduct of warfare to protect civilians and combatants who are hors de combat (out of combat). Compliance requires programming AWS to respect principles like distinction, proportionality, and military necessity.

10.8.2.2 Weapons review and use of force

States are obligated to review new weapons to ensure they comply with international law before deployment. AWS raise questions about the adequacy of current legal frameworks in assessing their compliance.

10.8.2.3 State responsibility

States deploying AWS are responsible for ensuring their compliance with international law, including IHL. This includes both operational use and development phases.

10.8.2.4 Ethics of development and research

Researchers and developers have ethical responsibilities to consider the implications of their work on AWS, including potential misuse or unintended consequences.

10.8.3 Addressing the concerns

Addressing ethical and legal concerns in AWS is crucial due to their potential implications for warfare, human rights, and international law. Here are some key considerations [22, 25, 26]:

10.8.3.1 Policy and regulation

Establish clear policies and regulations governing the development, deployment, and use of AWS. This includes international agreements, national laws, and industry standards.

10.8.3.2 Transparency and accountability

Promote transparency in the development and deployment of AWS, ensuring clear accountability mechanisms for both developers and deploying states.

10.8.3.3 Ethics boards and review

Establish independent ethics boards or review processes to evaluate the ethical implications of AWS projects and ensure compliance with ethical standards.

10.8.3.4 Public and stakeholder engagement

Engage with civil society, academia, and other stakeholders to ensure diverse perspectives are considered in the development and governance of AWS.

10.8.3.5 International cooperation

Foster international dialogue and cooperation to address global concerns related to AWS, including through forums like the United Nations and other international organizations.

10.9 CONCLUSION

Weapon autonomy, the capability of weapons to operate independently or semi-independently of human control, stands at the forefront of contemporary debates on military ethics, technology, and international security. This chapter presents AI techniques driving weapon autonomy and examines the far-reaching implications of their integration into military contexts. Understanding weapon autonomy requires navigating complex intersections of technology, morality, and policy, shaping the future landscape of armed conflict and global stability.

As advancements in AI, ML, and robotics accelerate, so too do the possibilities and concerns surround. These technologies are driving the evolution of AWS, enhancing their capabilities and operational effectiveness. As the integration of AI and ML techniques into weapon systems proliferates, it becomes increasingly apparent that this technological advancement presents a dual-edged sword. While offering unprecedented capabilities, autonomous weapons also usher in ethical dilemmas that demand careful consideration. It is imperative to address the ethical, legal, and strategic implications to ensure its responsible and ethical use in military operations and global security contexts.

The evolution of autonomous weapons signifies a paradigm shift in military operations and promises to reshape the landscape of modern warfare by enhanced strategic advantage and operational efficiency. However, alongside these advancements come significant challenges and risks, necessitating a holistic approach to address the ethical, legal, and societal implications of AWS. As AI-powered weapons continue to mature, it is imperative to establish regulatory frameworks that promote accountability, transparency, and adherence to IHL. By doing so, society can harness the potential of autonomous weapons to contribute to peace and security while mitigating potential ethical pitfalls and ensuring responsible usage in the pursuit of national defense objectives.

It is incumbent upon policymakers, military strategists, and ethicists to engage in informed and collaborative discussions to navigate this complex landscape. By fostering dialogue and cooperation, stakeholders can work towards ensuring the responsible development and deployment of autonomous systems, thereby balancing technological innovation with ethical considerations to uphold principles of humanity and justice.

By addressing these complexities with foresight and diligence, stakeholders can navigate the ethical and strategic terrain of autonomous weapons, fostering a future where technological innovation aligns harmoniously with ethical imperatives to promote peace and security on a global scale.

REFERENCES

1. Scharre, P. (2018). *Army of None: Autonomous Weapons and the Future of War.* WW Norton & Company.
2. Pedron, S. M., & da Cruz, J. D. A. (2020). The future of wars: artificial intelligence (AI) and lethal autonomous weapon systems (laws). *International Journal of Security Studies*, 2(1), 2.
3. Longpre, S., Storm, M., & Shah, R. (2022). Lethal autonomous weapons systems & artificial intelligence: trends, challenges, and policies. *Edited by Kevin McDermott. MIT Science Policy Review*, 3, 47–56.
4. Davison, N. (2018). A legal perspective: autonomous weapon systems under international humanitarian law. *United Nations Office of Disarmament Affairs (UNODA) Occasional Papers*, 30, pp. 5–18, https://doi.org/10.18356/29a571ba-en

5. Dremliuga, R. (2020). General legal limits of the application of the lethal autonomous weapons systems within the purview of international humanitarian law. *Journal of Politics and Law, 13*, 115.

6. Schwarz, E. (2021). Autonomous weapons systems, artificial intelligence, and the problem of meaningful human control. *Philosophical Journal of Conflict and Violence.*

7. Abaimov, S., & Martellini, M. (2020). Artificial intelligence in autonomous weapon systems. In *21st Century Prometheus: Managing CBRN Safety and Security Affected by Cutting-Edge Technologies*, 141–177.

8. Horowitz, M. C. (2021). When speed kills: lethal autonomous weapon systems, deterrence and stability. In *Emerging Technologies and International Stability* (pp. 144–168). Routledge.

9. Gunawan, Y., Aulawi, M. H., Anggriawan, R., & Putro, T. A. (2022). Command responsibility of autonomous weapons under international humanitarian law. *Cogent Social Sciences, 8*(1), 2139906.

10. Barrera, C., Padron, I., Luis, F. S., & Llinas, O. (2021). Trends and challenges in unmanned surface vehicles (USV): from survey to shipping. *TransNav: International Journal on Marine Navigation and Safety of Sea Transportation, 15*,135–142.

11. Bolbot, V., Sandru, A., Saarniniemi, T., Puolakka, O., Kujala, P., & Valdez Banda, O. A. (2023). Small unmanned surface vessels—a review and critical analysis of relations to safety and safety assurance of larger autonomous ships. *Journal of Marine Science and Engineering, 11*(12), 2387.

12. Parry, J., & Hubbard, S. (2023). Review of sensor technology to support automated air-to-air refueling of a probe configured uncrewed aircraft. *Sensors, 23*(2), 995.

13. Zhu, K., Yang, J., Zhang, Y., Nie, J., Lim, W. Y. B., Zhang, H., & Xiong, Z. (2022). Aerial refueling: scheduling wireless energy charging for UAV enabled data collection. *IEEE Transactions on Green Communications and Networking, 6*(3), 1494–1510.

14. Lv, C., & Cao, L. (2021). Target recognition algorithm based on optical sensor data fusion. *Journal of Sensors, 2021*(1), 1979523.

15. Bistron, M., & Piotrowski, Z. (2021). Artificial intelligence applications in military systems and their influence on sense of security of citizens. *Electronics, 10*(7), 871.

16. McNabb, M. (2023, June 22). Teal 2 Makes Blue UAS List: Red Cat and Athena Announce New AI and Computer Vision Abilities for Teal 2 Drone. *DRONE-LIFE.* https://dronelife.com/2023/06/22/red-cat-and-athena-announce-new-ai-and-computer-vision-abilities-for-teal-2-drone/

17. Lingel, S., Hagen, J., Hastings, E., Lee, M., Sargent, M., Walsh, M., & Blancett, D. (2020). *Joint All-Domain Command and Control for Modern Warfare. Santa Monica: RAND Corporation.*

18. Rivera, J. P., Mukobi, G., Reuel, A., Lamparth, M., Smith, C., & Schneider, J. (2024, June). Escalation risks from language models in military and diplomatic decision-making. In *The 2024 ACM Conference on Fairness, Accountability, and Transparency* (pp. 836–898).

19. Starken, A., Caulkins, B., Wu, A. S., & Mondesire, S. C. (2022, December). Trends in machine learning for adaptive automated forces. In *the Proceedings of the Interservice/Industry Training, Simulation and Education Conference (IT/TSEC'22).*

20. De Marco, A., D'Onza, P. M., & Manfredi, S. (2023). A deep reinforcement learning control approach for high-performance aircraft. *Nonlinear Dynamics, 111*(18), 17037–17077.

21. Mehrabi, N., Morstatter, F., Saxena, N., Lerman, K., & Galstyan, A. (2021). A survey on bias and fairness in machine learning. *ACM Computing Surveys (CSUR, 54*(6), 1–35.

22. Hynek, N., & Solovyeva, A. (2021). Operations of power in autonomous weapon systems: ethical conditions and socio-political prospects. *AI & Society*, *36*, 79–99.
23. Chengeta, T. (2022). Autonomous weapon systems: accountability gap and racial oppression. In S. Christopher (Ed.), Reclaiming Human Rights in a Changing World Order. Chatham House/Brookings Institution Press, (pp. 216–236)
24. Winter, E. (2022). The compatibility of autonomous weapons with the principles of international humanitarian law. *Journal of Conflict and Security Law*, *27*(1), 1–20.
25. Riebe, T., Schmid, S., & Reuter, C. (2020). Meaningful human control of lethal autonomous weapon systems: the CCW-debate and its implications for VSD. *IEEE Technology and Society Magazine*, *39*(4), 36–51.
26. Chengeta, T. (2022). Is the convention on conventional weapons the appropriate framework to produce a new law on autonomous weapon systems. In F. Viljoen, F. Charles, T. Dire, S. Ann, & K. Magnus (Eds.), A Life Interrupted: Essays in Honour of the Lives and Legacies of Christ of Heyns. Pretoria University Law Press, (pp. 379–397)

Chapter 11

Navigating the complexities

AI, security, and liberty at the border

Vijaypal Singh Rathor

11.1 INTRODUCTION

The actions a nation takes to manage and keep an eye on its borders are referred to as border security. The actions include preventing the admission of persons, products, and illegal imports (guns, narcotics, etc.) and controlling the lawful flow of goods and people for travel and trade. Such actions are needed to thwart risks to border security and cross-border terrorism. It is a complex and evolving field influenced by geopolitical factors such as Greece and Turkey share more than 350 nautical miles of maritime borders. Border security is made more difficult by illegal threats and activities such as terrorism, drug and weapon proliferation, people trafficking, and asylum seekers' obfuscation. Citizens of the state, who view foreign nationals as competitors and believe they would hurt their interests, wealth, and daily lives, are partly to blame for the need for border security.

Border security encompasses a range of strategies and technologies including physical barriers (such as fences and walls), surveillance systems (such as cameras and sensors), patrol agents (like border guards and police), and legal frameworks (such as immigration laws and customs regulations) [1–4].

Border security is increasingly incorporating artificial intelligence (AI) technologies to enhance monitoring, detection, and management capabilities. AI is utilized in several key areas of border security such as [1, 3, 5]:

1. **Surveillance and Monitoring:** AI-powered cameras, drones, and sensors can continuously monitor border areas, analyzing live feeds for anomalies or unauthorized crossings. Machine learning (ML) algorithms can detect patterns indicative of suspicious activities, alerting border agents in real time.
2. **Facial Recognition and Biometrics:** AI algorithms are used to match facial images against databases of known individuals, aiding in the identification of suspects, criminals, or persons of interest at border checkpoints.
3. **Predictive Analytics:** AI models can analyze historical data on border crossings, crime patterns, and immigration trends to predict potential threats or hotspots, allowing for proactive deployment of resources.

DOI: 10.1201/9781032657264-11

4. **Automation of Border Procedures:** AI streamlines routine tasks such as passport verification and visa processing, reducing processing times and human error.
5. **Risk Assessment:** AI-based systems assess the risk profile of travelers based on various factors like travel history, biometric data, and background checks, helping prioritize inspections and improve decision-making.
6. **Language Translation:** AI-powered translation tools facilitate communication between border officers and travelers who speak different languages, improving efficiency and accuracy during interactions.
7. **Data Analysis and Collaboration:** AI helps in analyzing vast amounts of data from multiple sources (such as intelligence databases, social media, and travel records) to identify trends, networks, and potential threats across borders. It also supports international collaboration by standardizing data formats and facilitating information sharing between agencies.

While AI enhances border security capabilities, challenges include ensuring data privacy, addressing algorithm biases, and navigating legal and ethical considerations. Effective deployment requires balancing technological advancement with ethical guidelines and public trust to maximize security benefits while safeguarding individual rights [4, 6].

11.1.1 Objectives of this chapter

The objectives of the chapter are:

1. To explore the transformative role of AI in enhancing border security and management and the benefits thereof.
2. To explain and evaluate the effectiveness of specific AI tools and techniques being used such as facial recognition, predictive analytics, automated decision-making, biometric identification, and enhanced surveillance capabilities.
3. To discuss ethical and privacy concerns regarding the use of AI in border security which demands the policy frameworks and regulations governing the use of AI in border security.
4. To discuss case studies or examples where AI has been successfully implemented in border control.

11.2 AI-ENABLED VEHICLES FOR BORDER SURVEILLANCE

11.2.1 Unmanned aerial vehicle

Drones, commonly referred to as unmanned aerial vehicles (UAVs) [7, 8], are flying machines that run without a human pilot present. A UAV is shown in Figure 11.1. UAVs can be operated remotely or independently, and their

Figure 11.1 Unmanned aerial vehicles (UAV).

diverse design, size, and set of features allow for a multitude of uses in both the military and commercial domains.

11.2.1.1 Uses in the border security

- **Reconnaissance and Surveillance:** UAVs offer troops on the ground an invaluable aerial perspective, facilitating instantaneous tracking of adversary locations, movements, and actions in hazardous or isolated regions. Drones equipped with AI algorithms can autonomously patrol borders, identify suspicious activities, and track movement patterns [7].
- **Target Acquisition and Strike:** Armed drones carrying bombs and missiles can attack targets more precisely than conventional airstrikes, which may result in fewer civilian casualties.
- **Intelligence Gathering:** By using high-resolution cameras and other sensors, UAVs may obtain invaluable intelligence that is essential for threat assessment and mission planning.
- **Search & Rescue:** By using thermal imaging sensors, UAVs can quickly discover survivors in disaster areas or look for missing people over large areas.
- **Delivery Services:** Drone delivery is becoming more and more popular. It has the potential to be used for delivering quick food and emergency medical supplies at geologically tough borders.

11.2.1.2 Benefits of unmanned aerial vehicles

- **Cost-Effectiveness:** UAVs are often less expensive to create, run, and maintain than manned aircraft.
- **Safety:** UAVs are perfect for high risk or dangerous operations since they remove the possibility of pilot injuries or fatalities.
- **Versatility:** UAVs can be outfitted with a range of sensors and equipment to enable them to carry out a variety of functions in a variety of settings.
- **Endurance:** Compared to human aircraft, many UAVs have longer flight periods, which allows them to conduct longer surveillance or data-collecting missions.

11.2.1.3 The difficulties with UAVs

- **Control and Regulation:** Securing the responsible and safe operation of UAVs in shared airspace is a continuous task that necessitates strict rules and air traffic management procedures.
- **Privacy Issues:** Since UAVs are able to take high-definition photos and videos, there should be clear *policies* in place regarding the gathering and use of data.
- **Security Flaws:** UAVs are vulnerable to hacking and hijacking by hostile entities, hence robust cybersecurity measures are required to stop unauthorized use. Some research studies have shown that commercial drones, such as those made by well-known electronics firms like DJI, Xiaomi, or even the Aerialtronics police's drone, have serious security flaws that would be easy for would-be hackers to take advantage of [9]
- **Battery Life:** Although getting better, battery life can still have an impact on the time and range of UAV operations.

11.2.1.4 UAVs' future

New developments in technology indicate that UAVs will have many more creative uses in the future. Anticipate extended flight durations, heightened payload capability, enhanced AI integration for self-navigating vehicles, and an expanded array of specialized features customized for certain missions. UAVs have the potential to revolutionize the way we collect data, transport commodities, and carry out operations across borders.

11.2.2 USV

Often referred to as autonomous surface vehicles or unmanned surface vessels (USVs) [7, 8, 10], USVs are robotic boats that run on the water without a human crew. A USV is shown in Figure 11.2. They are available in a variety of sizes, from tiny hobby drones to massive tanks. Different degrees

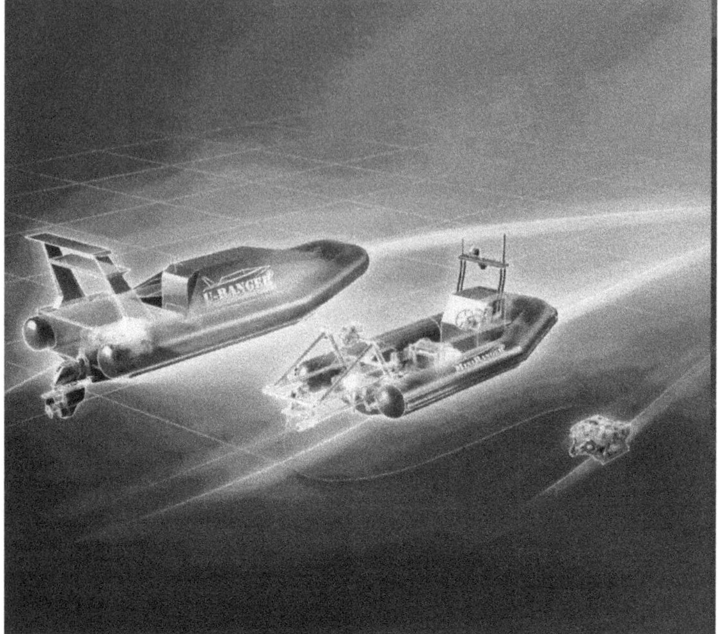

Figure 11.2 Unmanned surface vessels (USV).

of autonomy are used by USVs, ranging from totally autonomous navigation to remote control by a human operator.

11.2.2.1 Uses in military

USVs are employed in anti-submarine warfare, mine countermeasures, observation, and reconnaissance. They are perfect for covert missions because of their quiet operation and low profile. USVs can be used for search and rescue operations at nautical borders.

11.2.2.2 Advantages of USVs

- **Cost-Effective:** Because USVs don't need crew pay or supplies, they may be less expensive to run than manned vessels.
- **Safety:** They eliminate the possibility of exposing people to hazardous situations or unfavorable weather.
- **Endurance:** Because crew weariness does not limit USVs, they can stay at sea for longer periods of time than human vessels.
- **Data Collection:** By using advanced sensors, USVs can gather important information over wide regions.

11.2.2.3 The difficulties with USVs

1. **Regulation:** To guarantee the safe operation of USVs alongside manned vessels on shared waterways, clear regulations are required.
2. **Technology:** Reliable autonomous navigation requires advanced technology, such as strong communication networks and AI with advanced decision-making capabilities.
3. **Security:** To guarantee unauthorized control of USVs and stop hacking, cybersecurity measures are essential.

11.2.2.4 USVs' prospects

- USVs are anticipated to grow in sophistication and importance in a variety of maritime applications as technology develops.
- **Enhanced Autonomy:** AI developments will give USVs the ability to manage challenging jobs and make choices in unpredictable situations.
- **Better Integration:** Data exchange and coordinated operations will probably be made easier by USVs' more seamless integration with other marine systems and boats.
- **Increased Usage:** It is anticipated that USV technology will become less expensive, resulting in increased usage in a number of industries.

11.3 AI-POWERED SURVEILLANCE SYSTEMS

11.3.1 Key technological solutions

11.3.1.1 Automated Virtual Agent for Truth Assessments in Real-Time (AVATAR)

Automated Virtual Agent for Truth Assessments in Real-Time (AVATAR) [11–13] is a technology intended to automate the processes of screening, interviewing, and determining a person's credibility when they cross international borders. With assistance from FRONTEX and the US Department of Homeland Security, the BORDERS research Centre at the University of Arizona created the system. The AVATAR lie detector device uses changes in pupil dilation, voice quality, and eye movements to identify people who could pose a risk during automated screening and interviews. As per Frontex's own statement, the efficacy of AVATAR and other systems is questionable because of the possibility of algorithmic prejudices.

11.3.1.2 iBorderCtrl

In response to AVATAR, the European Union (EU) launched the iBorderCtrl [11] initiative, which was created by a partnership spanning the entire continent. Four border crossing sites in Hungary, Latvia, and Greece already have an AI-based lie detector deployed as part of this project. Additionally, the system would automatically alert fast-reaction teams in the event of an emergency requiring immediate action, such as when a boat carrying refugees or immigrants overturned, prompting them to rush toward the scene and greatly reducing the possibility of casualties. The iBorderCtrl [14, 15] comprises a pre-registration phase where an avatar asks a traveler who wants to cross EU borders a series of questions while the system authenticates the traveler's micro-gestures, or non-verbal behaviors, to create their complete profile. The information gathered during this phase, along with the traveler's answers, feeds the "automatic deception detection system" which in the end determines a score of the reliability of a particular traveler. The use of particular tools, such as the biometrics module, which uses fingerprints to validate the traveler's identity, or the face-matching tool, which receives and processes images from the travelers, helps to inform the border guard who has the final say in whether to accept or reject a traveler's application. The risk-based assessment tool is also crucial, since it gives the border guard information primarily influenced by the data gathered from the procedures, as previously described. Only if the traveler scores higher than a predetermined threshold based on the low-risk assessment and the findings of the deception detection is the standard border inspection conducted at border crossing locations.

11.4 AI-POWERED CUTTING EDGE TECHNOLOGIES FOR BORDER SECURITY

AI is revolutionizing border security at a rapid pace, providing a robust toolkit of techniques and features that extend beyond human identification. Here's a closer look at the application of AI in border security:

11.4.1 Intelligent video analytics and object detection

11.4.1.1 Beyond object recognition

AI systems are capable of classifying items according to their size, speed, and even behavior in addition to detecting objects like people and cars [16]. AI algorithms analyze live and recorded video feeds from cameras to detect anomalies, such as unauthorized border crossings or suspicious behavior. Consider an AI system that can tell the difference between an unauthorized individual entering the country at night and a harmless tumbleweed. By adding an additional layer of intelligence, border guards are able to focus on real threats and drastically minimize the number of false alarms.

11.4.1.2 Drone and sensor integration

To create a complete image of border activities, AI may combine video feeds from a network of drones, ground-based cameras, and different sensors [17]. AI can process data from various sensors (seismic, infrared, acoustic) [17] to detect and classify potential threats at the border. This makes it possible to monitor distant locations in real time and respond to crises more quickly.

11.4.1.3 Enhanced voice, facial recognition, and biometric identification

AI can analyze speech patterns and biometric characteristics to verify identities, enhancing border security screening processes. AI algorithms enable rapid and accurate identification of individuals at border crossings, comparing faces against databases of known suspects or watch lists [18]. AI's ability to distinguish between a real person and an image or video presentation is known as "liveness detection," which thwarts attempts to use false identification to get over border protection [18, 19].

11.4.1.4 Mobile facial matching

Static picture-based facial recognition technologies are becoming obsolete. AI can now identify people even when they are moving through a crowd by tracking and matching faces in real-time video streams [19].

11.4.1.5 Multimodal biometric integration

For more reliable identification, AI systems can combine fingerprint, iris, or voice recognition in addition to facial recognition. This is particularly useful in scenarios where low illumination or disguise may compromise facial detection [20].

11.4.2 Predictive analytics and risk assessment for advanced screening

ML algorithms analyze historical data and current trends to predict potential security threats, enabling proactive measures by border security agencies. AI models assess traveler profiles and behaviors to identify high-risk individuals or cargo, prioritizing inspections and resources accordingly [21, 22].

11.4.2.1 Trip pattern analysis

To spot questionable trends, AI can examine enormous volumes of trip data. This could include taking odd routes, visiting dangerous locations frequently, or having ties to notorious criminals.

11.4.2.2 Social media screening

Posts on social media can be screened for possible threats or extreme ideologies by employing Natural Language Processing text analysis. However, privacy considerations must be taken into account.

11.4.2.3 Dynamic risk scoring

Using a variety of criteria, AI is able to rate each traveler's risk. This makes it possible for border officials to expedite the process for low-risk travelers while giving priority to high-risk persons for more thorough screening.

11.4.3 Threat detection in cargo and luggage inspection

11.4.3.1 Automatic anomaly detection in scans

This feature allows AI to automatically identify hidden compartments or anomalies in X-ray or millimeter wave scans of luggage and cargo containers. This may involve recognizing items with peculiar densities, forms, or packing patterns [23, 24].

11.4.3.2 Material classification

AI may be trained to identify the materials that make up things in cargo and packages, which can be used to flag suspicious items like drugs or weapons [24].

11.4.4 Additional AI techniques on the horizon for border security

11.4.4.1 AI-powered threat assessment at checkpoints

When a traveler is being questioned, AI systems may examine their body language, facial expressions, and physiological reactions to spot signs of uneasiness or deceit. Nonetheless, here it is important to pay close attention to ethical issues of justice and potential bias.

11.4.4.2 Conversational AI for multilingual border control

By managing regular border control encounters, chat-bots driven by AI can free up human agents to work on more difficult tasks. The fact that these chat-bots may be built to speak several languages makes the process easier for visitors from other countries [25].

11.5 MERITS AND DEMERITS OF USING ARTIFICIAL INTELLIGENCE IN BORDER SECURITY

Autonomous vehicles such as UAVs, USVs, etc. heavily depend on the sensors and data collected by them, it is clear that there are significant risks associated with the use of small, quick, and most importantly efficient sensors. These days, UAVs are equipped with high-tech, high-resolution cameras, including

4K cameras, which can detect even the smallest objects in low light and darkness, like a man overboard. They also have laser radar to gain improved accuracy in range and thermal imaging equipment to detect moving objects in unfavorable environments.

They are also less vulnerable to noise-jamming, which could otherwise divert their attention from their goal [26]. However, there are several aspects that affect the video analytics process, such as the presence of light. The absence of modern sensors, such as 4K cameras, would lead to subpar surveillance, regardless of how complex the algorithms are which analyzes the captured video streams. Aside from UAVs, there are other important tools that can be used to strengthen the cohesiveness of borders, particularly when it comes to data collection: USVs.

The developed infrastructure may perform worse due to sporadic weather events. For instance, in a marine environment, particularly in the morning, it is not uncommon to have thick fog, which severely restricts the sensors' field of view. Situational awareness is made much more difficult by unfavorable local occurrences like winter storms with strong winds and torrential rain. Intense rain not only impairs the functionality of its onboard sensors but also lowers the quality of connection with the Rescue Coordination Centers, making any coordinated search and rescue operation difficult, if not impossible.

There is a concerning possibility that the implementation of biometric identification systems would exacerbate discriminatory practices [27]. When a refugee's life depends on the quality of the answers they provide, the stress of the circumstance may lead them to exhibit unanticipated emotions like crying, which an avatar would not be able to comprehend [28]. Rejecting an asylum request will lead to more complaints if not adequately addressed; for instance, tensions were observed in most of the Greek refugee camps due to the incredibly sluggish processing and review of the application. The burden of proof for the traveler will increase if the contents of the database, which served as the basis for the retrieval of questionable information, remain a secret from them. This misconception would therefore lead to complaints from travelers or asylum seekers, permanently harming the system's reputation.

A judgment based on false negatives would have the same, if not greater, consequences. The harm that will be done in this instance will compromise not only the system's credibility but also the basic idea upon which it was built and developed: the safety of the country. The public debate against automated decision-making systems frequently centers on the fact that these systems can produce false positives and false negatives, emphasizing the seriousness of any potential malfunction's consequences.

It is true that technologies in the field of AI-based border security have evolved significantly in recent years, leveraging advancements in various domains such as surveillance, communication, and data analytics. AI has immense potential to enhance security measures is immense, but it also comes with challenges. By acknowledging both the merits and demerits, border security agencies can implement AI in a responsible and effective manner, maximizing its benefits while minimizing its risks. Let's explore the merits and demerits of AI in border security and management.

11.5.1 Merits of AI in border security

11.5.1.1 Enhanced accuracy and efficiency

AI can automate a lot of laborious jobs, such as facial recognition, object detection, and anomaly detection in scans, freeing up border agents to work on more complicated duties and accelerating processing times overall [28].

11.5.1.2 Better threat detection

AI's ability to evaluate enormous volumes of data and spot trends that people might overlook makes it possible to detect suspicious activity, concealed items, and other threats more accurately [28].

11.5.1.3 24/7 Real-time monitoring

AI-powered systems are able to continuously monitor borders, even in remote places, which facilitates enhanced situational awareness and quicker incident response times.

11.5.1.4 Risk-based screening

AI can evaluate travelers' data and rate travelers according to risk factors, allowing border officials to give priority to high-risk passengers and expedite low-risk travelers [28].

11.5.1.5 Advanced threat detection in cargo

AI may examine scans and spot irregularities in luggage and cargo containers, which may result in the finding of weapons, drugs, or other contraband.

11.5.2 Demerits of AI in border security

Even though AI has a lot of promise to improve border security, ethical issues must be addressed, and responsible application must be ensured. Important areas of attention consist of:

11.5.2.1 Data security and privacy issues

Gathering, storing, and utilizing a large quantity of personal information for border security raises privacy concerns. To safeguard sensitive traveler information and stop misuse, strict data governance laws are essential to prevent abuse [27].

11.5.2.2 Algorithmic bias

AI systems have the potential to reinforce prejudices found in the data they are trained on, which could result in unfair profiling and treatment of particular populations unfairly [27]. To make sure AI systems don't uphold discrimination based on race, ethnicity, or other criteria, frequent audits and bias detection methods are crucial.

11.5.2.3 Ethical aspects of threat assessment driven by AI

It is unethical to analyze body language and physiological reactions to detect dishonesty; fairness and individual rights must be carefully taken into account [27].

11.5.2.4 Accountability and transparency

It might be difficult to hold sophisticated AI systems responsible for mistakes or biased results since it can be difficult to understand how they make judgments. To establish public confidence in the application of AI to border security, unambiguous policies and public awareness campaigns are required. There should always be human oversight and appeal procedures in place.

11.5.2.5 Over-reliance on technology

In border security operations, an over-reliance on AI may result in a lack of human judgment and critical thinking.

11.5.2.6 Potential for abuse

If AI border security systems fall into the wrong hands, they might be used to target particular groups or people for widespread surveillance or repression [27].

11.6 RETHINKING AI IN BORDER MANAGEMENT

11.6.1 Beyond facial profiling

AI, particularly ML, works with observable patterns in face images captured by surveillance cameras. But there is no such pattern that can establish a relationship between facial characteristics and criminality. Alternatively, you can say that there are no *physical features to criminality* function exist in nature. Having a particular type of face or facial features does not make someone a criminal.

Predictions that stem from the identification of relationships between criminal behavior and facial features are acknowledged as legitimate and understood as the result of knowledgeable and *objective* technological evaluations. In actuality, these *predictions* materially confuse criminal activity with the common societal circumstances of being unjustly over-policed. Such algorithmic recommendations for police produce more data, which is subsequently fed back into the system to reproduce *skewed outcomes* [29, 30]. Computer scientists can greatly benefit from ongoing methodological debates, insights, and research work in other disciplines such as anthropology, sociology, media, and communication studies. Researchers have worked for decades to develop more robust frameworks for video surveillance by considering the circumstances embedded in countless social and cultural contexts.

11.6.2 Emotional state profiling

AI polygraphs, also known as lie detectors and emotion recognition technologies, are a subset of systems that make the claim that they can infer a person's intention, emotional state, or mental state such as deception, trustworthiness, and truthfulness based on their biometric information or other information about their physiological, behavioral, or physical traits [31]. AI polygraphs use advanced algorithms to analyze various data points such as facial expressions, voice patterns, physiological responses (like heart rate and sweat gland activity), and even linguistic cues (word choice, sentence structure) to determine if someone is being truthful or deceptive. Unlike traditional polygraphs that rely on sensors attached to the body, AI polygraphs can potentially analyze data from video and audio recordings in real-time [32].

11.6.2.1 Expected applications of AI polygraphs in border security and migration

1. **Screening at Borders:** AI polygraphs could be used to screen individuals at border checkpoints or during visa interviews to detect deception regarding their identity, purpose of travel, or intentions. This could potentially identify individuals with malicious intent, such as terrorists or criminals trying to enter a country.
2. **Managing Asylum Claims:** In the context of migration, AI polygraphs might be used during asylum interviews to assess the credibility of applicants' claims about persecution or fear in their home countries. This could help immigration authorities make more informed decisions about granting asylum [4].
3. **Monitoring Immigration Compliance:** AI polygraphs could also be used for monitoring individuals who are on temporary visas or parole to ensure they are complying with the conditions of their stay. This might include verifying employment status, adherence to geographic restrictions, or compliance with reporting requirements [4, 33].

11.6.2.2 Challenges and concerns of AI polygraphs

1. **Accuracy and Reliability:** The effectiveness of AI polygraphs in real-world scenarios remains uncertain [27]. Factors such as cultural differences in communication styles, the ability of individuals to manipulate their behavior, and the accuracy of the underlying algorithms pose significant challenges.
2. **Privacy and Ethical Issues:** The use of AI polygraphs raises serious privacy concerns, as it involves continuous monitoring and analysis of personal data [27]. There are also ethical considerations regarding the fairness of using AI to make decisions that could have profound impacts on individuals' lives.
3. **Legal and Regulatory Frameworks:** Implementing AI polygraphs in border security and migration would require clear legal and regulatory frameworks to address issues such as data protection, consent, and the rights of individuals subjected to such technology.
4. **Bias and Discrimination:** Like any AI system, there is a risk of bias due to inappropriate data and algorithms biases and in AI polygraphs, which could disproportionately impact certain groups based on factors such as race, ethnicity, or language proficiency [27].

11.6.2.3 AI polygraphs are not trustworthy

Although *iBorderCtrl* and AVATAR are the systems that incorporate emotion recognition technology as part of its assessment process. Tech countries such as the United States and EU members are not showing trust in AI polygraphs. These countries are exploring and implementing advanced technologies for border control and migration management, including AI-driven systems. However, AI polygraphs specifically have not been widely adopted or publicly acknowledged even by these countries except pilot projects [34].

11.6.3 The algorithmic threat: discrimination and injustice in AI-powered border control

The UK Home Office is a ministerial department of the UK government responsible for immigration, security, and law and order. Its primary responsibilities include managing immigration and visas, ensuring national security, tackling crime, and overseeing policing and fire services in England and Wales. Visa streaming refers to the process of categorizing visa applications into different streams or categories based on various factors such as the applicant's nationality, purpose of visit, and risk assessment.

The UK Home Office employed a *visa streaming* algorithm to sift through visa applications. The UK Home Office itself stated that the algorithm *entrenched racism and bias into the visa system* and was suspended in 2020 as a result of a judicial challenge. Additionally, automated monitoring systems at border and refugee camps apply risk assessments [35]. The right to be free from discrimination is seriously threatened by the risky practice of using computerized risk assessment and profiling systems during migration procedures. This threat comes from both direct and indirect sources. Additionally, the European Court of Justice drew attention to the potential for discrimination when automated decision-making is used [36, 37].

Article 5(1) of the General Data Protection Regulation's talks [38, 39] regarding the quality of personal data is likely to be compromised by the use of AI systems for profiling and individual risk assessments. Concern should be given in particular to data that is of poor or erroneous quality when it comes to EU migration databases. It states that the data can only be stored for as long as it's needed for the specific reason it was collected. These systems should be acknowledged as posing an unacceptable risk to fundamental rights, rather than advocating for increased data gathering to improve the quality of the data which would further undermine the concepts of consent.

11.6.4 Address people concerns

In order to determine whether there is a match, remote biometric identification (RBI) systems and other AI-based border surveillance systems compare an individual in a surveillance stream (using a biometric template which may include facial features, fingerprints, iris scans, etc.) against a reference database or watch-list. The superpowers are likewise interested in this largely unexplored field of technology due to its potential since AI technologies will have far-reaching impacts on the national defense and border security industries. Any individual who is outside their country of nationality due to a well-founded fear of persecution because of their race, religion, nationality, membership in a particular social group, or political opinion is considered a refugee if they are unable or unwilling to seek protection from their country's government. Conversely, immigrants are those who make the decision to leave their home nation not because of political or ethnic unrest, state persecution, or other issues; rather, they have other motivations, such as the desire to improve their families' quality of life.

Border regions experiencing refugee flows may be exploited as testing grounds for new AI technologies, sometimes without the affected communities' consent or with little openness. This gives rise to worries over possible injury and exploitation of weaker demographics. Cristina Galdon Cavell, a researcher and advocate in the field of technology and human rights, has raised several criticisms regarding AI border management and security. He contends that sensitive and personal information about travelers obtained by automated border procedures should be subject to checks and balances, is more widely applicable. If these automated systems are successful in meeting the basic standards of respect and human rights, citizens' faith in them should increase [3].

Ethics related to a person's race, ethnicity, or religion come up during the process of determining whether or not they qualify as a valid refugee and are unquestionably crucial in determining the outcome [40]. Any human-managing tool or service that categorizes people based on their traits or skills must be transparent in order to allay concerns about people being mistreated because of their race, color, or religion. The Schengen Agreement is a treaty that was signed in 1985 near Schengen, Luxembourg, originally signed by five members of the European Economic Community [41] but now includes 26 countries. The main purpose of the Schengen Agreement is to abolish physical borders among European countries and to establish a common visa policy among member states. The Schengen Agreement was incorporated into European Union law in 1999.

Therefore, border guard training should emphasize both upholding humanitarian standards in accordance with the Schengen Agreement and effectively supervising a sophisticated automated decision-making system. To create a reliable AI, other crucial principles like human monitoring, responsibility, etc. are needed.

China, who is working very hard to become the AI industry leader, has a very interesting and potential example of to grab the attention of civic society. In an apparent attempt to spy on the Muslim residents of Xinjiang, a province in northwest China, the Chinese government has already put in place an AI-based monitoring network [42]. The system has been widely criticized for its potential to infringe on human rights and perpetuate discrimination against minority groups. Reports have emerged of widespread surveillance, arbitrary detention, and human rights abuses in Xinjiang. However, the relevant authorities contend that despite the growing number of claims of grave human rights violations, the system is so effective that the frequency of terrorist acts has nearly decreased. A decision maker who is situationally aware need not be inundated with unreliable and unassessed information, which can be just as detrimental as one who has neither reliable nor any information at all. The supply of a significant information flow to the relevant fusion center is questionable.

The European Commission (EC), the executive arm of the European Union (EU) [43, 44], proposes and enforces legislation, implements policies, and manages the EU budget. The European Commission has expressed concerns that the use of AI-powered border control systems could lead to racial profiling and discriminatory targeting of certain individuals, particularly in border areas. This is because AI systems can perpetuate existing biases and stereotypes if they are trained on biased data, leading to unfair outcomes and discrimination.

11.7 CASE STUDIES: DEPLOYED AI IN BORDER SECURITY

11.7.1 SARI real-time

SARI (Sistema Automatico di Riconoscimento Immagini) [45] stands for Automatic Image Recognition System. The *Real-Time* part refers to its ability to analyze faces captured by cameras instantly. A facial recognition system with RBI capabilities, the SARI Real-Time was purchased by the Italian Interior

Ministry in 2017. *A tactical system to monitor disembarkation operations and the various types of related illegal activities, filming them and identifying the people involved* was the intended application. SARI Real-Time was never actually implemented since the implementation of SARI Real-Time was declared illegal by the Italian Data Protection Authority in 2021 because it lacked a legal foundation for the automatic analysis of facial images and amounted to indiscriminate mass monitoring.

11.7.2 Dialect identification recognition system (DIAS)

DIAS (Dialekt Identification and Analysis System) [46] is a dialect recognition system. A Dialect Recognition System (DIAS) was implemented in 2017 by the German Federal Office for Migration and Refugees (BAMF) to review asylum applications. The goal of the dialect recognition systems was to help the asylum authorities evaluate the accuracy of the applicants' assertions about their place of origin. A two-minute voice recording of the subject explaining a picture in their native tongue is captured by the dialect identification system. After processing the voice data, it determines, as a percentage, how close the speech is to a particular dialect. Despite the inherent prejudice and technological shortcomings of these systems, there is evidence that in certain instances, the decision about an asylum claim was made using DIAS.

11.7.3 Visitor and Immigrant Status Indicator Technology

The Visitor and Immigrant Status Indicator Technology (US-VISIT) [47] is a program run by the Department of Homeland Security (DHS) in the United States. US-VISIT collects biometric data from individuals crossing borders with the intention of enhancing the pertinent databases, border security devices are already widely employed at the U.S.-Mexico borders. The ultimate idea would be a completely automated biometric entry-exit system that would eliminate the need for passports for travelers.

11.7.4 Lock-on Naval Electro-optic/infrared Sensor System (LockNESS)

This is a real-world defense system created by BAE systems. It's a shipboard defense system designed to protect vessels from various threats [48]. It is used on ships to detect and track threats like drones, missiles, and small boats. Unlike radar, *LockNESS* is particularly adept at detecting and tracking low radar cross-section threats. This includes things like UAVs (drones), anti-ship missiles, and small, fast attack craft. It uses electro-optical and infrared sensors to provide a 360-degree view of the ship. Advanced algorithms are used to analyze the visual data to identify and track potential threats. AI can be used to train systems on vast datasets of various threats, allowing them to recognize patterns and classify objects in real time. LockNESS offers automated detection and tracking, potentially reducing the workload for human operators. Although its primary purpose is to notify ship platforms, it can also easily carry out tasks like coastal patrol. The primary advancement over conventional command and control systems is the improvement of tracking algorithms for locating and interacting with subterranean, surface, and aerial targets.

11.7.5 Reconfigurable Surveillance System with Communicating Smart Sensors (RECONSURVE)

The ever-growing need for security and border control has driven innovation in surveillance technology. Reconfigurable Surveillance Systems with Communicating Smart Sensors (RECONSURVE) [49] represent a significant advancement in this field. This project, developed through a collaborative effort between Turkey, France, and South Korea, aimed to create a dynamic and adaptable maritime surveillance system to address contemporary security challenges.

RECONSURVE is a cutting-edge technology designed to revolutionize the field of surveillance. Traditional surveillance systems often suffer from limitations in adaptability and interoperability. RECONSURVE sought to overcome these limitations by introducing a key feature: reconfigurability. This allows the system to be customized based on specific needs and threats. Imagine a modular system where various sensor types, like radar, thermal cameras, and UAVs, can be integrated or removed depending on the situation. This flexibility enables focused surveillance efforts on areas of high risk, while optimizing resource allocation. Due to reconfigurability feature enables RECONSURVE to be easily adapted to different surveillance scenarios, allowing it to be used in a variety of applications, from border security to law enforcement.

Another crucial aspect of RECONSURVE is the concept of communicating smart sensors. These sensors are not merely data collection points; they actively communicate with each other and with a central processing unit. This communication network allows for real-time data sharing, analysis, and decision-making. Imagine a scenario where a radar detects a suspicious vessel. This information can be instantly relayed to thermal cameras for visual identification, while UAVs are dispatched to investigate further. This coordinated response would be significantly faster and more efficient compared to traditional systems.

RECONSURVE also utilizes advanced analytics and AI to enhance its surveillance capabilities. The system can analyze data from various sensors and sources, providing valuable insights and patterns that can be used to identify potential threats. The AI algorithms can also predict potential threats, allowing for proactive measures to be taken.

While RECONSURVE presents a promising vision for future surveillance systems, some challenges need to be addressed. Data security and privacy concerns are paramount. Robust data encryption and user access control measures are essential to ensure responsible use of the technology. Additionally, the ethical implications of such comprehensive surveillance systems require careful consideration.

RECONSURVE represents a significant leap forward in surveillance technology. Its emphasis on reconfigurability, communication, and adaptability paves the way for more effective and responsive security solutions.

11.8 AI SECURITY GATES: THE FUTURE OF BORDER CONTROL

AI security gates are revolutionizing the way borders are protected, making them more secure, efficient, and effective. Traditional security gates rely on human judgment and manual checks, which can be time-consuming and prone to errors. AI security gates [50] on the other hand, utilize ML algorithms and advanced sensors to detect and identify potential threats in real-time. These gates are equipped with facial recognition software, biometric scanners, and object detection systems, making them more accurate and reliable than their traditional counterparts.

One of the significant advantages of AI security gates is their ability to process large amounts of data quickly and efficiently. They can analyze traveler information, flight data, and security alerts in real-time, enabling them to make informed decisions about who to let in and who to flag for further screening. This not only enhances security but also reduces wait times and improves the overall traveler experience.

AI security gates are also highly adaptable and can be integrated with existing security systems, making them a cost-effective solution for border control agencies. They can be easily updated with new algorithms and software, ensuring that they remain effective against evolving threats.

Moreover, AI security gates can help reduce the risk of human error, which is a significant concern in traditional security checks. They can detect and prevent illegal activities, such as drug trafficking and terrorism, more effectively than human security personnel. However, it is important to address the ethical and privacy concerns surrounding their use to ensure that they are deployed responsibly and ethically [6].

AI security gates have been implemented as pilot projects at the following border crossings [51]:

- Israel, the United Arab Emirates (UAE), and Saudi Arabia: biometric civil registries for citizens and expatriates
- West Bank: Israeli authorities have incorporated facial recognition technologies into the checkpoints and Closed Circuit Television/Internet Protocol (CCTV/IP) video surveillance systems
- Iraq: the Kurdistan province of Sulaymaniyah introduced a similar system

- Egypt: biometric gateways or biometric access systems have been installed at the airports of Cairo, Hurghada, and Sharm El Sheikh
- Tunisia: similar projects have been implemented, including at the Tunisian-Libyan border

11.9 CONCLUSION

AI is increasingly shaping modern approaches to border security and management, revolutionizing how nations monitor and control their borders. The integration of AI in border security and management has transformed the way borders are protected and managed. AI-powered systems enable more efficient and effective surveillance, detection, and response to potential threats. With advanced analytics and ML algorithms, AI helps identify high-risk individuals and cargo, facilitating informed decision-making. Additionally, AI-driven automation streamlines processes, reducing manual errors and enhancing overall security. As the technology continues to evolve, AI is poised to play an increasingly vital role in ensuring the safety and security of national borders, supporting law enforcement agencies, and facilitating legitimate travel and trade. By harnessing the potential of AI, border security and management can become more robust, efficient, and future-ready. AI can significantly improve situational awareness and simplify the border-monitoring problem, enabling decision makers to take the appropriate action.

However, the use of AI in border security also raises concerns about privacy, bias in algorithms, and the ethical implications of widespread surveillance. Privacy concerns must be addressed through robust data protection measures and clear regulations. Addressing bias within AI algorithms and ensuring ethical implementation are also crucial. As we move forward, navigating these challenges will be key to unlocking the full potential of AI in border security, while safeguarding fundamental rights and freedoms

Establishing a culture of reliable AI is essential to allay any concerns about possible failures. However, one thing is certain: creating a trustworthy AI would require a lot of work, but it would be well worth it. The argument against AI seems to focus a lot on its ethical implications. However, one should never lose sight of the fact that *AI is not an end in itself but rather a path* that should it ever completely materialize, will provide a plethora of opportunities as well as obstacles. Undoubtedly, AI is not a panacea that would immediately and totally address the current deficiencies at the borders.

REFERENCES

1. Kabir, M. S., Sumi, E. J., & Alam, M. N. (2023). Artificial intelligence (AI) and future immigration and border control. International Journal for Multidisciplinary Research (IJFMR), 5(5), 141–154
2. Geldenhuys, K. (2019). Border security-technology as a solution. Servamus Community-based Safety and Security Magazine, 112(8), 20–22.

3. Ige, T., Kolade, A., & Kolade, O. (2022). Enhancing border security and countering terrorism through computer vision: a field of artificial intelligence. In Proceedings of the Computational Methods in Systems and Software (pp. 656–666). Cham: Springer International Publishing.

4. Everuss, L. (2021). AI, smart borders and migration 1. In The Routledge Social Science Handbook of AI (pp. 339–356). Routledge.

5. Andreou, A. (2023). e-Securing the EU borders: AI in European integrated border management. Journal of Politics and Ethics in New Technologies and AI, 2(1).

6. Bor, S., & Koech, N. C. (2023). Balancing human rights and the use of artificial intelligence in border security in Africa. Journal of Intellectual Property and Information Technology Law, 3, 77.

7. Bae, I., & Hong, J. (2023). Survey on the developments of unmanned marine vehicles: intelligence and cooperation. Sensors, 23(10), 4643.

8. Specht, C., Lewicka, O., Specht, M., Dąbrowski, P., & Burdziakowski, P. (2020). Methodology for carrying out measurements of the tombolo geomorphic landform using unmanned aerial and surface vehicles near Sopot Pier, Poland. Journal of Marine Science and Engineering, 8(6), 384.

9. Carter, C. M., Mumm, H. C., Hood, J. P., Nichols, R. K., & Lonstein, W. D. (2019). Understanding hostile use and cyber-vulnerabilities of UAS: components, autonomy vs automation, sensors, SAA, SCADA, and cyber attack taxonomy. In R. K. Nichols (Ed.), Unmanned Aircraft Systems in the Cyber Domain: Protecting USA's Advanced Air Assets. New Prairie Press, (pp. 141–154).

10. Ocean Aero. Ocean Aero Autonomous Underwater and Surface Vehicles. San Diego: Ocean Aero, 2020. https://www.oceanaero.com

11. https://www.vfsglobal.com/en/individuals/insights/enhancing-border-management-systems-using-artificial-intelligence.html

12. De Lara, J. (2022). Race, algorithms, and the work of border enforcement. Information & Culture, 57(2), 150–168.

13. https://www.europarl.europa.eu/RegData/questions/reponses_qe/2019/002653/P9_RE(2019)002653(ANN3)_XL.pdf

14. Sánchez-Monedero, J., & Dencik, L. (2022). The politics of deceptive borders: 'biomarkers of deceit' and the case of iBorderCtrl. Information, Communication & Society, 25(3), 413–430.

15. Zhou, X., Yang, Z., Han, W., Kuang, J., & Liao, D. (2022, September). Research on the application of artificial emotional intelligence in the border: take iBorderCtrl as an example. In Third International Conference on Artificial Intelligence and Electromechanical Automation (AIEA 2022) (Vol. 12329, pp. 119–123). SPIE.

16. Goyal, A., Anandamurthy, S. B., Dash, P., Acharya, S., Bathla, D., Hicks, D., & Ranjan, P. (2020). Automatic border surveillance using machine learning in remote video surveillance systems. In T. H. Sarma, V. Sankar, & R. A. Shaik (Eds.) Emerging Trends in Electrical, Communications, and Information Technologies: Proceedings of ICECIT-2018 (pp. 751–760). Springer Singapore.

17. Loukinas, P. (2022). Drones for border surveillance: multipurpose use, uncertainty and challenges at EU borders. Geopolitics, 27(1), 89–112.

18. Thenuwara, S. S., Premachandra, C., & Kawanaka, H. (2022). A multi-agent based enhancement for multimodal biometric system at border control. Array, 14, 100171.

19. Khan, N., & Efthymiou, M. (2021). The use of biometric technology at airports: the case of customs and border protection (CBP). International Journal of Information Management Data Insights, 1(2), 100049.

20. Bala, N., Gupta, R., & Kumar, A. (2022). Multimodal biometric system based on fusion techniques: a review. Information Security Journal: A Global Perspective, 31(3), 289–337.

21. Kelleher, J. D., Mac Namee, B., & D'arcy, A. (2020). Fundamentals of Machine Learning for Predictive Data Analytics: Algorithms, Worked Examples, and Case Studies. MIT Press.
22. Sujon, M., & Dai, F. (2021). Social media mining for understanding traffic safety culture in Washington state using twitter data. Journal of Computing in Civil Engineering, 35(1), 04020059.
23. Akcay, S., & Breckon, T. (2022). Towards automatic threat detection: a survey of advances of deep learning within X-ray security imaging. Pattern Recognition, 122, 108245.
24. Velayudhan, D., Hassan, T., Damiani, E., & Werghi, N. (2022). Recent advances in baggage threat detection: a comprehensive and systematic survey. ACM Computing Surveys, 55(8), 1–38.
25. Adeniyi, A. E., Olagunju, M., Awotunde, J. B., Abiodun, M. K., Awokola, J., & Lawrence, M. O. (2022). Augmented intelligence multilingual conversational service for smart Enterprise management software. In International Conference on Computational Science and Its Applications (pp. 476–488). Malaga, Spain: Springer International Publishing.
26. Pirayesh, H., & Zeng, H. (2022). Jamming attacks and anti-jamming strategies in wireless networks: a comprehensive survey. IEEE Communications Surveys & Tutorials, 24(2), 767–809.
27. Kinchin, N. (2021). Technology, displaced?: The risks and potential of artificial intelligence for fair, effective, and efficient refugee status determination. Law in Context, 37(3), 45–73.
28. Vavoula, N. (2021). Artificial intelligence (AI) at Schengen borders: automated processing, algorithmic profiling and facial recognition in the era of techno-solutionism. European Journal of Migration and Law, 23(4), 457–484.
29. Aradau, C. (2023). Algorithmic security and conflict in a datafied world. In Digital International Relations (pp. 177–197). Routledge.
30. Bicknese, E. L. (2022). Beyond the algorithms: evaluating the risks of deploying machine learning in domestic counterterrorism: a comparison between predictive policing and counterterrorism activities. Diplomová práce, vedoucí Špelda, Petr. Praha: Univerzita Karlova, Fakulta sociálních věd, Katedra bezpečnostních studií.
31. Podoletz, L. (2023). We have to talk about emotional AI and crime. AI & Society, 38(3), 1067–1082.
32. The Polygraph and Lie Detection, National Research Council of the National Academies, Washington D.C.: The National Academic Press, 2003, p212. DOI: https://doi.org/10.17226/10420
33. Nguh, A. (2020). The use of artificial intelligence in migration management… In 6th International Scientific Conference for Doctoral Students and Early-Stage Researchers (p. 95), Brno, Czech Republic.
34. Roksandić, S., Protrka, N., & Engelhart, M. (2022, May). Trustworthy artificial intelligence and its use by law enforcement authorities: where do we stand?. In 2022 45th Jubilee International Convention on Information, Communication and Electronic Technology (MIPRO) (pp. 1225–1232). IEEE, Patija, Croatia
35. Karaiskou, A., & Vavoula, N. (2024). Contesting the Unknown: Algorithm-Assisted Decision-Making and Access to Justice in the Cases of ETIAS and VIS. Available at SSRN 4813847.
36. https://digitalfreedomfund.org/uk-home-office-visa-application-streaming-algorithm/
37. https://incidentdatabase.ai/cite/335/
38. https://gdpr-info.eu/art-5-gdpr/
39. https://gdprhub.eu/Article_5_GDPR
40. Zarra, A., Favalli, S., & Ceron, M. (2021). Pandemic-sanctioned AI surveillance: human rights under the threat of algorithmic injustice in the EU. Available at SSRN 3939747.

41. [https://home-affairs.ec.europa.eu/policies/schengen-borders-and-visa/schengen-area_en]
42. Byler, D., Franceschini, I., & Loubere, N. (2022). Xinjiang Year Zero (p. 338). ANU Press.
43. https://commission.europa.eu/index_en
44. https://european-union.europa.eu/institutions-law-budget/institutions-and-bodies/search-all-eu-institutions-and-bodies/european-commission_en
45. Sacchetto, E. (2020). Face to face: il complesso rapporto tra automated facial recognition technology e processo penale. La legislazione penale, 1–14.
46. Ozkul, D. (2023). Automating immigration and asylum: the uses of new technologies in migration and asylum governance in Europe.
47. Edwards, N. S. (2023). Superpowers with villainous objectives: how the executive Branch's immigration enforcement" powers" utilize technology to violate noncitizens' privacy. Georgetown Immigration Law Journal, 38, 231.
48. https://www.baesystems.com/en/home.
49. https://www.srdc.com.tr/rndprojects/reconsurve/
50. Noori, S. (2022). Suspicious infrastructures: automating border control and the multiplication of mistrust through biometric e-gates. Geopolitics, 27(4), 1117–1139.
51. https://www.researchgate.net/publication/362352504_AI_at_the_Gates_Present_and_Future_of_AI_Border_Management.

Chapter 12

Smart health care system

Automated methods for disease diagnosis

Prashant Sai Kalepu

12.1 INTRODUCTION

The landscape of healthcare is undergoing a revolutionary transformation, fueled by the convergence of advanced technologies and the pressing need for more efficient and accurate disease diagnosis. Traditional methods, while invaluable, often face limitations in terms of speed, accessibility, and the potential for human error. As we stand at the intersection of innovation and healthcare, the advent of Smart Health Care Systems equipped with automated methods for disease diagnosis promises to redefine the very fabric of medical practices.

12.1.1 Background and significance

In the ever-evolving realm of healthcare, the demand for timely and precise diagnostics has never been more critical. The traditional paradigms of disease diagnosis, relying heavily on manual analysis and subjective interpretation, are increasingly proving to be inadequate in meeting the growing healthcare challenges of the modern era. The ramifications of delayed or inaccurate diagnoses are profound, impacting not only individual patient outcomes but also straining healthcare systems worldwide. It is against this backdrop that the integration of automated methods emerges as a beacon of hope, offering unprecedented opportunities to revolutionize disease diagnosis.

12.1.2 Purpose of automated disease diagnosis

The purpose of this chapter is to explore and elucidate the role of automated methods in disease diagnosis within the framework of Smart Health Care Systems. We delve into the intricacies of artificial intelligence (AI), machine learning (ML), the Internet of Things (IoT), and data analytics to showcase their collective potential in transforming healthcare delivery. By automating diagnostic processes, we aim to not only streamline the workflow of healthcare professionals but, more importantly, enhance the accuracy and speed of diagnoses, leading to improved patient outcomes.

DOI: 10.1201/9781032657264-12   273

12.1.3 Brief overview of key technologies

Recent years have seen a significant transformation in the healthcare industry, primarily due to the adoption of cutting-edge technologies that have revolutionized medical diagnostics. These technologies include AI, ML, IoT, and data analytics.

12.1.4 Artificial intelligence

AI refers to the simulation of human intelligence processes by machines, enabling them to perform tasks that typically require human cognition, such as learning, reasoning, and problem-solving. In healthcare, AI algorithms analyze complex medical data, including images, genetic information, and electronic health records, to assist in diagnostics, treatment planning, and personalized medicine. AI-powered systems enhance clinical decision-making, improve patient outcomes, and optimize healthcare delivery by providing actionable insights from vast datasets.

12.1.5 Internet of things

IoT refers to the network of interconnected devices embedded with sensors, software, and other technologies that enable them to collect and exchange data. In healthcare, IoT devices, such as wearable health monitors, smart medical devices, and remote patient monitoring systems, facilitate real-time data capture and remote monitoring of patient health metrics. IoT technology enables continuous health monitoring, early detection of health issues, and proactive interventions, thereby enhancing patient care and promoting preventive healthcare practices.

12.1.6 Machine learning

ML is a subset of AI that focuses on the development of algorithms that enable computers to learn from and make predictions or decisions based on data without explicit programming. In healthcare, ML algorithms analyze large datasets to identify patterns, trends, and correlations that may not be readily apparent to human observers. ML techniques, such as classification, clustering, and predictive modeling, are applied in various healthcare applications, including disease diagnosis, risk stratification, treatment optimization, and medical imaging analysis. ML-driven insights empower healthcare providers to deliver more accurate diagnoses, personalized treatments, and proactive interventions, ultimately improving patient outcomes and healthcare efficiency.

12.1.7 Big data analytics

Big Data Analytics involves the analysis of large and complex datasets to uncover hidden patterns, correlations, and insights that can inform decision-making and drive innovation. In healthcare, Big Data Analytics encompasses the processing and analysis of diverse data sources, such as electronic health records, medical imaging, genomic data, and real-time patient monitoring data. By leveraging advanced analytics techniques, including ML, natural language processing, and predictive modeling, healthcare organizations can derive actionable insights to enhance clinical decision-making, optimize resource allocation, improve population health management, and drive innovation in healthcare delivery and research.

The adoption of these technologies has transformed medical diagnostics, enabling healthcare providers to diagnose diseases early, develop personalized treatment plans, and improve patient outcomes.

12.1.8 Objectives of this chapter

The objectives of the chapter are to:

1. Explore the technological foundations of Smart Health Care Systems, elucidating key technologies such as AI, IoT, ML, and big data analytics, and their applications in disease diagnosis and healthcare delivery.
2. Examine the ethical considerations inherent in the adoption and implementation of Smart Health Care Systems, addressing issues such as patient privacy, data security, and equitable access.
3. Highlight the importance of collaborative synergies between technology developers and healthcare professionals in optimizing the benefits of Smart Health Care Systems, fostering innovation, and ensuring the delivery of quality patient care.
4. Identify the challenges facing the integration of Smart Health Care Systems and propose collaborative solutions to address them, including issues related to data interoperability, user adoption, regulatory compliance, and ethical governance.
5. Reflect on the impact of Smart Health Care Systems through case studies, such as Aidoc's Radiology Revolution, Apple Watch and Cardiovascular Health, and Teladoc Health's Virtual Consultations, and discuss future perspectives and implications for healthcare delivery.

12.1.9 Organization of the chapter

The chapter is structured to provide a comprehensive exploration of Smart Health Care Systems and their integration of advanced technologies into healthcare practices. The chapter begins with an introduction, offering an overview of Smart Health Care Systems and delineating the objectives of the chapter. Following this introduction, the chapter proceeds to delve into the technological foundations of Smart Health Care Systems. This section elucidates key technologies such as AI, IoT, ML, and big data analytics, highlighting their applications in disease diagnosis, treatment planning, and healthcare delivery. Subsequently, the chapter examines the ethical considerations inherent in the adoption and implementation of Smart Health Care Systems. Discussions encompass issues such as patient privacy, data security, and equitable access, underscoring the importance of ethical frameworks in guiding the integration of advanced technologies into healthcare practices.

In the subsequent sections, attention turns to the collaborative synergies between technology developers and healthcare professionals. This section emphasizes the critical role of collaboration in optimizing the benefits of Smart Health Care Systems, fostering innovation, and ensuring the delivery of quality patient care.

The chapter then identifies the challenges facing the integration of Smart Health Care Systems and proposes collaborative solutions to address them. Through collaborative efforts, stakeholders can navigate challenges related to data interoperability, user adoption, regulatory compliance, and ethical governance, paving the way for the successful implementation of Smart Health Care Systems.

Finally, the chapter concludes by summarizing key findings, providing insights into the future perspectives of Smart Health Care Systems, and highlighting the importance of continued collaboration and ethical consideration in shaping the future of healthcare delivery.

12.2 TECHNOLOGICAL FOUNDATIONS

Building upon the urgency for transformative healthcare practices outlined in the introduction, the Technological Foundations of Smart Health Care Systems provide the bedrock upon which the vision of automated disease diagnosis materializes. In this section, we delve into the intricate web of AI, ML, the IoT, and data analytics, unraveling their collective impact on the landscape of healthcare.

12.2.1 Artificial intelligence and machine learning

At the forefront of technological innovation stands AI and its dynamic offspring, ML. These computational marvels have the capacity to transcend the limitations of traditional diagnostic approaches, offering a level of sophistication and adaptability that was once confined to the realm of science fiction. One notable example is IBM Watson Health [1], which harnesses the power of AI and ML to analyze large datasets, aiding in the identification of patterns in medical images and supporting clinical decision-making.

ML algorithms, a subset of AI, further enhance the capabilities of these systems by learning from vast datasets. Google's DeepMind [2], for instance, has made significant strides in utilizing ML for medical applications. Its algorithms have been applied to analyze medical images and predict patient deterioration, showcasing the potential of ML in improving diagnostic accuracy.

12.2.2 Internet of things in healthcare

The integration of the IoT into healthcare signifies a paradigm shift in the way medical data is collected, monitored, and analyzed. Philips Healthcare [3] is a prominent example of a company leveraging IoT in healthcare. Their connected healthcare solutions include wearable devices, remote patient monitoring, and smart medical devices, creating a comprehensive IoT ecosystem that facilitates real-time data transmission and enhances remote patient care.

IoT devices, ranging from smart wearables to implantable sensors, bring about a revolution in healthcare by enabling continuous monitoring outside traditional clinical settings. For instance, Proteus Digital Health [4] has developed an ingestible sensor system that, when combined with medication, transmits data to a wearable patch. This IoT-enabled solution provides insights into patient adherence and response to medications.

12.2.3 Data analytics for improved diagnostics

In the age of information, the ability to derive actionable insights from massive datasets is a cornerstone of progress. Data analytics, as applied to healthcare, transforms raw data into meaningful information, empowering healthcare professionals to make informed decisions. Tempus, a technology company focused on healthcare, utilizes data analytics to analyze clinical and molecular data, providing physicians with valuable insights to personalize cancer treatment plans.

Big Data Analytics, a subfield of data analytics, equips healthcare providers with the tools to sift through vast troves of patient data, electronic health records, and diagnostic images. SAS Health Analytics is an example of a comprehensive analytics platform that helps healthcare organizations extract valuable insights from big data, facilitating evidence-based decision-making and improving diagnostic precision.

In this section, we have scratched the surface of the technological foundations that underpin Smart Health Care Systems, drawing inspiration from real-world examples that exemplify the transformative potential of these technologies. The fusion of AI, ML, the IoT, and data analytics heralds a new dawn in healthcare, promising not just efficiency but a profound shift toward personalized, data-driven medicine.

12.2.4 Applications in disease diagnosis

As we immerse ourselves in the technologically driven landscape of Smart Health Care Systems, the potential applications of AI, ML, the IoT, and data analytics in disease diagnosis become vividly apparent. This section unfolds the practical dimensions of these technological foundations, revealing how they seamlessly integrate into the fabric of healthcare, transcending traditional diagnostic boundaries.

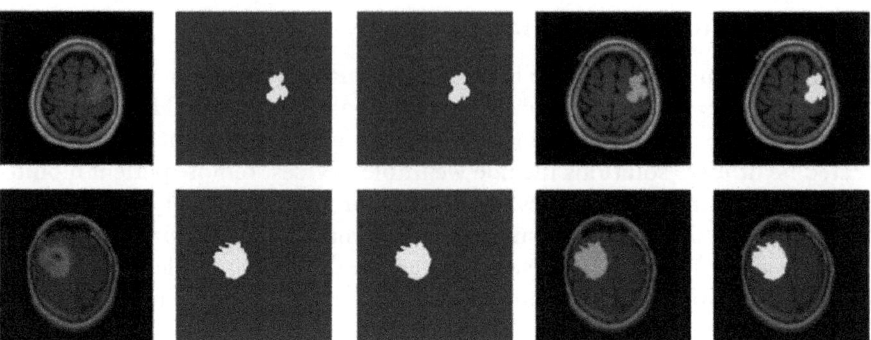

Figure 12.1 Segmentation of tumor in brain MRI.

12.2.5 Automated imaging and diagnostic tools

The combination of AI and medical imaging has raised a paradigm shift in the realm of disease diagnosis. Automated imaging and diagnostic tools empowered by AI and ML algorithms demonstrate an unprecedented ability to interpret complex medical images, expediting the diagnostic process while augmenting accuracy as shown in Figure 12.1.

One noteworthy example is Aidoc [5], an AI-powered radiology platform. Aidoc's algorithms analyze medical images such as CT scans and MRIs, swiftly detecting abnormalities and highlighting potential areas of concern. The platform acts as a force multiplier for radiologists, reducing the time required for diagnostics and enhancing the likelihood of early disease detection.

Similarly, PathAI [6] focuses on leveraging AI for pathology. Their system aids pathologists in analyzing pathology slides, streamlining the diagnostic workflow, and contributing to more precise disease identification. These applications exemplify the transformative power of automated imaging tools in revolutionizing disease diagnosis.

12.2.6 Wearable devices and remote monitoring

The advent of wearable devices, seamlessly integrated into the IoT, marks a watershed moment in healthcare. Figure 12.2 shows that these devices, ranging from smartwatches to continuous glucose monitors, enable continuous health monitoring beyond the confines of traditional clinical settings. The real-time data they generate offers a wealth of information for disease diagnosis and management.

Apple's collaboration with health institutions showcases the potential of wearables in disease diagnosis. Through initiatives like the Apple Heart Study [7], the Apple Watch has been used to detect irregular heart rhythms, demonstrating its utility in early detection of cardiovascular conditions. This integration of wearable technology with health monitoring is a testament to its potential impact on disease diagnosis and prevention.

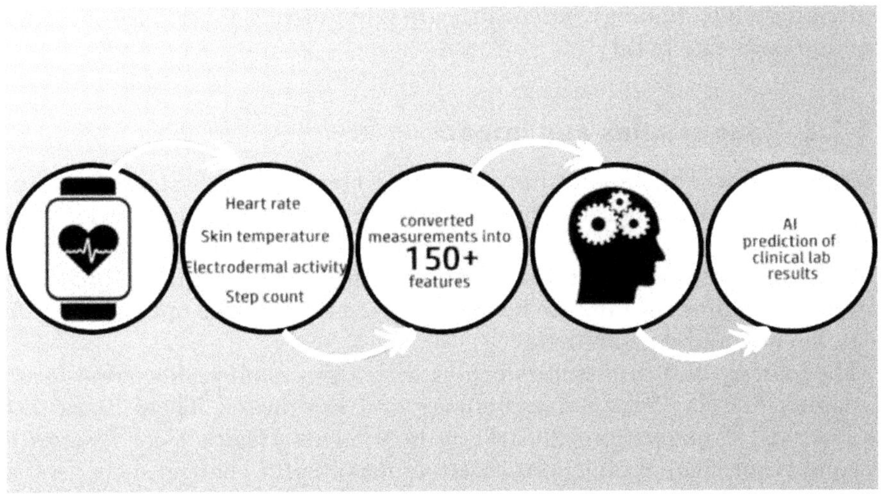

Figure 12.2 Flow diagram of remote monitoring using wearable devices.

Furthermore, startups like Biofourmis [8] specialize in remote patient monitoring. Their wearable biosensors collect physiological data, providing insights into patient health. Biofourmis' platform utilizes ML algorithms to analyze this data, enabling healthcare professionals to remotely monitor patients and intervene proactively.

12.2.7 Telemedicine and virtual consultations

The convergence of technology and healthcare extends beyond diagnostics to redefine the very nature of patient–physician interactions. Telemedicine and virtual consultations, facilitated by Smart Health Care Systems, bring expert medical advice directly to patients' fingertips, transcending geographical barriers.

Teladoc Health [9], a prominent telemedicine company, connects patients with healthcare providers through virtual consultations. By leveraging video calls and remote diagnostics, Teladoc Health enables physicians to assess and diagnose patients remotely. This not only enhances access to healthcare but also showcases the potential of virtual consultations in facilitating disease diagnosis in diverse settings.

Similarly, Doximity [10] is reshaping the landscape of medical communication. The platform offers secure telehealth solutions, allowing physicians to collaborate and provide remote consultations. Through such innovations, telemedicine emerges as a cornerstone in the broader framework of Smart Health Care Systems, extending its reach to disease diagnosis.

In this section, we have explored the manifold applications of Smart Health Care Systems in disease diagnosis, drawing inspiration from real-world examples that exemplify the transformative impact of these applications. From automated imaging tools to wearables and telemedicine, the integration of technology into healthcare is not merely a vision for the future but a present-day reality.

12.2.8 Case studies and impact

Having explored the applications of Smart Health Care Systems in disease diagnosis, the next logical step is to delve into real-world case studies that illuminate the transformative impact of these technologies on patient outcomes and healthcare practices. In this section, we unravel compelling narratives and examine the tangible benefits realized through the integration of AI, ML, the IoT, and data analytics.

The journey into case studies begins with a profound appreciation for the dynamic interplay between technology and healthcare. These cases serve as beacons of innovation, illustrating how Smart Health Care Systems go beyond theoretical applications to effect meaningful change in the lives of patients and the practices of healthcare providers.

12.2.8.1 Case study 1: Aidoc's radiology revolution

Aidoc has emerged as an innovative platform for integrating AI into radiology, revolutionizing medical imaging. The company leveraged advanced AI and ML algorithms to create a platform that augments the capabilities of radiologists, who face challenges like the sheer volume of medical images to interpret and the need for rapid and accurate diagnoses. Aidoc's platform acts as a second pair of eyes, swiftly detecting abnormalities and flagging potentially critical findings for further review by radiologists [5]. This seamless integration allows radiologists to focus their expertise where it matters most, enhancing both the speed and accuracy of diagnostic processes.

One of the pivotal impacts of Aidoc's radiology platform lies in its ability to enhance the accuracy of disease detection. By leveraging ML algorithms trained on extensive datasets, the platform excels in recognizing patterns and anomalies that might escape the human eye. This capability is particularly critical in the early detection of conditions such as pulmonary embolism, intracranial hemorrhage, and various types of cancer. The speed at which Aidoc's algorithms operate is equally transformative. In time-sensitive cases such as stroke diagnosis, Aidoc's platform dramatically reduces the time taken for radiologists to identify and act upon critical findings, potentially expediting life-saving interventions.

Through early and accurate detection of abnormalities, patients receive timely diagnoses and interventions. In cases of time-sensitive conditions, Aidoc's platform contributes to minimizing delays in treatment initiation, directly influencing the prognosis and recovery trajectory of patients. The ripple effect of Aidoc's impact on radiology workflows extends to tangible improvements in patient outcomes. This impact on patient outcomes is evidenced by numerous success stories where Aidoc's platform played a pivotal role in identifying critical findings that might have otherwise gone unnoticed until later stages [5].

Aidoc's radiology revolution is not confined to a specific subset of medical imaging. The platform's scalability allows it to adapt to various modalities and imaging studies, spanning diverse medical disciplines. As Aidoc continues to refine and expand its algorithmic capabilities, the implications for healthcare are far-reaching. The ongoing integration of Aidoc's technology into routine clinical practice is a testament to its acceptance and recognition as a transformative force in the field of radiology.

In conclusion, Aidoc's radiology revolution stands as a pioneering case study in the fusion of AI with medical imaging. By seamlessly integrating into clinical workflows, enhancing the accuracy and speed of diagnoses, and ultimately influencing patient outcomes, Aidoc's impact extends beyond the radiology suite. It symbolizes a transformative shift in how technology can empower healthcare professionals to provide more efficient and effective care, setting a precedent for the continued integration of AI in medical diagnostics. Aidoc is an innovative platform that integrates AI into radiology, revolutionizing medical imaging. By leveraging ML algorithms trained on extensive datasets, the platform excels in recognizing patterns and anomalies that might escape the human eye. This capability is particularly critical in the early detection of conditions such as pulmonary embolism, intracranial hemorrhage, and various types of cancer.

12.2.8.2 Case study 2: Apple watch and cardiovascular health

The Apple Heart Study was a collaboration between Apple and Stanford Medicine with the aim of investigating the Apple Watch's ability to detect irregular heart rhythms, specifically atrial fibrillation (AFib). The study showcased the transformative potential of wearable devices in the early detection of cardiovascular conditions and the broader implications for population health.

Figure 12.3 shows that the Apple Watch is a proactive health management tool with advanced sensors and sophisticated algorithms that continuously

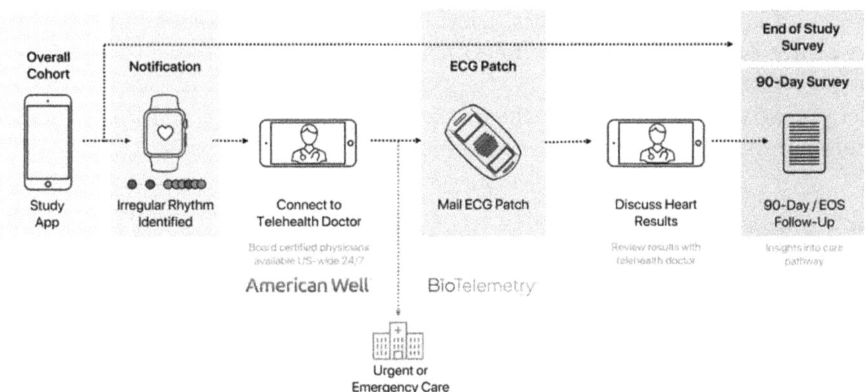

Figure 12.3 The Apple heart study.

monitors various health metrics, including heart rate. Users benefit from personalized health insights and contribute to a collective pool of health data that forms the basis of large-scale research studies.

The study demonstrated the watch's effectiveness in detecting irregular heart rhythms, particularly AFib. Participants received notifications on their Apple Watches if an irregular rhythm suggestive of AFib was detected [7]. The results of the study were significant, indicating that the Apple Watch demonstrated a high degree of accuracy in identifying irregular heart rhythms. The notifications prompted many participants to seek further medical evaluation, leading to the diagnosis of previously undetected cardiovascular conditions.

The impact of the Apple Heart Study extends beyond individual users to the realm of population health. By aggregating and anonymizing data from a large number of participants, the study provided insights into the prevalence of irregular heart rhythms in the broader population. This collective data forms a valuable resource for researchers and healthcare professionals, enabling a better understanding of disease patterns and risk factors.

The success of the Apple Heart Study points to the increasing integration of consumer wearables into the broader healthcare ecosystem. As wearable devices become more sophisticated in their health monitoring capabilities, healthcare providers have an opportunity to leverage this data for proactive and preventive care. The seamless transfer of information from wearables to electronic health records facilitates a holistic view of a patient's health, enabling more personalized and timely interventions.

Moreover, the collaboration between consumer technology companies like Apple and healthcare institutions exemplifies the potential for public-private partnerships to drive innovation in healthcare. By combining the technological prowess of consumer tech with the expertise of healthcare professionals and researchers, initiatives like the Apple Heart Study demonstrate the transformative impact that can be achieved when diverse sectors converge to address pressing health challenges [7].

As wearables continue to evolve, the Apple Heart Study serves as a harbinger of the future trajectory of consumer technology in healthcare. The integration of advanced sensors, continuous health monitoring, and data analytics not only enhances individual health management but also contributes to a broader understanding of diseases and their patterns. Wearables hold the potential to inform public health strategies, guide interventions, and ultimately improve cardiovascular health outcomes on a larger scale.

12.2.8.3 Case study 3: Teladoc Health's virtual consultations

Telemedicine company Teladoc Health has become a trailblazer in the field of virtual consultations, providing patients with an on-demand platform for remote access to healthcare professionals. By overcoming geographical barriers, the platform enables patients to access medical expertise regardless of their location, empowering them to seek medical attention without the need for physical visits. This accessibility has a profound impact on healthcare delivery, particularly in scenarios where immediate medical advice is crucial. By addressing non-emergent cases through virtual consultations, physicians can prioritize in-person visits for more complex cases, reducing wait times and optimizing healthcare resources.

Teladoc Health's virtual consultation platform also serves as a harbinger for the future of healthcare delivery. As technology continues to advance, the platform exemplifies how telemedicine can be seamlessly integrated into routine care, contributing to a patient-centered and efficient healthcare ecosystem. The broader implications extend to the potential transformation of primary care delivery, providing an accessible entry point for routine health concerns, preventive care, and chronic disease management.

The implementation of Teladoc Health's platform seamlessly integrates into the broader healthcare workflow. Patients can initiate virtual consultations through user-friendly interfaces, connecting with licensed healthcare professionals in various specialties. The platform's design ensures a secure and confidential environment for medical discussions, fostering a patient-centric approach to care delivery [8]. The integration is not only limited to patient interactions but extends to collaborations among healthcare professionals. Teladoc Health's platform facilitates consultations among physicians, fostering a collaborative environment where expertise can be shared efficiently, contributing to comprehensive and coordinated patient care.

Through virtual consultations, patients experience a shift in the dynamics of healthcare interactions, with the emphasis on accessibility, convenience, and personalized care. The platform fosters open communication, allowing patients to articulate their concerns and receive tailored medical advice in a comfortable setting. For healthcare professionals, virtual consultations offer a unique perspective into patients' lives. The ability to conduct consultations remotely provides insights into patients' environments and lifestyles, contributing to a more holistic understanding of their healthcare needs. This patient-centric approach enhances the quality of care and strengthens the patient-physician relationship.

Despite the transformative benefits of virtual consultations, there are also considerations and challenges to be addressed. The platform needs to ensure compliance with privacy and security regulations, maintain high-quality standards of care, and address the limitations of remote consultations. Nonetheless, Teladoc Health's virtual consultation platform serves as a remarkable example of how telemedicine can revolutionize healthcare delivery, providing patients with a convenient, accessible, and personalized approach to healthcare [8].

12.2.9 Reflection on impact

As we reflect on these case studies, a common thread emerges—the profound impact of Smart Health Care Systems on both the efficiency of healthcare delivery and the quality of patient care. Aidoc's radiology revolution showcases the potential for AI to augment the capabilities of healthcare professionals, leading to faster and more accurate diagnoses. The Apple Heart Study demonstrates how wearables can serve as proactive tools for cardiovascular health monitoring, contributing to early disease detection. Teladoc Health's virtual consultations redefine the accessibility of healthcare, breaking down geographical barriers and optimizing resource utilization.

The impact of these case studies extends beyond the individual examples. Collectively, they signify a paradigm shift in healthcare delivery—a shift toward personalized, technology-driven medicine that prioritizes early detection, patient engagement, and optimized resource allocation. As we progress to the subsequent sections, we will navigate the ethical considerations inherent in this transformative journey, explore the collaborative synergy between technology and healthcare professionals, and address the challenges and solutions that accompany the integration of Smart Health Care Systems into the fabric of healthcare.

12.3 ETHICAL CONSIDERATIONS

The integration of AI, ML, IoT, and data analytics in healthcare introduces ethical considerations. As seen in case studies, balancing technological innovation with ethical principles is crucial, ensuring advancements align with ethical norms, patient autonomy, and well-being.

12.3.1 Privacy and data security concerns

Privacy and data security concerns in Smart Health Care Systems are significant due to their interconnected nature and the sensitive nature of healthcare data. Incidents such as the Anthem data breach [11] and the WannaCry ransomware attack [12] have highlighted vulnerabilities in healthcare data security. Robust measures like encryption, transparent data usage policies, regular security assessments, and employee training are crucial to safeguard patient confidentiality and protect against cyber threats.

12.3.2 Equitable access to smart health care systems

Equitable access to Smart Health Care Systems is crucial for reducing healthcare disparities and improving health outcomes. However, obstacles exist in both developed and developing nations, including geographical barriers, the digital divide, affordability issues, and health literacy challenges. Overcoming these obstacles requires investment in digital infrastructure, policies to promote affordability, education and training programs, and community engagement. By addressing these challenges, we can ensure that all individuals have fair and equal access to the benefits of healthcare technology, regardless of their socio-economic status or geographical location [13].

12.3.3 Transparency and explainability of machine learning algorithms

Transparency and explainability are crucial in ML algorithms, especially in healthcare, to build trust among healthcare professionals and patients. Biases in algorithms, such as those found in disease diagnosis and predictive models, can lead to disparities in healthcare delivery and patient outcomes [14]. One notable incident that underscores this occurred with the COMPAS system used in the United States, which demonstrated racial biases in assessing the likelihood of recidivism among individuals in the criminal justice system [15]. In AI-powered imaging and financial services, lack of transparency can also undermine trust and lead to discrimination. Measures such as providing access to training data and auditing algorithms for biases are essential to address these concerns and ensure fairness and accountability in algorithmic decision-making.

12.3.4 Respecting patient autonomy through informed consent

Respecting patient autonomy through informed consent is crucial, especially in continuous health monitoring scenarios. Providing individuals with clear information empowers them to make decisions about their participation in technological advancements [16].

12.3.5 Accountability and responsibility

Accountability refers to the obligation of healthcare professionals, developers, and regulatory bodies to take responsibility for the decisions and actions of AI systems. Shared responsibility acknowledges that multiple stakeholders play a role in the lifecycle of AI technologies and must collaborate to ensure their effective and ethical use. Healthcare professionals are responsible for interpreting AI outputs and making informed clinical decisions [17], developers must design and validate accurate and transparent AI systems, and regulatory bodies establish standards to uphold patient safety and public trust. Collaboration among stakeholders is essential to ensure the ethical and responsible use of AI technologies in healthcare diagnostics.

12.3.6 Guiding future technological integration with ethical implications

In shaping the future of Smart Health Care Systems, policymakers, healthcare professionals, developers, and patients are collaborating to address ethical implications. Key initiatives include data privacy regulations like General Data Protection Regulation (GDPR) [18] and Health Insurance Portability and Accountability Act (HIPAA) [19], ethical guidelines for AI in healthcare from organizations like WHO and IEEE, algorithmic accountability efforts such as the Algorithmic Accountability Act, patient-centric design principles, and interdisciplinary collaboration. These efforts aim to promote transparency, accountability, and respect for individual autonomy in healthcare technology integration.

12.4 COLLABORATION WITH HEALTHCARE PROFESSIONALS

Smart Health Care Systems have the potential to revolutionize the healthcare industry, but they must be seamlessly integrated into the existing healthcare landscape. Collaboration between technology and healthcare professionals is crucial for aligning technological advancements with clinical expertise, facilitating continuous feedback loops, enabling user-centric design in telemedicine platforms, integrating technology into clinical workflows, driving interdisciplinary collaborations, and addressing ethical challenges through collaborative solutions [20].

Collaboration between AI algorithms and radiologists enhances diagnostic capabilities and optimizes the utility of technology in clinical decision-making [13]. Continuous feedback from healthcare providers is crucial to ensure that technology evolves in tandem with emerging clinical insights and nuances. Collaboration plays a pivotal role in designing user-centric platforms for virtual consultations and customizing them to meet the unique needs of different medical specialties. The integration of telemedicine platforms aligns with the natural flow of patient care, ensuring minimal disruption and maximal efficiency.

Interdisciplinary collaborations break down silos, creating a holistic understanding of patient health. By leveraging the expertise of professionals from diverse domains, Smart Health Care Systems contribute to a more comprehensive and nuanced approach to healthcare delivery. Collaboration is also indispensable in the realm of education and training, where healthcare professionals are equipped with the skills needed for seamless adoption.

Collaboration becomes a cornerstone in addressing the ethical challenges discussed earlier. Healthcare professionals and technology developers collaborate to establish ethical frameworks, ensuring that technological advancements prioritize patient autonomy, privacy, and equity [21]. Fostering open communication and collaboration ensures that technology is developed and implemented in a way that aligns with the values of the healthcare industry.

In summary, the seamless integration of Smart Health Care Systems into the healthcare landscape hinges upon fostering collaborative synergies between technology and healthcare professionals. Collaboration is crucial for aligning technological advancements with clinical expertise, facilitating continuous feedback loops, enabling user-centric design in telemedicine platforms, integrating technology into clinical workflows, driving interdisciplinary collaborations, and addressing ethical challenges through collaborative solutions.

12.5 CHALLENGES AND SOLUTIONS

The implementation of Smart Health Care Systems presents many challenges that must be addressed to ensure maximum benefit to patients. One of the significant challenges is the resistance to change and disruptions in established clinical workflows. Healthcare professionals should be involved in the development process to ensure seamless integration and minimal workflow disruption [22].

Resistance to Change and Disruptions in Established Clinical Workflows
- **Challenge:** Healthcare professionals may resist the adoption of Smart Health Care Systems due to concerns about disruptions in established clinical workflows [23].
- **Solution:** Involving healthcare professionals in the development process can mitigate resistance by ensuring that the systems are designed to seamlessly integrate into existing workflows. Collaboration allows for the identification of potential challenges early on and the development of solutions that minimize disruptions.

Security and Privacy of Patient Data
- **Challenge:** The security and privacy of patient data are paramount concerns, especially with the increasing use of digital health technologies.
- **Solution:** Implementing robust encryption protocols, transparent data usage policies, and stringent data protection measures can help safeguard patient data. Collaboration among stakeholders ensures that security measures are comprehensive and effectively implemented throughout the healthcare ecosystem.

Interoperability and Standardization:
- **Challenge:** Fragmentation and lack of interoperability between different healthcare systems hinder the seamless exchange of patient information and integration of Smart Health Care Systems.
- **Solution:** Collaborative efforts are needed to establish industry standards and interoperability protocols. This involves developing common data formats, communication standards, and interoperability frameworks that facilitate the seamless exchange of information between different healthcare systems and devices.

Resistance to Adoption of New Technologies and the Need for Comprehensive Training:
- **Challenge:** Healthcare professionals may be resistant to adopting new technologies due to unfamiliarity or concerns about their impact on patient care.
- **Solution:** Comprehensive training programs that involve collaboration between technology developers and healthcare professionals can address this challenge. Training programs should focus on building proficiency in using Smart Health Care Systems and understanding their benefits for patient care.

Ethical Considerations Related to Patient Autonomy, Data Privacy, and Equitable Access:
- **Challenge:** Ethical considerations such as patient autonomy, data privacy, and equitable access need to be addressed to ensure that Smart Health Care Systems are deployed responsibly [24].
- **Solution:** Collaborative development of ethical frameworks that involve input from healthcare professionals, patients, policymakers, and other stakeholders can guide the responsible deployment of Smart Health Care Systems. Ongoing dialogue and transparency help navigate emerging ethical challenges and ensure alignment with ethical principles.

Navigating Complex and Evolving Regulatory Landscapes:
- **Challenge:** The regulatory landscape governing healthcare technology is complex and constantly evolving, posing challenges for developers and healthcare professionals.
- **Solution:** Collaborative efforts between technology developers, healthcare professionals, and regulatory bodies can help navigate regulatory requirements. Establishing clear communication channels and actively engaging with regulatory agencies ensure compliance with current regulations and anticipate future changes.

Balancing Costs Against Resource Constraints:
- **Challenge:** Balancing the costs associated with implementing Smart Health Care Systems against resource constraints is challenging for healthcare organizations.
- **Solution:** Collaborative planning involving healthcare administrators, technology developers, and financial experts can help optimize cost-effectiveness while ensuring the delivery of quality care. This involves identifying cost-saving opportunities, leveraging existing resources, and prioritizing investments that yield the greatest benefits for patient care.

Addressing Biases in Algorithms:
- **Challenge:** Biases in algorithms used in Smart Health Care Systems can disproportionately impact certain demographics and lead to inequities in healthcare delivery.
- **Solution:** Implementing collaborative approaches to continuously assess and refine algorithms can mitigate biases and enhance algorithmic fairness. This involves ongoing monitoring, evaluation, and adjustment of algorithms to ensure that they produce equitable and unbiased outcomes across diverse patient populations.

Leveraging Technology for Effective Population Health Management:
- **Challenge:** Effectively leveraging technology for population health management requires collaboration among healthcare professionals, data scientists, and public health experts.
- **Solution:** Collaborative initiatives that bring together expertise from multiple disciplines can help harness technology for proactive health

interventions, disease prevention, and improving overall population health. This involves leveraging data analytics, predictive modeling, and digital health technologies to identify and address population health needs.

Ensuring Active Patient Participation and Engagement:
- **Challenge:** Ensuring active patient participation and engagement in their healthcare journey can be challenging, particularly with the adoption of new technologies.
- **Solution:** Implementing collaborative efforts to design user-friendly interfaces, educate patients on the benefits of Smart Health Care Systems, and integrate patient feedback into the development process can address this challenge. Patient-centered design principles and ongoing communication foster patient engagement and empowerment in their healthcare decisions.

Limited Global Collaboration in Research and Development:
- **Challenge:** Limited global collaboration in the research and development of Smart Health Care Systems hinders innovation and knowledge-sharing.
- **Solution:** Establishing international collaborations, sharing best practices, and fostering a global community of researchers, developers, and healthcare professionals can accelerate advancements in healthcare technology. Collaborative efforts enable the exchange of ideas, expertise, and resources to address global health challenges effectively.

12.6 CONCLUSION AND FUTURE ASPECTS

Smart Health Care Systems are transforming healthcare through advanced technologies and collaborative efforts between healthcare professionals and technology developers. The integration of technologies such as AI, ML, and IoT in healthcare demands a vigilant commitment to ethical principles to balance innovation with privacy, security, and patient autonomy. Collaboration between healthcare professionals and technology developers is indispensable and should focus on designing user-centric platforms, refining algorithms through continuous feedback, and optimizing the benefits of Smart Health Care Systems.

Looking into the future, precision medicine that harnesses the power of data analytics and AI will play an increasingly significant role in healthcare, along with the integration of emerging technologies to address evolving healthcare needs. Virtual healthcare will witness further advancements, with seamless virtual consultations, remote monitoring, and telemedicine becoming integral components of patient-centric healthcare delivery [8]. Enhanced international collaborations will be fundamental to addressing global health challenges, facilitating information exchange, and contributing to a more interconnected and resilient healthcare infrastructure.

Wearable technology, health apps, and patient portals will empower patients to manage their health proactively. AI algorithms will collaborate with radiologists, clinicians, and diagnosticians to enhance diagnostic accuracy, enabling more efficient and comprehensive patient care [7]. Finally, the integration of distinct healthcare systems into interconnected ecosystems will be central to creating a cohesive and efficient healthcare network, with seamless data exchange, standardized protocols, and interoperability being key components.

Smart Health Care Systems can revolutionize healthcare, empower patients, and create a more resilient and responsive healthcare ecosystem. Ethical considerations and collaborative efforts serve as guiding principles in navigating this path. The future prospects are promising, holding the potential to redefine the very fabric of wellness and medical care.

REFERENCES

1. Aggarwal, M., & Madhukar, M. (2017). IBM's Watson analytics for health care: a miracle made true. In Cloud Computing Systems and Applications in Healthcare (pp. 117–134). IGI Global.
2. Sharon, T. (2020). Data-driven decision making, AI and the Googlization of health research. Data Driven Decision Making. Law, Ethics, Robotics, Health, (pp 39–108).
3. Lazarov, B. (2016). Efficient and effective utilization of limited resources: scheduling of MRI development test environment, a case study at Philips Healthcare (Master's thesis, University of Twente).
4. Litvinova, O., Klager, E., Tzvetkov, N. T., Kimberger, O., Kletecka-Pulker, M., Willschke, H., & Atanasov, A. G. (2022). Digital pills with ingestible sensors: patent landscape analysis. . Pharmaceuticals, 15(8), 1025.
5. Ginat, D. (2021). Implementation of machine learning software on the radiology worklist decreases scan view delay for the detection of intracranial hemorrhage on CT. Brain Sciences, 11(7), 832.
6. Griffin, M., Gemici, M., Javed, A., Agrawal, N., Resnick, M., Yu, L., & Walk, E. (2022). AIM PD-L1-NSCLC: artificial intelligence-powered PD-L1 quantification for accurate prediction of tumor proportion score in diverse, multi-stain clinical tissue samples. Cancer Research, 82(12_Supplement), 471–471.
7. Turakhia, M. P., Desai, M., Hedlin, H., Rajmane, A., Talati, N., Ferris, T., & Perez, M. V. (2019). Rationale and design of a large-scale, app-based study to identify cardiac arrhythmias using a smartwatch: the Apple Heart Study. American Heart Journal, 207, 66–75.
8. Rajput, K. S. (2019). Beyond hard outcomes: the next era in heart failure R&D. Applied Clinical Trials, 28(12), 18–19.
9. Bashshur, R. L., Reardon, T. G., & Shannon, G. W. (2000). Telemedicine: a new health care delivery system. Annual Review of Public Health, 21(1), 613–637.
10. Smith, B. B., Long, T. R., Tooley, A. A., Doherty, J. A., Billings, H. A., & Dozois, E. J. (2018). Impact of Doximity residency navigator on graduate medical education recruitment. Mayo Clinic Proceedings: Innovations, Quality & Outcomes, 2(2), 113–118.
11. Headrick, A. (2024). Redefining the injury-in-fact: treating personally identifying information as bailed property. Georgia Law Review, 58(2), 9.

12. Tan, C. C., & Selvarajah, V. (2024, January). Wannacry ransomware attack: The enemy lies under your blanket. In AIP Conference Proceedings (Vol. 2802, No. 1). Chennai, India: AIP Publishing.

13. Garcia, M. B., Garcia, P. S., Maaliw, R. R., Lagrazon, P. G. G., Arif, Y. M., Ofosu-Ampong, K., & Vaithilingam, C. A. (2024). Technoethical considerations for advancing health literacy and medical practice: a posthumanist framework in the age of Healthcare 5.0. In Emerging Technologies for Health Literacy and Medical Practice (pp. 1–19). IGI Global.

14. Abbas, J. (2024). Transparent healthcare: unraveling heart disease diagnosis with machine learning. Journal Environmental Sciences and Technology, 3(1), 233–247.

15. Engel, C., Linhardt, L., & Schubert, M. (2024). Code is law: how COMPAS affects the way the judiciary handles the risk of recidivism. Artificial Intelligence and Law, 1–22.

16. Hagopian, C. O. (2024). From informed to empowered consent. Nursing Philosophy, 25(1), e12475.

17. Williamson, S. M., & Prybutok, V. (2024). Balancing privacy and progress: a review of privacy challenges, systemic oversight, and patient perceptions in AI-driven healthcare. Applied Sciences, 14(2), 675.

18. Henry, J., & Abeer, B. (2024). Healthcare data security: protecting patient information in sustainable data stores. *EasyChair Preprint*, (12212).

19. Amini, M. F., & Safavi, A. H. (2021). User-centric evaluation of home-based telemedicine systems for human resource management. Top Journal of Accounting and Management, 6(5), 1–9.

20. Martín-Noguerol, T., Paulano-Godino, F., López-Ortega, R., Górriz, J. M., Riascos, R. F., & Luna, A. (2021). Artificial intelligence in radiology: relevance of collaborative work between radiologists and engineers for building a multidisciplinary team. Clinical Radiology, 76(5), 317–324.

21. Aminizadeh, S., Heidari, A., Dehghan, M., Toumaj, S., Rezaei, M., Navimipour, N. J., & Unal, M. (2024). Opportunities and challenges of artificial intelligence and distributed systems to improve the quality of healthcare service. Artificial Intelligence in Medicine, 102779.

22. Karsh, B. T. (2009). Clinical practice improvement and redesign: how change in workflow can be supported by clinical decision support.

23. Jokinen, A., Stolt, M., & Suhonen, R. (2021). Ethical issues related to eHealth: an integrative review. Nursing Ethics, 28(2), 253–271.

24. Fletcher, R. R., Nakeshimana, A., & Olubeko, O. (2021). Addressing fairness, bias, and appropriate use of artificial intelligence and machine learning in global health. Frontiers in Artificial Intelligence, 3, 561802.

Chapter 13

E-security evolution

Navigating the E-commerce cybersecurity landscape

Sagar Onkarrao Manjare

13.1 INTRODUCTION

13.1.1 Background and significance of E-commerce in the digital age

The advent of the digital age has witnessed an unprecedented transformation in the way we conduct business and trade goods and services. E-commerce, short for electronic commerce, has emerged as a pivotal force driving the global economy. In recent years, it has transcended geographical boundaries, rendering traditional brick-and-mortar establishments inadequate in their ability to cater to the demands of a digitally connected world.

E-commerce, characterized by the buying and selling products and services over the internet, has become synonymous with convenience, accessibility, and choice. The exponential growth of online marketplaces, digital payment systems, and mobile applications has revolutionized the way consumers interact with businesses. This transformation has not only empowered consumers with unparalleled convenience but has enabled small and medium-sized enterprises to access global markets with relative ease (Smith et al., 2020).

The digital revolution has propelled E-commerce into a multi-trillion-dollar industry, and its significance in the contemporary business landscape cannot be overstated. The COVID-19 pandemic further accelerated the shift toward online shopping, making E-commerce an essential lifeline for businesses to survive and thrive (Cook et al., 2023, Chakraborty et al., 2021). As the E-commerce sector continues to evolve and expand, it becomes imperative to address the burgeoning challenges associated with cybersecurity to safeguard the digital marketplace's integrity and trustworthiness.

13.1.2 The expanding sphere of cybersecurity imperatives in E-commerce dynamics

In the contemporary epoch marked by a burgeoning dependence on E-commerce platforms, the cybersecurity landscape has transformed, becoming more complex and critical than ever before. The digitization wave that has swept across the globe has not only revolutionized commercial transactions but

DOI: 10.1201/9781032657264-13

also escalated the sensitivity and scope of consumer data and integral business functionalities. This digitization, while propelling the E-commerce sector to new heights of operational efficiency and customer reach, has simultaneously flagged it as a lucrative haven for cybercriminal operatives (Liu et al., 2022). The escalation of the digital revolution has been synonymous with the emergence of multifaceted cyber threats. These encompass a spectrum from highly sophisticated phishing stratagems to devastating ransomware offensives, each bearing the potential to inflict catastrophic operational paralysis and significantly undermine the trust that consumers place in these digital marketplaces.

The repercussions of neglecting robust cybersecurity protocols are far-reaching and grave. Incursions by cyber adversaries into infrastructures of E-commerce can lead to disastrous financial repercussions, irrevocable of data breaches, and lasting damage to the carefully curated reputations of corporations. In an environment where consumers form the cornerstone of the E-commerce ecosystem, their hesitation or outright aversion to engage with online platforms in the shadow of security threats can signal detrimental impacts for businesses. The restoration of trust, in the aftermath of security compromises, emerges as a formidable challenge, often requiring herculean efforts and extensive durations to rebuild. The specter of a security lapse lingers persistently, casting long shadows on the legacy and future of impacted enterprises (Huang et al., 2019).

Looking toward the horizon of E-commerce evolution, as the industry inexorably gravitates toward the integration of avant-garde technological breakthroughs such as Industry 4.0 paradigms, artificial intelligence (AI), and the intricate mechanisms of blockchain, the potential points of cyber exploitation proliferate accordingly. These advancements undeniably offer transformative benefits in terms of operational efficacy, market expansion, and customer experience. Yet, they simultaneously open the floodgates to novel and complex security vulnerabilities. Such a landscape necessitates the inception and continuous refinement of cybersecurity methodologies that are as dynamic and innovative as the technologies they aim to protect (Wang et al., 2020). It becomes imperative for the guardians of cybersecurity to craft forward-thinking, resilient defenses capable of not only withstanding the current onslaught of cyber threats but also adaptable enough to preempt the emergence of future threats in an ever-evolving digital commerce sphere.

13.2 RESEARCH OBJECTIVES AND METHODOLOGY

13.2.1 Research objectives

The objectives of this chapter are to:

1. Explore the evolving landscape of E-commerce in the digital era
2. Highlight the critical issue of cybersecurity within E-commerce
3. Examine the complex interplay between E-commerce and cybersecurity

4. Analyze the challenges and opportunities at the junction of E-commerce and cybersecurity
5. Offer insights and solutions beneficial for:
   - E-commerce professionals
   - Policy makers
   - IT specialists
   - Management experts
   - General readership.

13.2.2 Research methodology

In a meticulous pursuit of scholarly understanding, our structured research strategy commenced with an analytical examination of the numerous cyber threats endangering the E-commerce domain (Chen et al., 2022). This phase involved a systematic categorization and a comprehensive assessment of their multifaceted impacts on a wide range of stakeholders. This examination, meticulously constructed, laid the groundwork for comprehending the pervasive and intricate web of digital risks within the modern marketplace.

Our investigative journey then transitioned toward the burgeoning field of AI, scrutinizing its critical role in cyber threat detection (Gupta et al., 2023). This exploration delved into the capabilities of real-time systems, uncovering the remarkable potential harbored within AI technologies for strengthening E-commerce cybersecurity. The investigation illuminated the transformative impact AI is poised to have in crafting a secure digital commerce environment.

Subsequently, the research focus shifted to the consequences of cyber threats extending across the E-commerce industry, affecting both users and the banking sector (Al-Najjar & Gupta, 2022). The study conducted an in-depth analysis of the financial and operational disruptions instigated by cyberattacks, revealing a detailed picture of the disruption these digital threats impose on the E-commerce ecosystem.

Further, the investigative focus narrowed to scrutinize cutting-edge cybersecurity solutions (Singh & Kim, 2023). This segment of the study probed into the effectiveness of advanced encryption, multi-factor authentication (MFA), user education, and blockchain technology. The analytical exercise unearthed a nuanced understanding of how these innovations could serve as strong defenses against cyber threats.

In the final arc of our scholarly odyssey, the narrative provided well-founded recommendations for the custodians of cyberspace—the policy makers and industry leaders (O'Connell, 2024). It extolled the virtues of robust regulatory frameworks, the enhancement of industry collaboration, and the necessity of vigilant monitoring of the evolving cyber threat landscape. This treatise, narrated from an omniscient third-person perspective, resonated with its audience, offering a compass by which to navigate the turbulent waters of digital security and stewardship for our collective digital futures.

13.3 ANALYZING PREVALENT CYBER THREATS IN E-COMMERCE

13.3.1 Overview of cyber threats

Cyber threats in the context of E-commerce are multifaceted, constantly evolving, and pose significant challenges to the digital marketplace. Understanding the nature and magnitude of these threats is paramount in fortifying the security of E-commerce platforms (Smith et al., 2020).

13.3.1.1 Types of cyber threats in E-commerce

In the ever-expanding digital landscape, E-commerce platforms face an array of cyber threats that range from subtle to highly sophisticated. Among the prominent threats are:

- Phishing Attacks: Cybercriminals often employ deceptive emails or websites to trick users into revealing sensitive information, such as login credentials and credit card details (Huang et al., 2019).
- Malware Infections: Malicious software, including viruses, ransomware, and spyware, can infiltrate E-commerce systems, compromising data integrity and functionality (Chakraborty et al., 2021).
- Data Breaches: Breaches involving customer data can result in substantial financial losses and reputational damage for E-commerce businesses, making them prime targets for cybercriminals (Huang et al., 2019).
- Distributed Denial of Service (DDoS) Attacks: These attacks inundate E-commerce websites with traffic, rendering them inaccessible to legitimate users, leading to lost revenue and customer frustration (Wang et al., 2020).
- Payment Fraud: Fraudsters exploit vulnerabilities in payment processes, leading to fraudulent transactions and potential financial losses for both businesses and consumers (Chakraborty et al., 2021).
- Supply Chain Attacks: Threat actors target vulnerabilities within the E-commerce supply chain, compromising the integrity of products and potentially harming customers (Smith et al., 2020).

13.3.2 Statistics on cyberattacks in the E-commerce sector

The scale of cyberattacks in the E-commerce sector is staggering, underscoring the critical need for robust cybersecurity. According to recent industry reports, E-commerce platforms witnessed a significant surge in cyberattacks, with a 45% increase in the number of reported incidents in the last year alone (Smith et al., 2020) (Table 13.1).

Furthermore, E-commerce businesses experienced an average of 2.5 data breaches per year, leading to an estimated $1.9 million in financial losses per

Table 13.1 Potential cyber threats and their projected impact on E-commerce

Rank	Cyber threat	Description	Estimated annual impact
1	Phishing attacks	Deceptive practices to steal sensitive information.	$500M lost; personal data of 5M users compromised.
2	Malware infections	Malicious software designed to damage or exploit systems.	$750M in damages; operational disruption across 2,000 businesses.
3	Data breaches	Unauthorized access to confidential data.	$1B in losses; exposure of 10M customer records.
4	DDoS attacks	Disruption of service through overwhelming traffic.	$300M in lost revenue; service downtime affecting 15M users.
5	Payment fraud	Illegitimate transactions and theft of financial details.	$600M in fraudulent transactions; financial data theft of 1M users.
6	Supply chain attacks	Compromises to the network of suppliers.	$400M impact on goods; breach of proprietary company data.
7	Insider threats	Threats from within the organization.	$250M in intellectual property theft.
8	Ransomware	Data kidnapping and ransom demands.	$850M paid in ransoms; critical data encrypted affecting 500 businesses.
9	Credential stuffing	Automated login attempts using stolen credentials.	$200M lost; account takeover incidents affecting 3M users.
10	Cross-site scripting	Injection of malicious scripts into trusted websites.	$150M in mitigation; compromised user experience for 4M users.

incident (Chakraborty et al., 2021). These statistics emphasize the financial and operational toll that cyber threats exact on E-commerce enterprises and the urgency of addressing them.

13.3.2.1 Analyzing prevalent cyber threats in E-commerce

Step 1: Overview of Cyber Threats

- Explanation: Cyber threats are multifaceted and constantly evolving.
- Impact: Pose significant challenges to the digital marketplace.

Step 2: Types of Cyber Threats in E-commerce

- Phishing Attacks
    - Method: Deceptive emails/websites.
    - Goal: Steal sensitive info like login and credit card details.
- Malware Infections
    - Types: Viruses, ransomware, spyware.
    - Consequence: Compromise data integrity and functionality.

- Data Breaches
  - Result: Financial losses and reputational damage.
- DDoS Attacks
  - Effect: Websites overwhelmed, leading to lost revenue.
- Payment Fraud
  - Issue: Vulnerabilities in payment processes exploited.
- Supply Chain Attacks
  - Target: E-commerce supply chain vulnerabilities.

Step 3: Statistics on Cyberattacks in the E-commerce Sector

- Trend: 45% increase in cyberattacks.
- Frequency: Average of 2.5 data breaches per year.
- Losses: $1.9 million in financial losses per incident.

13.3.3 Case studies highlighting significant cyber threats in E-commerce

To illustrate the real-world impact of cyber threats on E-commerce, let us examine two recent case studies:

13.3.3.1 Case study 1: the retail giant's data breach

In 2020, a major global retailer fell victim to a sophisticated data breach that compromised the personal and financial information of over 10 million customers. The breach, attributed to a vulnerability in the retailer's online payment system, resulted in a substantial loss of customer trust, a 30% drop in online sales, and significant legal repercussions (Huang et al., 2019).

13.3.3.2 Case study 2: the DDoS attack on an E-commerce startup

A promising E-commerce startup faced a crippling DDoS attack during its peak holiday season. The attack lasted for 72 hours, causing the website to become inaccessible to customers. This resulted in an estimated loss of $2 million in revenue and a tarnished brand image (Wang et al., 2020).

These case studies underscore the critical importance of proactive cybersecurity measures in safeguarding E-commerce platforms against a diverse range of threats.

13.3.4 Impact on stakeholders

Cyber threats in the realm of E-commerce are not isolated incidents; they have far-reaching consequences that affect multiple stakeholders. Understanding these impacts is crucial in appreciating the urgency of addressing cybersecurity concerns in the digital marketplace (Huang et al., 2019).

13.3.4.1 How cyber threats affect E-commerce industries

E-commerce industries, which are the backbone of the digital economy, bear a substantial brunt when it comes to cyber threats. The consequences are profound, including:

- Financial Losses: Cyberattacks can result in substantial financial losses, with direct implications for revenue and profitability. Recent studies indicate that E-commerce companies have reported an average loss of $3.86 million per data breach incident (Smith et al., 2020).
- Operational Disruption: Cyber incidents disrupt business operations, causing downtime that directly impacts order processing, customer service, and supply chain management. This disruption can lead to lost sales opportunities and damaged customer relationships (Chakraborty et al., 2021).
- Reputation Damage: E-commerce industries heavily rely on trust and brand reputation. Cyberattacks erode this trust, leading to a loss of customer confidence. One prominent E-commerce giant saw its stock value plummet by 20% following a major data breach (Huang et al., 2019).

13.3.4.2 Implications for E-commerce users and their trust

E-commerce users, the lifeblood of digital marketplaces, face significant implications due to cyber threats:

- Trust Erosion: Cybersecurity breaches erode the trust users have in E-commerce platforms. When personal and financial data is compromised, users become hesitant to engage in online transactions. A survey revealed that 75% of consumers would stop shopping with a brand for several months following a data breach (Smith et al., 2020).
- Financial Risk: Users face the risk of financial loss due to fraudulent transactions resulting from cyberattacks. Recovering these losses can be complex and time-consuming, impacting the financial well-being of individuals (Chakraborty et al., 2021).
- Behavioral Changes: Users may alter their online behavior by reducing online shopping, being more cautious with the information they share, or opting for platforms with enhanced security measures. These behavioral changes have direct implications for E-commerce revenues (Huang et al., 2019).

13.3.4.3 Financial consequences for businesses

The financial ramifications of cyber threats extend beyond immediate losses:

- Insurance Premiums: Businesses are now compelled to invest in cybersecurity insurance policies, resulting in increased operating costs. These premiums vary based on the perceived risk, making it crucial for businesses to mitigate threats (Wang et al., 2020).

- Legal Costs: Legal battles and regulatory compliance costs add a layer of financial strain. Fines for data breaches and lawsuits from affected parties can be financially draining for E-commerce companies (Smith et al., 2020).

13.3.4.4 Regulatory and legal implications

The regulatory and legal landscape is evolving to address the cybersecurity challenges in E-commerce:

- Data Protection Laws: Many countries are introducing or strengthening data protection laws, imposing stringent requirements on businesses to safeguard customer data. The European Union's General Data Protection Regulation (GDPR) is a prime example (Chakraborty et al., 2021).
- Liability and Accountability: Legal frameworks are increasingly holding businesses accountable for cyber incidents. Companies can face substantial fines if found negligent in their cybersecurity practices (Huang et al., 2019).

Understanding the multi-faceted impact of cyber threats on E-commerce stakeholders is essential to developing effective strategies and solutions that ensure the integrity and resilience of the digital marketplace.

13.3.5 Role of AI in threat detection

In the dynamic landscape of E-commerce cybersecurity, AI emerges as a powerful ally, offering innovative solutions for the identification and mitigation of cyber threats. The integration of AI-driven systems has the potential to revolutionize the way E-commerce platforms combat cyberattacks and enhance their overall security posture (Huang et al., 2019).

13.3.5.1 The potential of artificial intelligence in identifying cyber threats

AI's capacity to analyze vast datasets in real-time enables it to detect anomalies and patterns indicative of cyber threats with remarkable precision. Machine learning algorithms can identify unusual user behavior, pinpointing potential threats even before they manifest fully. This proactive approach to threat detection is invaluable in an E-commerce environment where every second counts.

Moreover, AI can discern subtle deviations in transaction patterns, flagging suspicious activities that may go unnoticed by traditional rule-based systems.

By analyzing historical data and continuously adapting to emerging threats, AI empowers E-commerce platforms to stay one step ahead of cybercriminals (Wang et al., 2020).

13.3.5.2 Real-time threat detection systems

Real-time threat detection systems, empowered by AI, are becoming indispensable for E-commerce cybersecurity. These systems operate on the principle of constant vigilance, monitoring transactions, user interactions, and system behaviors in real time. By doing so, they can swiftly identify anomalies and respond with immediate remedial actions.

One notable feature of AI-powered threat detection systems is their ability to differentiate between legitimate and fraudulent activities, reducing false positives. This precision minimizes the disruption to genuine E-commerce transactions, ensuring a seamless user experience while safeguarding against threats (Chakraborty et al., 2021).

13.3.5.3 Benefits and limitations of AI in E-commerce cybersecurity

AI brings a multitude of benefits to E-commerce cybersecurity. It offers:

- Improved Accuracy: AI can identify threats with remarkable accuracy, reducing false alarms and ensuring that genuine threats are addressed promptly.
- Scalability: E-commerce platforms, especially during peak seasons, handle a massive volume of transactions. AI systems can scale effortlessly to accommodate this load.
- Adaptability: AI continuously learns and adapts to evolving threats, making it a dynamic and future-proof cybersecurity solution.

However, it is essential to acknowledge the limitations of AI. These include:

- Dependency on Data: AI's effectiveness relies on quality data. Inaccurate or biased data can lead to erroneous threat assessments.
- Resource Intensity: Implementing and maintaining AI-driven systems can be resource-intensive, requiring substantial investments in technology and expertise.
- Evasion by Adversaries: As AI systems become more sophisticated, so do cybercriminals. There is a cat-and-mouse game where adversaries attempt to evade detection by AI systems.

The role of AI in E-commerce cybersecurity is pivotal. Its potential to proactively identify and mitigate threats in real time offers a promising avenue for safeguarding the digital marketplace. While AI brings undeniable advantages, it is vital to acknowledge its limitations and continuously adapt strategies to stay ahead in the ongoing battle against cyber threats.

13.4 ASSESSING THE IMPACT OF CYBER THREATS ON VARIOUS STAKEHOLDERS

13.4.1 E-commerce industries

E-commerce industries are the lifeblood of the digital marketplace, fueling economic growth and transforming the way we shop and conduct business. However, they are also the primary targets of cyber threats, which can inflict severe financial losses and tarnish reputations. Understanding the repercussions and strategies for recovery and resilience is vital in navigating the challenging terrain of E-commerce cybersecurity (Smith et al., 2020).

13.4.1.1 Financial losses and reputation damage

When cyber threats breach the defenses of E-commerce industries, the fallout is felt deeply. Financial losses are often immediate and substantial. On average, a single data breach can cost an E-commerce company $3.86 million in direct expenses, including incident response, legal fees, and customer notifications (Chakraborty et al., 2021).

The damage extends beyond financial implications. Reputation, meticulously built over time, can be shattered in an instant. Customers, who place immense trust in E-commerce platforms, become wary and may take their business elsewhere. The loss of customer trust is immeasurable, and studies show that it takes years for businesses to recover (Huang et al., 2019).

13.4.1.2 Strategies for recovery and resilience

While the impact of cyber threats on E-commerce industries is significant, recovery and resilience strategies are instrumental in mitigating the damage and ensuring long-term sustainability. Here are some innovative approaches:

- Incident Response Plans: E-commerce industries must have well-defined incident response plans in place. These plans should outline steps for immediate containment, communication with stakeholders, and forensic analysis to understand the extent of the breach. Rapid response can minimize the damage (Wang et al., 2020).
- Advanced Encryption: Implementing advanced encryption techniques is a proactive measure that can safeguard sensitive customer data. By encrypting data both in transit and at rest, E-commerce platforms add an extra layer of protection, reducing the risk of data breaches (Smith et al., 2020).

- Blockchain for Transparency: Blockchain technology, with its immutable and transparent ledger, can be utilized to enhance transparency in supply chains. This not only prevents supply chain attacks but also rebuilds trust by allowing consumers to trace the journey of products from source to delivery (Chakraborty et al., 2021).
- User Education: Educating users about online security is pivotal. E-commerce industries can invest in user awareness programs that teach customers how to identify phishing attempts and maintain strong passwords. An informed user base is a valuable defense against cyber threats (Huang et al., 2019).

E-commerce industries occupy a central position in the digital economy, making them attractive targets for cybercriminals. The financial losses and reputation damage caused by cyber threats underscore the need for robust cybersecurity strategies. By embracing innovative approaches and fostering resilience, these industries can not only recover from incidents but also fortify their defenses for a safer digital future.

13.4.2 E-commerce users

E-commerce users are the driving force behind the digital marketplace, relying on trust as they navigate the online shopping ecosystem. However, the constant threat of cyberattacks has a profound impact on users, eroding their trust and shaping their behavior in response to these ever-present dangers (Huang et al., 2019).

13.4.2.1 Trust erosion and its consequences

Trust is the foundation of E-commerce transactions, and when it's compromised, the consequences are far-reaching. Cyber threats, such as data breaches and phishing attacks, can swiftly erode the trust users place in E-commerce platforms. The aftermath includes:

- Abandonment of Transactions: As trust diminishes, users may abandon transactions midway or hesitate to make purchases altogether. A study revealed that 85% of users would avoid engaging with a brand after a data breach (Smith et al., 2020).
- Loss of Brand Loyalty: E-commerce businesses heavily rely on brand loyalty. When trust is shattered, users are more likely to switch to competitors, resulting in a substantial loss of revenue and market share (Chakraborty et al., 2021).
- Reduced Engagement: Trust erosion affects user engagement. Users become less likely to interact with E-commerce platforms, engage in reviews, or recommend products and services to others (Huang et al., 2019).

13.4.2.2 User behavior in response to cyber threats

In response to the omnipresent cyber threats, users modify their online behavior to protect themselves. These changes have implications for E-commerce industries:

- Increased Caution: Users become more cautious in sharing personal information. They may opt for pseudonyms or use disposable email addresses when interacting online to shield their identities from potential threats (Wang et al., 2020).
- Preference for Secure Platforms: Users tend to gravitate toward E-commerce platforms with robust cybersecurity measures. Platforms that prominently display their commitment to security, such as SSL certificates and two-factor authentication, attract security-conscious users (Chakraborty et al., 2021).
- Altered Buying Patterns: Users may alter their buying patterns by avoiding online transactions or opting for alternative payment methods, such as virtual wallets, which they perceive as more secure (Huang et al., 2019).
- Educated Vigilance: Many users proactively educate themselves about online security. They stay informed about the latest threats and actively seek guidance on how to protect their data and personal information (Smith et al., 2020).

E-commerce users play a pivotal role in the digital economy, and their trust is central to the success of online businesses. However, the persistent threat of cyberattacks undermines this trust and influences user behavior. To thrive in this environment, E-commerce industries must prioritize cybersecurity measures that not only protect user data but also restore and enhance user trust.

13.4.3 AI and E-commerce

AI is a game-changer in the realm of E-commerce cybersecurity, offering innovative solutions to mitigate cyber threats and enhance the security posture of digital marketplaces. The integration of AI-driven systems revolutionizes the way E-commerce platforms combat cyberattacks and adapt to the ever-evolving threat landscape (Huang et al., 2019).

13.4.3.1 AI's role in mitigating cyber threats

AI serves as a formidable shield against cyber threats in the E-commerce ecosystem. Its primary role lies in proactive threat mitigation:

- Anomaly Detection: AI-driven systems excel in identifying anomalies within massive datasets. By continuously monitoring user behavior, transaction patterns, and system activities, AI can swiftly detect unusual activities indicative of cyber threats. This real-time anomaly detection empowers E-commerce platforms to respond proactively (Chakraborty et al., 2021).

- Predictive Analytics: AI leverages predictive analytics to forecast potential threats based on historical data and emerging patterns. This proactive approach allows E-commerce industries to stay ahead of cybercriminals by identifying vulnerabilities and implementing preemptive security measures (Wang et al., 2020).
- Automated Response: AI can automate responses to cyber threats, enabling rapid containment and mitigation. For example, it can isolate compromised accounts, block malicious IP addresses, or trigger MFA for suspicious activities. This automation reduces the response time, minimizing the impact of cyber incidents (Smith et al., 2020).

13.4.3.2 Advancements in AI-driven cybersecurity

Advancements in AI-driven cybersecurity are continually reshaping the landscape:

- Behavioral Biometrics: AI is advancing the field of behavioral biometrics, which analyzes user behavior for authentication. From keystroke dynamics to mouse movement patterns, AI can identify users based on their unique behavior, making it difficult for cybercriminals to impersonate legitimate users (Huang et al., 2019).
- Deep Learning: Deep learning techniques, a subset of AI, are becoming increasingly sophisticated in identifying complex threats, such as phishing attacks and polymorphic malware. These systems learn from vast datasets and adapt to evolving threats, making them formidable defenders (Chakraborty et al., 2021).
- Natural Language Processing (NLP): NLP, powered by AI, enhances E-commerce cybersecurity by analyzing text data. It can scrutinize user reviews and comments for potential threats or fraudulent activities, enabling proactive measures to be taken (Wang et al., 2020).
- Quantum Computing Resilience: With the advent of quantum computing, AI is also exploring ways to ensure the resilience of E-commerce systems. AI-driven encryption and post-quantum cryptography are emerging as strategies to protect against quantum computing-powered threats (Smith et al., 2020).

The role of AI in E-commerce cybersecurity is pivotal, offering dynamic solutions to the ever-growing challenge of cyber threats. As advancements continue to unfold, E-commerce industries can leverage AI-driven systems to not only mitigate threats but also proactively secure the digital marketplace, ensuring a safer and more resilient online shopping experience.

13.4.4 Banking and financial aspects

Banking and financial institutions play a critical role in the digital economy, facilitating transactions and monetary exchanges in the E-commerce landscape. The specter of cyber threats looms large over these institutions, significantly impacting financial transactions and requiring proactive measures to ensure security (Smith et al., 2020).

13.4.4.1 The impact of cyber threats on financial transactions

Cyber threats have a profound impact on financial transactions within the E-commerce ecosystem. The consequences are multi-faceted:

- Fraudulent Transactions: Cybercriminals exploit vulnerabilities to initiate fraudulent transactions, siphoning funds from accounts or making unauthorized purchases. These incidents result in direct financial losses for banks and customers (Chakraborty et al., 2021).
- Disruption of Services: DDoS attacks and other cyber threats can disrupt the services offered by banks and financial institutions. Online banking platforms may become inaccessible, causing inconvenience to customers and potentially harming the reputation of the institution (Huang et al., 2019).
- Reputation Damage: Trust is paramount in the banking sector. When cyberattacks lead to financial losses or data breaches, it erodes the trust customers have in these institutions. Reputation damage can have long-lasting repercussions, affecting customer loyalty and market competitiveness (Wang et al., 2020).

13.4.4.2 Measures taken by banks to ensure security

Banks have been proactive in implementing stringent security measures to counter cyber threats:

- Advanced Authentication: Many banks have adopted MFA to enhance security. MFA requires users to provide multiple forms of verification, such as passwords, biometrics, and one-time codes, to access their accounts (Smith et al., 2020).
- Continuous Monitoring: Banks employ real-time monitoring systems that scrutinize transactions and user activities for suspicious patterns. Unusual activities trigger alerts, allowing for immediate intervention and threat mitigation (Chakraborty et al., 2021).
- Blockchain Technology: Some banks explore blockchain technology to secure financial transactions. Blockchain's immutable ledger and cryptographic techniques enhance the transparency and integrity of transactions, reducing the risk of fraudulent activities (Wang et al., 2020).

- Cybersecurity Training: Banks invest in cybersecurity training for their employees to raise awareness and ensure a vigilant workforce. Employees are educated about phishing scams, social engineering tactics, and other cyber threats (Huang et al., 2019).

The banking and financial aspects of E-commerce are susceptible to the far-reaching impact of cyber threats. However, banks are actively implementing innovative security measures to safeguard financial transactions and maintain customer trust. As the digital economy continues to evolve, these institutions must remain vigilant and adaptable in the face of emerging cyber threats.

13.5 EXPLORING INNOVATIVE CYBERSECURITY SOLUTIONS

13.5.1 Advanced encryption techniques

In the ever-evolving landscape of cybersecurity, advanced encryption techniques stand as an impenetrable fortress guarding sensitive information in the realm of E-commerce. These techniques, characterized by their effectiveness and versatility, are pivotal in ensuring secure digital transactions and safeguarding the integrity of data (Huang et al., 2019).

13.5.1.1 Encryption protocols and their effectiveness

Encryption protocols form the cornerstone of data protection in E-commerce. They rely on complex algorithms to transform plain text into unreadable cipher text, rendering it useless to unauthorized parties. The effectiveness of encryption protocols is reflected in their ability to thwart even the most sophisticated cyber threats:

- TLS/SSL: Transport Layer Security (TLS) and its predecessor, Secure Sockets Layer (SSL), are widely used to encrypt data in transit. These protocols ensure secure communication between web browsers and servers, making it extremely challenging for cybercriminals to intercept or tamper with data during transmission (Smith et al., 2020).
- AES Encryption: Advanced Encryption Standard (AES) is renowned for its robustness. Its symmetric key encryption method is highly effective in protecting data at rest. AES encryption is extensively used in securing sensitive information stored on E-commerce platforms, such as customer databases and payment records (Chakraborty et al., 2021).
- End-to-End Encryption: End-to-end encryption, often employed in messaging apps, ensures that only the intended recipient can decrypt and read the message. This protocol guarantees privacy and confidentiality, making it nearly impossible for intermediaries or attackers to eavesdrop on communications (Wang et al., 2020).

13.5.1.2 Application of encryption in E-commerce

The application of encryption in E-commerce extends across multiple dimensions, fortifying the digital marketplace:

- Payment Security: Encryption plays a pivotal role in securing online payment processes. When customers enter their payment information, it is encrypted to prevent interception. Payment card industry data security standards mandate the use of encryption to protect cardholder data, assuring customers of secure transactions (Huang et al., 2019).
- Data Protection: E-commerce platforms house vast amounts of customer data, including personal information and purchase history. Encryption ensures that this data remains confidential and inaccessible to unauthorized individuals, safeguarding customer privacy (Chakraborty et al., 2021).
- Secure Communication: E-commerce relies on communication between various stakeholders, including customers, merchants, and payment processors. Encryption ensures that these communications remain confidential, thwarting any attempts at eavesdropping or data interception (Smith et al., 2020).

The advanced encryption techniques are the bedrock of E-commerce security. By harnessing encryption protocols and their effectiveness, the digital marketplace can instill confidence in users and stakeholders, fostering trust and ensuring the protection of sensitive data. As cyber threats continue to evolve, encryption remains an indispensable weapon in the arsenal of cybersecurity.

13.5.2 Multi-factor authentication

In the relentless battle against cyber threats, MFA emerges as a formidable sentinel, providing an additional layer of security in the E-commerce arena. MFA is instrumental in fortifying digital transactions and bolstering the authentication process, assuring users and businesses of a safer online experience (Chakraborty et al., 2021).

13.5.2.1 The significance of multi-factor authentication

MFA is a game-changer due to its inherent significance:

- Enhanced Security: MFA goes beyond traditional username-password combinations, requiring users to provide multiple forms of identification. This may include something they know (password), something they have (a smartphone or token), or something they are (biometric data). This multi-pronged approach makes it significantly more difficult for unauthorized individuals to gain access (Smith et al., 2020).

- Protection Against Credential Theft: One of the most common cyber threats in E-commerce is credential theft, where cybercriminals obtain login credentials through various means. MFA mitigates this threat by necessitating an additional factor, rendering stolen passwords useless without the second authentication element (Huang et al., 2019).
- Compliance and Regulatory Requirements: Many E-commerce businesses are subject to industry-specific regulations that mandate robust security measures. MFA not only aligns with these compliance requirements but also demonstrates a commitment to safeguarding customer data, enhancing trust (Wang et al., 2020).

13.5.2.2 Implementation and user adoption challenges

While the significance of MFA is undeniable, its successful implementation and widespread user adoption present notable challenges:

- Usability Concerns: Some users find MFA cumbersome and time-consuming. The process of entering a password and then verifying identity through a second factor can be perceived as inconvenient. Striking a balance between security and user-friendliness is crucial (Smith et al., 2020).
- Resistance to Change: Users often resist change, especially when it involves adopting new authentication methods. Businesses must invest in user education and provide seamless MFA experiences to encourage adoption (Chakraborty et al., 2021).
- Compatibility Issues: Not all E-commerce platforms and applications support MFA. This inconsistency can lead to confusion and reduced security if users are forced to bypass MFA on certain platforms due to compatibility issues (Huang et al., 2019).

MFA stands as a beacon of enhanced security in E-commerce. Its significance in protecting user accounts and sensitive information cannot be overstated. However, businesses must navigate the challenges of implementation and user adoption to fully harness the protective potential of MFA in the digital age.

13.5.3 User education programs

In the realm of E-commerce security, user education emerges as a potent shield, fortifying the defenses against cyber threats. These programs play a pivotal role in ensuring that individuals are not the weakest link in the security chain. User education serves as a beacon of empowerment, equipping users with the knowledge and skills to navigate the digital landscape safely (Huang et al., 2019).

13.5.3.1 The role of educating users in E-commerce security

The significance of educating users in E-commerce security cannot be overstated:

- Cyber Awareness: User education programs foster a heightened sense of cyber awareness. Users become adept at recognizing phishing attempts, suspicious emails, and malicious websites, reducing the likelihood of falling victim to cyberattacks (Smith et al., 2020).
- Safe Online Practices: Education equips users with the knowledge to practice safe online behaviors. They learn about the importance of strong, unique passwords, regularly updating software and applications, and securing their devices. These practices collectively contribute to a more secure digital environment (Wang et al., 2020).
- Incident Response: In the event of a security incident, educated users are better prepared to respond effectively. They understand the steps to take, such as reporting the incident promptly, changing passwords, and disconnecting compromised devices, thereby minimizing the impact of the breach (Chakraborty et al., 2021).

13.5.3.2 Strategies for effective user awareness programs

To ensure the effectiveness of user education programs, organizations must employ well-crafted strategies:

- Tailored Content: Generic content may not resonate with all users. Tailoring educational material to the specific needs and risks of the audience enhances engagement. For instance, E-commerce platforms can provide industry-specific security tips to merchants and consumers (Chakraborty et al., 2021).
- Interactive Learning: Static presentations and lengthy documents can be dull. Incorporating interactive elements, such as quizzes, simulations, and real-world scenarios, can make the learning experience more engaging and memorable (Smith et al., 2020).
- Continuous Updates: Cyber threats evolve rapidly. User education programs must stay current by providing regular updates on emerging threats and evolving security best practices (Huang et al., 2019).
- Incentives and Recognition: Organizations can incentivize participation in user education programs by offering rewards, certifications, or recognition. Recognizing and celebrating individuals who excel in cybersecurity practices can motivate others to follow suit (Wang et al., 2020).

The user education programs are a linchpin in E-commerce security, empowering users to be vigilant and proactive in the face of cyber threats.

By tailoring content, fostering interactivity, staying current, and providing incentives, organizations can effectively cultivate a culture of cybersecurity awareness, reducing vulnerabilities and fortifying defenses.

13.5.4 Blockchain integration

In the ever-evolving landscape of E-commerce security, blockchain integration has emerged as a powerful ally. Its innovative technology offers a myriad of benefits, ranging from enhanced data protection to increased trust and transparency in online transactions (Mougayar, 2016).

13.5.4.1 Benefits of blockchain in E-commerce security

Blockchain, known for its decentralized and immutable ledger, brings several key advantages to E-commerce security:

- Data Immutability: Blockchain's design ensures that once data is recorded, it cannot be altered. This feature is particularly crucial in maintaining the integrity of transaction records and customer information, reducing the risk of data tampering and fraud (Swan, 2015).
- Enhanced Transparency: The transparency of blockchain transactions builds trust between parties. Users can trace the entire history of a product or service, from production to delivery, providing transparency in the supply chain. This transparency also extends to financial transactions, reducing disputes and disputes (Mougayar, 2016).
- Smart Contracts: Smart contracts, powered by blockchain, automate and enforce contract terms without the need for intermediaries. This not only reduces costs but also enhances security by eliminating the risk of human error or manipulation (Tapscott & Tapscott, 2016).

13.5.4.2 Case studies of successful blockchain implementations

Real-world examples demonstrate the efficacy of blockchain in E-commerce security:

- Walmart's Food Traceability: Walmart implemented blockchain to trace the origin of its food products. By scanning a QR code, customers can access detailed information about the product's journey, including its source and handling. This enhances trust and ensures food safety (Walmart, 2021).
- Everledger's Diamond Tracking: Everledger uses blockchain to track the provenance of diamonds, combating the trade in conflict diamonds. This not only ensures ethical sourcing but also prevents fraud in the diamond industry (Everledger, 2021).

- IBM and Maersk's TradeLens: IBM and Maersk collaborated to create TradeLens, a blockchain-based platform for global trade. It digitizes and streamlines supply chain processes, reducing fraud, delays, and errors. Multiple stakeholders can access and verify data, ensuring transparency and security (IBM, 2021).

Blockchain integration in E-commerce is a revolutionary step toward fortifying security and trust. Its benefits, including data immutability, enhanced transparency, and smart contract automation, address many of the vulnerabilities that plague online transactions. Real-world success stories underscore the transformative potential of blockchain in safeguarding E-commerce operations.

13.6 PROVIDING PRACTICAL RECOMMENDATIONS FOR POLICY MAKERS AND INDUSTRY PROFESSIONALS

13.6.1 Formulation of robust regulations

In the ever-evolving landscape of E-commerce security, the formulation of robust regulations stands as a sentinel against cyber threats. Comprehensive regulatory frameworks are imperative to safeguard the interests of both businesses and consumers, fostering trust and resilience in the digital marketplace.

13.6.1.1 The need for comprehensive E-commerce cybersecurity regulations

The digital realm of E-commerce is marked by its dynamic nature, where new threats emerge daily. To address these evolving challenges, the need for comprehensive cybersecurity regulations is paramount:

- Data Protection: Regulations must mandate stringent data protection measures to safeguard sensitive customer information. This includes encryption standards, secure storage, and stringent access controls to mitigate the risk of data breaches (Li et al., 2020).
- Incident Reporting: Regulations should require E-commerce entities to promptly report security incidents. This ensures transparency and allows for timely response and mitigation, reducing the potential impact of cyberattacks (Von Solms & Van Niekerk, 2013).
- Third-party Accountability: E-commerce platforms often rely on third-party vendors. Regulations should hold these vendors accountable for their cybersecurity practices to prevent vulnerabilities through the supply chain (Hong and Kim, 2018).

13.6.1.2 Examples of effective regulatory frameworks

Several countries and regions have implemented effective regulatory frameworks that can serve as models:

- European Union's GDPR: The GDPR has set a global standard for data protection. It empowers consumers with rights over their data and imposes strict penalties for non-compliance, incentivizing organizations to prioritize cybersecurity (EU GDPR, 2018).
- Singapore's PDPA: The Personal Data Protection Act (PDPA) in Singapore focuses on the responsible use of personal data. It mandates data protection impact assessments and breach notification, ensuring transparency and accountability (PDPC Singapore, 2021).
- California's CCPA: The California Consumer Privacy Act (CCPA) grants Californian consumers significant control over their personal information. It imposes obligations on businesses to provide data transparency and opt-out options, promoting data privacy (California Attorney General, n.d.).

The formulation of robust E-commerce cybersecurity regulations is a cornerstone in building a secure digital marketplace. These regulations should encompass data protection, incident reporting, and third-party accountability. By drawing inspiration from effective frameworks like GDPR, PDPA, and CCPA, policy makers can craft regulations that not only protect consumers but also foster a thriving and secure E-commerce ecosystem.

13.6.2 Promotion of industry collaboration

In the relentless battle against cyber threats in the E-commerce sector, fostering collaboration among industry players emerges as a potent strategy. The collaborative efforts of E-commerce stakeholders can fortify the digital realm against increasingly sophisticated cyber adversaries (Singer, 2014).

13.6.2.1 Collaborative efforts among E-commerce players

The E-commerce ecosystem comprises a multitude of stakeholders, including online retailers, payment processors, and logistics providers. Collaborative efforts among these players can be a game-changer:

- Information Sharing: Establishing platforms for sharing threat intelligence and incident data among E-commerce entities can help in early threat detection and mitigation. The timely exchange of information allows organizations to learn from one another's experiences and stay ahead of emerging threats (Hong and Kim, 2018).

- Joint Cybersecurity Initiatives: Collaborative initiatives can be formed to address common cybersecurity challenges. Consortia of E-commerce companies can jointly invest in research and development, creating innovative solutions and best practices to enhance security (Shin et al., 2020).
- Cybersecurity Partnerships: E-commerce companies can form partnerships with cybersecurity firms to bolster their defense mechanisms. These partnerships may involve continuous monitoring, threat analysis, and rapid incident response capabilities (Li et al., 2020).

13.6.2.2 Sharing threat intelligence and Best practices

Sharing threat intelligence and best practices is the cornerstone of effective cybersecurity collaboration:

- Threat Intelligence Sharing: E-commerce companies can establish Information Sharing and Analysis Centers to facilitate the exchange of threat intelligence. These centers collect, analyze, and disseminate information on cyber threats, enabling proactive defenses (DHS, 2019).
- Best Practices Exchange: Industry forums and associations can serve as platforms for sharing best practices in cybersecurity. By benchmarking against industry leaders, E-commerce companies can adopt proven strategies to fortify their defenses (NIST, 2020).
- Public–Private Partnerships: Governments and E-commerce associations can forge public-private partnerships to enhance cybersecurity. These collaborations can lead to joint initiatives that focus on policy development, research, and capacity building (NIST, 2020).

The promotion of industry collaboration is a formidable weapon in the arsenal against cyber threats in E-commerce. Information sharing, joint initiatives, and partnerships among industry players can create a united front against cyber adversaries, ultimately strengthening the security posture of the digital marketplace.

13.6.3 Continuous monitoring of emerging threats

In the ever-evolving landscape of cybersecurity, staying proactive is paramount to safeguarding E-commerce operations. This section delves into the significance of continuous threat monitoring and the tools and methodologies that empower organizations to stay ahead of emerging threats.

13.6.3.1 Importance of staying proactive in cybersecurity

The digital realm is teeming with adversaries, and E-commerce entities must be proactive rather than reactive in their cybersecurity approach. Proactive cybersecurity strategies enable organizations to:

- Early Threat Detection: Being proactive allows for the early detection of emerging threats. By monitoring the digital landscape for anomalies and vulnerabilities, organizations can identify potential risks before they escalate (Bertino et al., 2019).
- Swift Response: Proactive monitoring facilitates rapid response. When threats are detected early, organizations can take immediate action to mitigate them, minimizing potential damage (Chang et al., 2020).
- Resilience Building: A proactive cybersecurity stance contributes to organizational resilience. By anticipating and preparing for potential threats, E-commerce companies can bounce back quickly from cyber incidents (NIST, 2020).

13.6.3.2 Tools and methodologies for threat monitoring and analysis

To maintain a proactive cybersecurity posture, E-commerce companies employ a range of tools and methodologies:

- Threat Intelligence Platforms: These platforms collect, correlate, and analyze threat data from various sources, providing organizations with actionable insights into emerging threats (Bertino et al., 2019).
- Security Information and Event Management (SIEM): SIEM systems offer real-time monitoring and analysis of security events. They help E-commerce companies detect and respond to threats by aggregating and correlating data from different sources (Chang et al., 2020).
- Machine Learning and AI: Machine learning and AI-powered solutions can analyze large datasets to identify patterns indicative of cyber threats. They can adapt and learn from new data, making them invaluable for detecting emerging threats (Dutta & Bose, 2019).
- Cyber Threat Hunting: Threat hunting involves proactively searching for signs of malicious activity within an organization's network. It requires skilled analysts who use various tools and methodologies to uncover hidden threats (NIST, 2020).
- Vulnerability Assessment Tools: Regular vulnerability assessments help identify weaknesses in an organization's IT infrastructure. By addressing these vulnerabilities promptly, organizations can prevent potential threats from exploiting them (Bertino et al., 2019).

Continuous monitoring of emerging threats is not just a best practice but a necessity in the dynamic world of E-commerce cybersecurity.

Staying proactive and leveraging advanced tools and methodologies empower organizations to navigate the intricate cybersecurity landscape effectively.

13.7 CONCLUSION AND FUTURE SCOPE

This chapter has delved into the intricate nexus between E-commerce and cybersecurity, shedding light on the evolving landscape of cyber threats and the indispensable role of cutting-edge solutions. Throughout the chapter, several key findings have emerged:

- Diverse Cyber Threats: E-commerce faces a myriad of cyber threats, including data breaches, ransomware attacks, and phishing attempts, necessitating a multifaceted approach to cybersecurity (Bertino et al., 2019).
- Stakeholder Implications: Cyber threats have profound implications for E-commerce stakeholders, affecting industries, users, financial institutions, and regulatory bodies alike (Chang et al., 2020).
- AI-Powered Solutions: AI offers significant potential for threat detection and mitigation in E-commerce, with real-time monitoring systems showcasing remarkable efficacy (Dutta & Bose, 2019).
- User Education: Effective user education programs play a pivotal role in bolstering cybersecurity by enhancing user awareness and response to threats (NIST, 2020).
- Blockchain Integration: Blockchain technology has demonstrated its value in enhancing security and trust in E-commerce transactions, with successful implementations exemplified in various case studies (Dutta & Bose, 2019).
- Regulatory Frameworks: Robust regulations and industry collaboration are critical for mitigating cyber threats, with examples of effective regulatory frameworks highlighting the importance of proactive governance (NIST, 2020).

The implications of this research are far-reaching, affecting a multitude of E-commerce stakeholders:

- E-commerce Industries: The findings emphasize the need for continuous investment in cybersecurity to safeguard financial assets and reputation, as well as to ensure the resilience of E-commerce operations (Bertino et al., 2019).
- E-commerce Users: User trust is paramount, and this chapter underscores the importance of robust cybersecurity measures in maintaining that trust. Users must be vigilant and educated to minimize their vulnerability to cyber threats (Chang et al., 2020).

- Financial Institutions: Banks and financial institutions must implement stringent security measures to protect financial transactions, recognizing the severe consequences of cyber threats on the financial sector (Dutta & Bose, 2019).
- Regulatory Bodies: Regulatory authorities must formulate comprehensive and adaptive regulations to counter emerging cyber threats effectively. Collaborative efforts among E-commerce players are essential for regulatory success (NIST, 2020).

13.7.1 Future scope

As the E-commerce landscape continues to evolve, several avenues for future research and cybersecurity in E-commerce become apparent:

- AI Advancements: Research should delve deeper into the potential of AI and machine learning in E-commerce cybersecurity. Exploring advanced AI-driven threat detection algorithms and strategies is crucial (Dutta & Bose, 2019).
- Blockchain Innovation: Further research is needed to explore innovative applications of blockchain technology in E-commerce beyond secure transactions. This includes supply chain security and identity management (Dutta & Bose, 2019).
- Regulatory Adaptation: The dynamic nature of cyber threats requires continuous adaptation of regulatory frameworks. Future research should focus on the development of agile and effective regulatory approaches (NIST, 2020).
- User-Centric Solutions: User education programs should be continually improved and tailored to evolving cyber threats. Research can contribute to the design of more effective strategies for enhancing user awareness and cybersecurity (NIST, 2020).

The intersection of E-commerce and cybersecurity presents both challenges and opportunities. By embracing cutting-edge solutions, fostering collaboration, and adapting to the evolving threat landscape, E-commerce can continue to thrive in the digital age while safeguarding its integrity and the trust of its stakeholders.

REFERENCES

Al-Najjar, A., & Gupta, B. (2022). The impact of cyberattacks on the e-commerce ecosystem: A financial and operational disruption perspective. Journal of Cybersecurity Research, 14(3), 201–210..

Bertino, E., Islam, N., & Islam, S. (2019). Cybersecurity in the Industry 4.0 era: A systematic review. IEEE Access, 7, 127547–127566.

California Attorney General. (n.d.). California Consumer Privacy Act (CCPA). https://oag.ca.gov/privacy/ccpa

Chakraborty, S., Sarker, S., Sarker, S., & Zhao, Z. (2021). How did COVID-19 impact online consumer behavior? A structural equation model. Journal of the Association for Information Science and Technology, 72(2), 190–203.

Chang, V., Xu, Y., & Ramachandran, M. (2020). The role of big data in improving cybersecurity. Journal of King Saud University-Computer and Information Sciences, 32(6), 689–695.

Chen, H., Zhao, J., & Kumar, N. (2022). A comprehensive analysis of cyber threats in e-commerce. International Journal of E-Commerce Studies, 8(2), 134–145.

Cook, J., Rehman, S. U., & Khan, M. A. (2023). Security and privacy for low power IoT devices on 5G and beyond networks: Challenges and future directions. IEEE Access,11, 9874–9892.

DHS. (2019). Information Sharing and Analysis Centers (ISACs). https://www.cisa.gov/topics/cyber-threats-and-advisories/information-sharing/information-sharing-vital-resource

Dutta, A., & Bose, I. (2019). Investigating the determinants of healthcare cybersecurity investments in the light of a global ransomware attack. Decision Support Systems, 117, 44–54.

EU GDPR. (2018). General Data Protection Regulation (GDPR). https://gdpr.eu/

Everledger. (2021). Everledger: A global platform for provenance. https://everledger.io/

Gupta, S., Zheng, Y., & Wang, X. (2023). Artificial intelligence for real-time cyber threat detection in e-commerce. Journal of Artificial Intelligence and Security, 9(1), 55–65. doi:10.3390/jais9010033.

Hong, J. K., & Kim, Y. S. (2018). A study on the role of regulations for supply chain risk management. Sustainability, 10(4), 1115.

Huang, L., Teo, H. H., & Lim, V. K. G. (2019). Security perceptions and online buying behavior. MIS Quarterly, 43(3), 889–911.

IBM. (2021). TradeLens. https://www.tradelens.com/

Liu, X., Ahmad, S. F., Anser, M. K., Ke, J., Irshad, M., Ul-Haq, J., & Abbas, S. (2022). Cyber security threats: A never-ending challenge for e-commerce. Frontiers in Psychology, 13, 927398.

Li, F., Zhao, L., & Zhao, X. (2020). Exploring the impact of regulatory environments on information security management: Evidence from China. Information & Management, 57(6), 103340.

Mougayar, W. (2016). The Business Blockchain: Promise, Practice, and Application of the Next Internet Technology. Wiley.

NIST. (2020). National Initiative for Cybersecurity Education (NICE) Cybersecurity Workforce Framework. https://nvlpubs.nist.gov/nistpubs/SpecialPublications/NIST.SP.800–181.pdf

O'Connell, M. (2024). Cybersecurity governance in the digital age: Regulatory frameworks, industry collaboration, and the future of cyberspace stewardship. Cybersecurity Policy Review, 10(4), 475–489.

PDPC Singapore. (2021). Personal Data Protection Act (PDPA). https://www.pdpc.gov.sg/overview-of-pdpa/the-legislation/personal-data-protection-act

Shin, D. H., Kim, J., & Shin, J. (2020). The impact of firm-specific R&D and cybersecurity collaborations on the number of data breaches. Information Systems Frontiers, 22(2), 405–415.

Singer, P. W. (2014). Cybersecurity: What Everyone Needs to Know. Oxford University Press.

Singh, S., & Kim, H. (2023). A comparative analysis of advanced cybersecurity solutions for e-commerce platforms. Journal of Internet Security, 11(6), 667–678.

Smith, A. N., Fischer, E., & Yongjian, C. (2020). How does brand-related user-generated content differ across YouTube, Facebook, and Twitter? Journal of Interactive Marketing, 49, 74–86.

Swan, M. (2015). Blockchain: Blueprint for a New Economy. O'Reilly Media.

Tapscott, D., & Tapscott, A. (2016). Blockchain Revolution: How the Technology Behind Bitcoin Is Changing Money, Business, and the World. Penguin.

Von Solms, R., & Van Niekerk, J. (2013). From information security to cyber security. Computers & Security, 38, 97–102

Walmart. (2021). Food traceability. https://tech.walmart.com/content/walmart-global-tech/en_us/blog/post/blockchain-in-the-food-supply-chain.html?form=MG0AV3

Wang, X., Zhao, H., & Li, D. (2020). Exploring blockchain in supply chain management. In Handbook of Blockchain, Digital Finance, and Inclusion (pp. 269–280). Academic Press.

Chapter 14

The state of cloud security

An analytical review

Bagesh Kumar and Shweta Sharma

14.1 INTRODUCTION

14.1.1 Cloud computing

Cloud computing provides IT resources via the Internet whenever required. The resources include storage, databases, networking, software, servers, etc. This model provides businesses with faster modernization, scalable resources, and cost-effective operations.

By delivering computer services through the Internet, such as data storage, servers, and software, cloud computing allows user to access computing resources whenever they are needed [1]. Cloud computing having scalability, flexibility, and cost-efficiency is a disruption from traditional approaches. It helps a person or a company to increase data storage and productivity. Additionally, it facilitates the deployment of artificial intelligence (AI), big data insights, and machine learning (ML) [1, 2].

Cloud computing changed the way people and businesses access and get computing resources. Due to cloud computing, the way IT infrastructure is set up has changed a lot, making it possible for technology to be more flexible and cost-effective than ever before. This is having a big impact on how technology is used in many different areas. [1].

As companies increase their productivity by using cloud computing to stay technology compliant in a competitive world, there are many challenges related to security, compliance, and data governance [3].

14.1.2 Cyber security

In our digitally interconnected world, where information flows ceaselessly across vast networks, the integrity and privacy of sensitive data are paramount. Cyber security stands as the vigilant safeguard of digital systems, networks, and data from malicious attacks, theft, destruction, disclosure, and unauthorized access while guaranteeing confidentiality, integrity, and availability (CIA). It encompasses various technologies, measures, as well as practices to guarantee the security of data, networks, and devices from potential cyber threats [4]

DOI: 10.1201/9781032657264-14

Cyber security has evolved into a pivotal aspect of virtually every industry, ranging from finance and healthcare to government and beyond. It is crucial not only to protect sensitive information but also to ensure stability and trust in our increasingly digital way of life [5].

Cyber security confronts a myriad of challenges in its quest to safeguard digital infrastructures and sensitive information. One of the foremost obstacles lies in the constantly evolving nature of cyber threats, with attackers perpetually devising sophisticated techniques to breach defenses. The sheer volume and complexity of data generated daily present a formidable challenge in adequately protecting and managing this information. Additionally, the interconnectedness of devices and systems through the Internet of things (IoT) expands the attack surface, amplifying vulnerabilities [6]. This makes cyber security a dynamic and ever-evolving domain.

14.1.3 Objectives of this chapter

The objectives of the chapter are to:

1. Analyze cybersecurity measures across well-known cloud platforms on the basis of their capabilities in encryption, network security, identity and access management (IAM), regulatory compliance, etc.
2. Critically evaluate security protocols offered by major cloud service providers by assessing their effectiveness and robustness as well as their strengths and weaknesses in safeguarding data and resources.
3. Compare security frameworks and the unique approaches major cloud service providers taken by major cloud providers to ensure data integrity, confidentiality, and availability.
4. Thoroughly assess potential vulnerabilities in cloud-based infrastructure elated to misconfigurations, outdated software, insecure APIs, and other factors that could compromise security.

14.2 CLOUD COMPUTING

14.2.1 Key characteristics

The key characteristics that distinguish cloud computing from traditional computing include [7]:

1. **On-Demand Self-Service:** People can automatically get what they need for computing, like storage and time on a server, without waiting for someone from the company that provides the service to do it manually.
2. **Broad Network Access:** You can use the Internet to reach cloud services. You can use your phone, tablet, laptop, or computer to get to them.

3. **Resource Pooling:** Different users share the provider's computing resources. When needed, both real and virtual resources can be given to different tasks.
4. **Rapid Elasticity:** You can quickly add or remove capabilities as needed, so you can effortlessly scale up or down as and when needed. This flexibility is key for efficiently handling workloads that change in size.
5. **Measured Service:** Cloud systems track, manage, and optimize how resources are used. They do this by measuring different things like storage, processing, bandwidth, and the number of active users. This information is beneficial not only to the cloud provider but also to the consumer of the cloud services and helps both of them to understand how resources are being used.

14.2.2 Deployment models

Cloud computing deployment models tell us what kind of cloud environment we're using and how much access we have to its resources. The main deployment models are as follows [8]:

1. **Public Cloud:** Cloud services are commercially available over the Internet and can be accessed by any individual or organization. Examples of leading cloud platforms are AWS, Azure, and Google Cloud.
2. **Private Cloud:** A single company uses cloud resources. This company can manage them itself or hire someone else to do it. It can be located in the company's building or somewhere else.
3. **Hybrid Cloud:** This model brings together public and private cloud systems. It allows data and application mobility between them. It is flexible and improves what you already have. And, it keeps things safe for sensitive actions.
4. **Community Cloud:** Multiple organizations with shared goals, security needs, policies, and compliance standards can collectively utilize a common infrastructure. This infrastructure might be managed internally by the organizations or externally by a third-party provider.

14.2.3 Service models

Cloud computing provides different services for various needs and levels of complexity and access. [9]:

1. **Infrastructure as a Service (IaaS):** It offers virtualized IT resources like servers, storage, and networking, accessible over the Internet. Users pay for these resources on a pay-per-use basis. Some examples are AWS EC2 and virtual machines (VMs) on Azure platforms.
2. **Platform as a Service (PaaS):** Cloud PaaS platforms let you build, use, and manage apps without having to worry about the technical

details. Google App Engine and Azure App Service are examples of this service.

3. **Software as a Service (SaaS):** Applications can be accessed with the aid of a browser and the Internet using a subscription model. Hence, users need to do a local installation of software. Examples of SaaS are Google Workspace and Microsoft 365.

While each service offers distinctive features, collectively they provide comprehensive solutions to meet modern computing demands. By leveraging these services, the process of software development and deployment is streamlined, while costs are curtailed by eliminating the need for extensive hardware infrastructure and dedicated facilities. This underscores the overarching potential of cloud computing to optimize resource utilization and bolster operational efficiency across businesses of all sizes.

14.2.4 Use cases

Cloud platforms provide many opportunities that go beyond traditional IT infrastructures. They offer a wide range of applications and services to let businesses innovate, grow, work together, and operate more efficiently in today's digital world. This technology allows for the quick release of applications and services. This speeds up the time it takes for products to reach the market and encourages new ideas.

Some key uses of cloud platforms include [10]:

1. **Data Storage and Backup:** Cloud platforms provide scalable and secure storage solutions. Users can store large amounts of data, back up critical information, and access it from anywhere with an Internet connection.

2. **Application Development and Deployment:** Cloud platforms offer tools and environments for developers that facilitate the development, testing, and deployment of applications without thinking about the required development infrastructure. PaaS streamlines the development process by offering required platforms.

3. **Scalable Computing Power:** Cloud platforms provide "on-demand" access to IT resources, allowing businesses to scale up or down as per the actual demand. This scalability is central to handle fluctuating workloads resourcefully.

4. **Collaboration and Productivity:** Cloud-based collaboration tools enable teams to work together in real time, share documents, and communicate seamlessly from different locations, fostering productivity and innovation.

5. **Disaster Recovery:** Cloud platforms help businesses recover after disasters. They do this by storing multiple copies of important data. This way,

if something happens to the main copy, businesses can still access their data and keep running smoothly.

6. **Big Data Analytics:** Cloud platforms provide powerful tools and services for analyzing large datasets. Companies can leverage these capabilities for business intelligence, predictive analytics, and decision-making.

7. **IoT (IoT) Applications:** Cloud platforms support IoT devices by delivering the infrastructure needed to collect, store, and process data generated by these devices, enabling various IoT applications [7, 11, 12].

8. **Content Delivery and Streaming:** Cloud platforms offer content delivery networks (CDNs) that enable fast and reliable delivery of content, facilitating streaming services for videos, music, and other media.

9. **AI Services:** Cloud platforms provide AI and ML services which lets businesses use pre-built models or create custom ML algorithms to derive insights, automate processes, and enhance user experiences.

10. **Cost Efficiency:** Cloud services let users pay only for what they need. This way, they don't have to spend a lot of money on infrastructure right away. [13].

14.3 SIGNIFICANCE OF CYBERSECURITY IN CLOUD COMPUTING

In cloud architecture, where a third party maintains, operates, and sells software, hardware, and infrastructure services and resources to users, security presents a significant challenge [14].

Cloud computing involves many different architectures, services, models, technologies, software, and applications. This makes cloud computing hard to secure. So, keeping cloud computing secure is a major job of the service provider [15, 16].

Cloud service providers shoulder the critical responsibility of upkeeping the confidentiality, integrity, reliability, and availability of data entrusted to them [17]. They must employ robust measures to prevent any form of compromise or unauthorized access, ensuring that the sensitive information of their clients remains secure and protected [18].

Maintaining user anonymity and data location privacy is another important aspect of security. Service providers must implement security mechanisms that provide adequate data and resource protection and mitigate malicious external threats.

The significance of cybersecurity in cloud computing can be understood through the following key aspects:

1. **Increased Attack Surface:** Cloud environments magnify the attack dimensions due to the distributed nature of data and services across multiple locations and shared infrastructure. This complexity requires robust cybersecurity measures [19].

2. **Dynamic Nature:** Cloud computing offers rapid scalability and flexibility. This dynamic nature poses threats in preserving steady security controls and visibility through the rapidly changing environments [20].
3. **Integration with IT Infrastructure:** Cloud services are integrated with existing IT infrastructure and applications, necessitating cohesive cybersecurity strategies to ensure seamless protection across hybrid and multi-cloud environments [19].
4. **Compliance and Regulatory Requirements:** Adhering to industry regulations such as "General Data Protection Regulation (GDPR)" and "Health Insurance Portability and Accountability Act (HIPAA)" and compliance standards requires robust cybersecurity controls and practices in cloud deployments [21].
5. **Business Continuity and Trust:** Identifying, assessing, and alleviating cybersecurity risks accompanying cloud adoption, including data breaches, insider threats, and vulnerabilities, are crucial for maintaining business continuity and trust by mitigating the impact of cyber incidents [22, 23].

14.4 SECURITY MANAGEMENT SYSTEMS IN CLOUD COMPUTING

Cybersecurity and cloud computing are intertwined in modern IT landscapes, where vigorous security procedures are crucial to guard data, software as well as infrastructure from evolving cyber threats. Security management systems (SMSs) in cloud computing play a critical role in safeguarding data, applications, and infrastructure from various threats and ensuring compliance with regulatory requirements. SMSs in cloud computing refer to the framework and practices put in place to guarantee the security of cloud-based assets, data, and operations [23]. Robust SMSs employ tools and technologies to aid organizations enhance their inclusive security posture in cloud computing, maintain trust with customers and stakeholders and effectively manage risks associated with cloud adoption. These systems help organizations:

1. **Ensure Compliance:** Meet regulatory requirements and industry standards for data protection and security.
2. **Mitigate Risks:** Identify and mitigate security threats and vulnerabilities proactively.
3. **Protect Data:** Encrypt sensitive data to prevent unauthorized access and data breaches.
4. **Monitor and Respond:** Continuously monitor cloud environments for security incidents and respond swiftly to threats.

14.4.1 Components of security management systems

The key components of a SMS tailored for cloud computing environments:

14.4.1.1 Identity and access management (IAM)

IAM systems manage user identities, permissions, and access to resources within cloud environments [24, 25]. They ensure that only authorized individuals and systems can access definite resources and accomplish defined actions. AWS IAM, Azure Active Directory (Azure AD), and Google Cloud IAM are the examples of examples of IAM tools.

14.4.1.2 Encryption and key management

These services protect data while it's stored or being sent. These services do so with the aid of robust key management for secure key storage and access [26, 27]. Examples of such services are Key Management Service (KMS) in AWS and Google Cloud and Azure Key Vault.

14.4.1.3 Network security

Azure Virtual Network (VNet), Virtual Private Cloud (VPC), Google Virtual Private Cloud (VPC), etc. are tools that provide network isolation, firewall controls, and network segmentation to secure communication between cloud resources [28].

14.4.1.4 Threat detection and prevention

These systems monitor cloud environments for suspicious activities, anomalies, and potential security threats. They use techniques like behavioral analysis and ML to identify and mitigate threats [29]. Examples are Azure Security Centre, AWS GuardDuty, and Google Cloud Security Command Centre (SCC).

14.4.1.5 Vulnerability management

Vulnerability management tools scan cloud resources for known vulnerabilities and misconfigurations [30, 31]. They help in identifying and remedying security weaknesses before they can be exploited. Examples of such tools are Google Cloud Security Scanner, Azure Security Centre and AWS Inspector, etc.

14.4.1.6 Security information and event management (SIEM)

These services provide monitoring, logging, and analysis of security events and incidents across cloud environments [32]. Examples of such tools are AWS CloudTrail, Azure Monitor, and SCC of Google Cloud.

14.4.1.7 Cloud security posture management (CSPM)

CSPM tools constantly monitor security configurations and compliance posture across cloud services and resources [33]. They ensure security best practices and adherence to regulatory requirements. Examples include Prisma Cloud (formerly known as RedLock), Dome9 (acquired by Check Point), and CloudHealth Secure State.

14.5 CYBERSECURITY ANALYSIS OF VARIOUS CLOUD PLATFORMS

Cybersecurity analysis for various cloud platforms typically involves evaluating several key components of SMSs such as:

- IAM
- Encryption and KMSs
- Threat Detection and Prevention Mechanisms
- Vulnerability Management Tools
- CSPM Tools

The next subsections analyze some popular cloud platforms on the basis of how effectively they implement the above components.

14.5.1 Identity and access management (IAM)

14.5.1.1 Amazon Web Services (AWS)

AWS-IAM is a foundational service offered by AWS. This service facilitates businesses to manage users and access permissions within their AWS cloud environment.

Features of AWS IAM:

1. **Centralized Identity Management:** With IAM, organizations can manage who can access what, all in one place. Administrators can create and control users and groups, assign permissions using policies, and grant access to AWS resources based on the principle of least privilege [34, 35].
2. **Granular Access Controls:** IAM lets admins control who can access specific AWS resources and what they can do. Policies are crafted using JSON and can be applied to users, groups, or roles. IAM also helps admins manage users, groups, and roles. Key features include [34]:
   a. **Users:** Individual identities with access keys and secret keys for authentication.
   b. **Groups:** Collections of users that can be managed as a single unit.
   c. **Roles:** Permissions that can be attached to users or groups, providing temporary access to resources.
   d. **Policies:** Define permissions and conditions for resource access.
   e. **IAM Identity Center:** Centralized identity management for AWS accounts.
3. **Multi-Factor Authentication (MFA):** MFA makes your AWS account extra safe. It asks you to give two or more ways to prove who you are

before you can use AWS resources. IAM supports virtual and hardware MFA devices, as well as SMS-based MFA [34, 35].

4. **Identity Federation:** IAM supports federated access, allowing users to sign in to AWS using existing corporate identities (via SAML 2.0) or social identities (via OpenID Connect). This simplifies user management and enhances user experience.

5. **Integration with AWS Services:** IAM integrates seamlessly with other AWS services, enabling secure access to services like Amazon S3, EC2, RDS, and others. Policies can be tailored to specific services and resource types.

6. **Audit and Monitoring:** IAM provides all-inclusive logging capabilities through AWS CloudTrail, which records IAM actions and API calls for auditing and compliance purposes. This helps organizations track user activity and identify security incidents.

7. **Scalability and Flexibility:** IAM scales with the organization's growth and allows flexible management of user access across multiple AWS accounts and regions [34].

8. **Integration and Automation:** IAM integrates well with DevOps processes and automation tools, enabling secure and streamlined management of AWS resources.

Cons and considerations:

1. **Complexity in Policy Management:** Writing and managing IAM policies can be complex, especially for large and complex environments.

2. **Learning Curve:** IAM's powerful features may require a learning curve for administrators and users unfamiliar with AWS security concepts and practices.

3. **Potential for Misconfigurations:** Misconfigurations in IAM policies can lead to security vulnerabilities or unintended exposure of resources. Regular audits and monitoring are essential to mitigate risks.

4. **Dependency on AWS Ecosystem:** Organizations heavily reliant on AWS services may find it challenging to integrate IAM with non-AWS environments or hybrid cloud setups.

5. **Cost of MFA Devices:** While MFA enhances security, organizations must consider the cost of hardware tokens or SMS-based MFA services, especially at scale.

14.5.1.2 *Azure Active Directory (AD)*

Azure Active Directory (Azure AD) is an all-inclusive IAM service that can be used for both cloud and in-house applications [36]. Built on a directory-based model, Azure AD integrates with on-premises AD. Its key features include [37]:

- **User Management:** Create, manage, and delete users and groups.
- **Application Management:** Integrate and manage applications used by your organization.

- **Conditional Access:** Implement gritty access controls established on user, location, device, and other conditions.
- **Azure AD B2B:** Team up with external users by allowing them to use your applications.
- **Azure AD Identity Protection:** Detect, investigate, and alleviate identity-based risks.
- **Single Sign-on (SSO)** across multiple applications.
- MFA for enhanced security.

14.5.1.3 Google Cloud IAM

Google Cloud IAM provides fine-grained access control for Google Cloud resources. It is based on a hierarchical structure of organizations, folders, and projects [38]. It offers:

- **Roles:** Predefined sets of permissions for different job functions.
- **Members:** Users, groups, or service accounts that can be assigned roles.
- **Policies:** Custom permissions for granular control.
- **Organization-wide IAM:** Manage access across multiple projects and folders.

14.5.2 Encryption

Cloud computing security relies on encryption to protect data from unapproved access. Different cloud platforms use encryption to keep data safe when it's stored, being moved, or being processed.

14.5.2.1 Amazon Web Services (AWS)

AWS provides comprehensive encryption capabilities across its services, allowing customers to encrypt data at multiple levels:

1. **Server-Side Encryption (SSE):**
    a. SSE-S3 [39]: Amazon S3 (Simple Storage Service) automatically encrypts data using AES-256 encryption. Customers can also administer their encryption keys using AWS KMS or bring their own keys (BYOK).
    b. SSE-KMS: Allows customers to use AWS KMS to manage encryption keys for various AWS services, ensuring granular control over data access and encryption [40].
    c. SSE-C: Customers can provide their own encryption keys for encryption of their data before storing it in S3.
2. **Client-Side Encryption:** AWS SDKs and CLI tools support client-side encryption, where customers encrypt data before uploading it to AWS services like S3, ensuring end-to-end encryption.

3. **Database Encryption:** AWS RDS (Relational Database Service) and other database services support encryption of data at rest using SSE or customer-managed keys through AWS KMS.
4. **Transit Encryption:** AWS ensures data transmitted between customers and AWS services is encrypted using industry-standard protocols like TLS (Transport Layer Security).

14.5.2.2 Microsoft Azure

Microsoft Azure offers robust encryption capabilities to protect data across its cloud services:

1. **Azure Storage Encryption:** Azure storage spontaneously encrypts stored data using AES-256 encryption [41]. Customers can manage encryption keys through Azure Key Vault, ensuring separation of duties and compliance.
2. **Azure Disk Encryption:** Azure Disk Encryption encrypts data stored on Azure VMs. The encryption of data is accomplished using BitLocker encryption on Windows VMs while Linux VM use DM-Crypt.
3. **Azure SQL Database Encryption [42]:** Azure SQL Database has a feature called transparent data encryption (TDE). This feature encrypts data that's stored in the database. This helps protect sensitive data from unauthorized access.

14.5.2.3 Google Cloud Platform (GCP)

Google Cloud Platform emphasizes strong encryption capabilities to protect customer data:

1. **Encryption at Rest:** GCP encrypts customer data at rest by default using AES-256 encryption. Encryption keys are managed and rotated automatically by Google, or customers can opt to manage their own keys using Cloud KMS.
2. **Encryption in Transit:** All data moving between users and GCP services is encrypted using TLS to prevent interception and eavesdropping.
3. **Database Encryption:** Google Cloud SQL and other database services support encryption at rest and in transit, ensuring data security for stored and transmitted data.
4. **Customer-Supplied Encryption Keys (CSEK):** GCP allows customers to provide their own encryption keys for additional control over data encryption and access.

14.5.3 Key management services

KMSs provided by leading cloud platforms such as AWS, Azure, and Google Cloud are essential tools for managing encryption keys securely [43]. These services enable organizations to create, control, and protect encryption keys used to encrypt data and ensure its confidentiality and integrity. When choosing a KMS, organizations must deal with factors such as integration capabilities, compliance requirements, scalability, and ease of use to meet their specific security needs in the cloud.

14.5.3.1 AWS key management service (KMS)

AWS KMS helps you make and control encryption keys for your data. It also helps you follow data protection regulations. Key features include:

1. **Centralized Key Management:** Allows you to create, import, and manage encryption keys that protect your data.
2. **Integration into AWS Services:** Unified integration with additional AWS services such as Amazon S3, EBS, RDS, and Redshift for encryption at rest.
3. **Granular Access Control:** Provides gritty control over who can use which keys and under what conditions using AWS IAM policies.
4. **Automatic Key Rotation:** Supports automatic rotation of keys to help you comply with security best practices and regulatory requirements.
5. **Hardware Security Modules (HSM) Integration:** Option to use AWS CloudHSM to generate and use your own encryption keys with dedicated hardware security modules.

14.5.3.2 Microsoft Azure key vault

Azure Key Vault helps safeguard cryptographic keys and secrets used by cloud applications and services. Key features include:

1. **Centralized Key Management:** Securely store and manage cryptographic keys, certificates, and secrets in Azure Key Vault.
2. **Key Usage Policies:** Define access policies and permissions to control who can access specific keys and secrets.
3. **Key Rotation and Versioning:** Supports automatic rotation of keys and versioning, ensuring that applications can seamlessly use new keys without downtime.
4. **Integration into Azure Services:** Integrates with Azure services like Azure Disk Encryption, Azure VMs, Azure SQL Database, and more for seamless encryption.
5. **Hardware Security Module (HSM) Integration:** Option to use Azure Dedicated HSM for FIPS 140-2 Level 3 validated hardware security modules.

14.5.3.3 Google Cloud

Google Cloud Key Management provides a scalable and centralized KMS for GCP. Key features include:

1. **Centralized Key Storage:** Store, use, and manage cryptographic keys and secrets securely in Google Cloud Key Management.
2. **Role-Based Access Control (RBAC):** Define and enforce fine-grained access controls using IAM roles and policies.
3. **Integration with Google Services:** Flawlessly integrates with additional Google Cloud services like Cloud Storage, Compute Engine, BigQuery, and more.
4. **Key Versioning and Rotation:** Supports automatic key rotation and versioning to enhance security and compliance.
5. **External Key Management:** Option to import your own keys and manage them with Google's Cloud External Key Manager.

14.5.4 Threat detection and prevention mechanisms in cloud platforms

Threat detection and prevention mechanisms are essential components of cloud security, aimed at identifying and mitigating potential security threats and vulnerabilities in real time. Leading cloud platforms provide specialized services for threat detection and prevention such as AWS GuardDuty, the Security Command Center (SCC) of Google Cloud, and Azure Security Center (ASC).

14.5.4.1 AWS GuardDuty

AWS GuardDuty is a managed threat detection service [44]. This service constantly keeps a watch on AWS accounts and workloads for malevolent activity and unapproved behavior. Key features include:

1. **Intelligent Threat Detection:** Utilizes ML algorithms and AWS' vast dataset to identify anomalies, malicious IPs, and unauthorized access patterns.
2. **Integration with AWS Services:** Spontaneously analyzes AWS CloudTrail logs, VPC Flow Logs, and DNS logs to detect threats without requiring additional agents or deployments.
3. **Actionable Insights:** Provides detailed findings and recommendations to remediate security issues, such as compromised EC2 instances or unauthorized access attempts.
4. **Scalability:** Scales automatically with AWS deployments and supports multi-account environments through centralized management.

5. **Native Integration:** Seamlessly integrates with AWS services, leveraging existing data sources for threat detection.
6. **Automated Remediation:** Offers automated responses and remediation recommendations to mitigate detected threats.

Limitations of AWS GuardDuty

1. **Limited to AWS Ecosystem:** Primarily focuses on threats within AWS environments, and may require additional tools for multi-cloud or hybrid environments.
2. **Complexity:** Requires configuration and tuning to reduce false positives and optimize threat detection accuracy.

14.5.4.2 Azure security center

ASC is a cohesive security management service. This service offers threat detection, policy enforcement, and security recommendations across Azure and hybrid cloud environments [45]. Its features are:

1. **Unified Security Dashboard:** Centralized view of security posture with insights into resource security hygiene, vulnerabilities, and detected threats.
2. **Threat Intelligence:** Utilizes Microsoft's global threat intelligence network to detect and alert on suspicious activities, malware infections, and potential data breaches.
3. **Integration with Azure Defender:** Enhances threat detection capabilities with advanced analytics and ML for proactive threat prevention.
4. **Compliance and Policy Management:** Assesses compliance with the supervisory standards and provides actionable recommendations for improving security posture.
5. **Hybrid Cloud Support:** Offers visibility and security management across Azure and hybrid environments.
6. **Automated Response:** Provides automated responses to security events and integrates with Azure Sentinel for advanced security orchestration.
7. **Integration with Microsoft Solutions:** Seamless integration with Microsoft services like Azure Active Directory and Office 365 for holistic security management.

Limitations of Azure security center

1. **Complex Pricing Model:** Pricing based on Azure Defender plans and usage, which can be complex to estimate and manage.
2. **Learning Curve:** Requires familiarity with Azure services and security configurations to maximize effectiveness.

14.5.4.3 Google Cloud security command center (SCC)

SCC in Google Cloud [46] is an inclusive security management service. The SCC provides insights into security threats across Google Cloud services. Key features of SCC are:

1. **Asset Inventory:** Automatically identifies and catalogs cloud assets to provide a comprehensive view of the environment.
2. **Threat Detection:** Uses built-in detectors and incorporation with third-party security tools to detect threats such as malware, vulnerabilities, and suspicious activity.
3. **Security Health Analytics:** Analyzes security health metrics and provides commendations for refining security posture and compliance.
4. **Integration with Google Services:** Integrates with Google Cloud services like Cloud Logging and Cloud Monitoring for centralized threat management and incident response.
5. **Comprehensive Visibility:** Offers a consolidated view of security posture across Google Cloud resources.
6. **Real-Time Insights:** Offers real-time alerts and notifications for detected threats and security incidents.
7. **Scalability:** Scales with Google Cloud deployments and supports multi-cloud environments with integration capabilities.

Limitations of Google Cloud SCC

1. **Integration Challenges:** Requires integration with Google Cloud services for full functionality, which may involve additional setup and configuration.
2. **Cost Considerations:** Pricing structure based on usage and features may require careful monitoring to manage costs effectively.

14.5.5 Vulnerability management tools in cloud platforms

Vulnerability management tools are crucial components of cloud security, enabling businesses to identify, assess, prioritize, and remediate vulnerabilities within their cloud environments. AWS Inspector [44], ASC, and Google Cloud Security Scanner are prominent tools offered by leading cloud providers for vulnerability management.

Organizations should carefully consider their specific cloud deployment requirements, including scalability, integration needs, and cost considerations when selecting and implementing these tools to effectively mitigate security risks and maintain compliance in the cloud. This section presents a survey of these tools, including their features, benefits, and limitations (considerations).

14.5.5.1 AWS inspector

AWS Inspector is a service that automatically checks the security of your apps on AWS. It helps you make your apps more secure. Key features include:

1. **Agent-Based Assessments:** Runs assessments using lightweight agents installed on EC2 instances, analyzing network, file system, and application configurations.
2. **Continuous Monitoring:** Provides continuous assessment and monitoring of applications for vulnerabilities and deviations from security best practices.
3. **Integration with AWS Services:** Integrates with AWS CloudTrail and AWS Config to provide comprehensive visibility and compliance monitoring.
4. **Actionable Insights:** Generates detailed findings and prioritizes vulnerabilities based on severity, providing recommendations for remediation.
5. **Native Integration:** Seamless integration with AWS services, leveraging AWS data sources for vulnerability assessment.
6. **Automated Assessments:** Automated scheduling and execution of assessments, reducing manual effort and ensuring continuous security monitoring.
7. **Scalability:** Scales automatically with AWS deployments, supporting large-scale environments and multi-account setups.

Limitations of AWS inspector

1. **Limited Scope:** Primarily focused on vulnerabilities within AWS environments, may require additional tools for comprehensive multi-cloud or hybrid environments.
2. **Complexity:** Requires configuration and tuning to decrease false positives and optimize vulnerability detection accuracy.

14.5.5.2 Azure security center

ASC [45, 47] includes vulnerability assessment capabilities to aid businesses identify and alleviate security weaknesses across Azure and hybrid cloud environments.

1. **Continuous Monitoring:** Automatically discovers and assesses vulnerabilities in VMs, containers, and other Azure resources.
2. **Integration with Azure Defender:** Enhances vulnerability management with advanced analytics and threat intelligence for proactive threat prevention.
3. **Security Recommendations:** Provides prioritized recommendations for mitigating vulnerabilities based on severity, compliance requirements, and industry standards.

4. **Compliance Dashboard:** Assesses compliance with regulatory standards and provides actionable insights for improving security posture.
5. **Hybrid Cloud Support:** Offers visibility and management across Azure and hybrid environments, providing a unified security view.
6. **Automated Remediation:** Provides automated recommendations and remediation steps to address identified vulnerabilities.
7. **Integration with Microsoft Solutions:** Seamless integration into Azure services like Azure Active Directory and Office 365 for all-inclusive security management.

Limitations of ASC

1. **Complex Pricing Model:** Pricing is based on Azure Defender plans and usage, which can be complex to estimate and manage.
2. **Dependency on Azure Ecosystem:** Requires familiarity with Azure services and configurations for effective use.

14.5.5.3 Google Cloud Security Scanner

Google Cloud Security Scanner is a tool. It helps find weaknesses in web applications on Google Cloud Platform. It performs the following functions:

Automatic Crawling and Testing: Automatically crawls web applications and tests for common vulnerabilities such as cross-site scripting (XSS) and SQL injection.

Integration with GCP Services: Integrates with Google Cloud services like Cloud Logging and Cloud Monitoring for centralized vulnerability management.

Detailed Reports: Provides detailed reports with identified vulnerabilities, including severity levels and recommended fixes.

Continuous Scanning: Supports continuous scanning to detect new vulnerabilities introduced during application updates and deployments.

Built-in Integration: Built-in integration with Google Cloud services for seamless vulnerability scanning and management.

Automated Testing: Automated scanning and testing processes reduce manual effort and ensure consistent security assessments.

Scalability: Scales with Google Cloud deployments, supporting large-scale applications and environments.

Limitations of Google Cloud Security Scanner

1. **Focus on Web Applications:** Primarily focused on vulnerabilities in web applications, may not cover vulnerabilities in other types of cloud resources.
2. **Limited to GCP:** Designed for vulnerabilities within Google Cloud Platform, may not integrate as seamlessly with non-GCP environments.

14.5.6 Cloud security posture management (CSPM) tools

CSPM [33] tools such as Prisma Cloud, Check Point CloudGuard (formerly Dome9), and CloudHealth Secure State are instrumental in helping organizations maintain a secure and compliant posture across their cloud environments. These tools provide visibility, automated remediation, and policy enforcement capabilities to mitigate security risks, ensure compliance with regulatory requirements, and optimize operational efficiency. A few tools are explored below, along with their features, benefits, and limitations.

14.5.6.1 Prisma Cloud (formerly RedLock)

Prisma Cloud by Palo Alto Networks [48] is an all-inclusive CSPM platform that offers visibility and control over multi-cloud environments which comprises AWS, Azure, Google Cloud, and Kubernetes clusters.

Features of Prisma Cloud

1. **Continuous Monitoring:** visibility of cloud resources, identifying potential security risks, misconfigurations, and compliance issuesProvides instantaneous.
2. **Compliance Assurance:** Enforces security best practices and regulatory compliance requirements (e.g., GDPR, PCI DSS) through predefined policies and custom checks.
3. **Threat Detection:** Uses ML to identify unusual behaviors, doubtful activities, and possible security incidents across cloud environments.
4. **Integration with DevOps:** Integrates seamlessly into CI/CD pipelines to ensure security and compliance throughout the software development lifecycle.
5. **Multi-Cloud Support:** Provides visibility and security management across AWS, Azure, Google Cloud, and Kubernetes environments.
6. **Automated Remediation:** Offers automated response actions and remediation for identified security issues, reducing manual intervention.
7. **Comprehensive Compliance:** Helps organizations meet compliance requirements through continuous monitoring and enforcement of security policies.

Limitations of Prisma Cloud

1. **Complexity:** Requires expertise and time to configure and fine-tune policies and alerts to reduce false positives and optimize security monitoring.
2. **Cost Considerations:** Pricing structure based on usage and features may require careful monitoring to manage costs effectively.

14.5.6.2 Dome9 (now Check Point CloudGuard)

Dome9, now integrated into Check Point CloudGuard [49, 50], is a CSPM solution that provides security posture management, compliance automation, and governance across cloud environments.

Features of Demo9

1. **Centralized Security Management:** Offers a unified dashboard for managing security policies, compliance checks, and remediation across AWS, Azure, and GCP.
2. **Automated Compliance Checks:** Automates security and compliance assessments based on industry standards and regulatory requirements.
3. **Network Security:** Provides network security group management, firewall policy management, and visualization of network traffic across cloud platforms.
4. **Integration with Check Point Solutions:** Integrates with Check Point's other security products for enhanced threat prevention and incident response capabilities.
5. **Multi-Cloud Visibility:** Supports comprehensive visibility and management across AWS, Azure, and GCP environments.
6. **Automated Remediation:** Automates remediation actions based on policy violations, reducing manual effort and improving response time.
7. Network Security Management: Provides robust network security management capabilities, including firewall policy orchestration and monitoring.

Limitations of Dmeo9

1. **Learning Curve:** Requires familiarity with Check Point's ecosystem and configuration settings to maximize effectiveness.
2. **Integration Challenges:** This may require additional setup and integration efforts for seamless operation with existing cloud deployments.

14.5.6.3 CloudHealth secure state

CloudHealth Secure State [51] by VMware is a CSPM platform that focuses on security and compliance posture management across cloud environments. CHSS collects data from cloud environments, analyzes it against security best practices and compliance standards, and generates reports and alerts to highlight potential risks. Users can then prioritize and address issues through automated or manual remediation actions.

Features of CloudHealth Secure State

1. **Continuous Monitoring:** Provides continuous visibility into cloud infrastructure, applications, and services to detect security risks and compliance issues.

2. **Automated Remediation:** Offers automated actions and remediation workflows to address identified security vulnerabilities and misconfigurations.
3. **Compliance management:** Ensures adherence to industry regulations (e.g., HIPAA and GDPR) through automated compliance checks.
4. **Customizable Policies:** Allows organizations to define custom security policies and compliance checks based on industry standards and regulatory requirements.
5. **Integration with CloudHealth Platform:** Integrates seamlessly with CloudHealth's cloud management platform for unified cloud cost and security management.
6. **Comprehensive Visibility:** Provides comprehensive visibility and monitoring capabilities across cloud environments.
7. **Automated Actions:** Automates security remediation actions based on predefined policies and thresholds, enhancing operational efficiency.
8. **Multi-cloud support:** CHSS offers unified visibility across major cloud providers (AWS, Azure, GCP), enabling consistent security policies and controls.

Limitations of CloudHealth Secure State

1. **Dependency on VMware Ecosystem:** Requires integration with VMware's CloudHealth platform for full functionality, which may involve additional costs and dependencies.
2. **Cost Considerations:** Pricing structure based on usage and features may require careful planning to optimize costs and ROI.

14.6 CONCLUSION

As cloud computing continues to dominate the technological landscape and provides exciting opportunities, ensuring robust cybersecurity in major cloud platforms is paramount. Ensuring robust cybersecurity in major cloud platforms is a multidimensional challenge that requires an all-inclusive approach.

IAM systems such as AWS IAM and Azure Active Directory are crucial in managing user identities, access controls, and authentication processes. Granular user access controls, implemented through MFA and least privilege philosophies, reduce the risk of unauthorized access. Continuously monitoring and auditing user activity help identify and address suspicious behavior. Additionally, integrating IAM with existing identity management systems fosters a cohesive security posture through on premise and cloud infrastructure

Encryption plays a pivotal role in safeguarding sensitive data at rest and in transit. Major cloud platforms offer robust encryption solutions and effective key management strategies. Utilizing KMS ensures secure key generation, storage, rotation, and access control protocols. Implementing encryption across all data layers, including databases, object storage, and backups, further strengthens the security posture of cloud environments.

Vulnerability management tools play a critical role in proactively identifying and addressing security weaknesses within cloud environments. These tools scan cloud resources for known vulnerabilities and misconfigurations, allowing organizations to prioritize patching efforts. Integrating vulnerability management tools with automated patching processes streamlines the remediation process, minimizing the window of opportunity for attackers.

CSPM tools provide a holistic view of an organization's cloud security posture. They offer centralized visibility into user access controls, encryption settings, network configurations, and potential vulnerabilities across all cloud resources. By analyzing security data and generating reports, CSPM tools enable organizations to identify and prioritize security gaps, fostering continuous improvement and proactive risk management.

By adopting a multi-layered approach that integrates IAM, encryption, threat detection, vulnerability management, and CSPM tools, businesses can successfully mitigate security risks and make sure CIA of their data and applications in the cloud.

REFERENCES

1. Varghese, B., & Buyya, R. (2018). Next generation cloud computing: new trends and research directions. *Future Generation Computer Systems*, 79, 849–861.
2. Liu, Y., Wang, L., Wang, X. V., Xu, X., & Zhang, L. (2019). Scheduling in cloud manufacturing: state-of-the-art and research challenges. *International Journal of Production Research*, 57(15–16), 4854–4879.
3. Tabrizchi, H., & Kuchaki Rafsanjani, M. (2020). A survey on security challenges in cloud computing: issues, threats, and solutions. *The Journal of Supercomputing*, 76(12), 9493–9532.
4. Kavak, H., Padilla, J. J., Vernon-Bido, D., Diallo, S. Y., Gore, R., & Shetty, S. (2021). Simulation for cybersecurity: state of the art and future directions. *Journal of Cybersecurity*, 7(1), tyab005.
5. Humayun, M., Niazi, M., Jhanjhi, N. Z., Alshayeb, M., & Mahmood, S. (2020). Cyber security threats and vulnerabilities: a systematic mapping study. *Arabian Journal for Science and Engineering*, 45, 3171–3189.
6. Gunduz, M. Z., & Das, R. (2020). Cyber-security on smart grid: threats and potential solutions. *Computer networks*, 169, 107094.
7. Sunyaev, A. Cloud computing. *Internet computing: Principles of distributed systems and emerging internet-based technologies*, Springer, 195–236.
8. Wurster, M., Breitenbücher, U., Falkenthal, M., Krieger, C., Leymann, F., Saatkamp, K., & Soldani, J. (2020). The essential deployment metamodel: a systematic review of deployment automation technologies. *SICS Software-Intensive Cyber-Physical Systems*, 35, 63–75.
9. Mohammed, C. M., & Zeebaree, S. R. (2021). Sufficient comparison among cloud computing services: IaaS, PaaS, and SaaS: a review. *International Journal of Science and Business*, 5(2), 17–30.

10. Marinescu, D. C. (2022). *Cloud computing: Theory and practice*. Morgan Kaufmann.
11. Ullah, F., Naeem, H., Jabbar, S., Khalid, S., Latif, M. A., Al-Turjman, F., & Mostarda, L. (2019). Cyber security threats detection in internet of things using deep learning approach. *IEEE Access, 7*, 124379–124389.
12. Kimani, K., Oduol, V., & Langat, K. (2019). Cyber security challenges for IoT-based smart grid networks. *International Journal of Critical Infrastructure Protection, 25*, 36–49.
13. Kavis, M. (2023). *Architecting the cloud*. Wiley.
14. Saha, M., Panda, S. K., & Panigrahi, S. (2019). Distributed computing security: issues and challenges. *Cyber security in parallel and distributed computing: Concepts, techniques, applications and case studies*, 129–138.
15. Wani, A. R., Rana, Q. P., & Pandey, N. (2019). Analysis and countermeasures for security and privacy issues in cloud computing. *System performance and management analytics*, 47–54.
16. Dey, H., Islam, R., & Arif, H. (2019, January). An integrated model to make cloud authentication and multi-tenancy more secure. In *2019 International Conference on Robotics, Electrical and Signal Processing Techniques (ICREST)* (pp. 502–506). Dhaka, Bangladesh: IEEE.
17. Pitchai, R., Babu, S., Supraja, P., & Anjanayya, S. (2019). Prediction of availability and integrity of cloud data using soft computing technique. *Soft Computing, 23*, 8555–8562.
18. Alhenaki, L., Alwatban, A., Alamri, B., & Alarifi, N. (2019, May). A survey on the security of cloud computing. In *2019 2nd International Conference on Computer Applications & Information Security (ICCAIS)* (pp. 1–7). Riyadh, Saudi Arabia: IEEE.
19. Dimitrov, W. (2020). The impact of the advanced technologies over the cyber attacks surface. In *Artificial Intelligence and Bioinspired Computational Methods: Proceedings of the 9th Computer Science On-line Conference 2020, Vol. 2 9* (pp. 509–518). Springer International Publishing.
20. Almutairy, N. M., Al-Shqeerat, K. H., & Al Hamad, H. A. (2019). A taxonomy of virtualization security issues in cloud computing environments. *Indian Journal of Science and Technology, 12*(3), 1–19.
21. Shah, V., & Konda, S. R. (2022). Cloud computing in healthcare: opportunities, risks, and compliance. *Revista Espanola de Documentacion Cientifica, 16*(3), 50–71.
22. Aljumah, A., & Ahanger, T. A. (2020). Cyber security threats, challenges and defence mechanisms in cloud computing. *IET communications, 14*(7), 1185–1191.
23. Tissir, N., El Kafhali, S., & Aboutabit, N. (2021). Cybersecurity management in cloud computing: semantic literature review and conceptual framework proposal. *Journal of Reliable Intelligent Environments, 7*(2), 69–84.
24. Mohammed, I. A. (2019). Cloud identity and access management–a model proposal. *International Journal of Innovations in Engineering Research and Technology, 6*(10), 1–8.
25. Alsirhani, A., Ezz, M., & Mostafa, A. M. (2022). Advanced Authentication Mechanisms for Identity and Access Management in Cloud Computing. *Computer Systems Science & Engineering, 43*(3), 967–984.
26. Ahmad, S., Mehfuz, S., & Beg, J. (2023). Hybrid cryptographic approach to enhance the mode of key management system in cloud environment. *The Journal of Supercomputing, 79*(7), 7377–7413.
27. Zhou, Y., Li, N., Tian, Y., An, D., & Wang, L. (2020). Public key encryption with keyword search in cloud: a survey. *Entropy, 22*(4), 421.
28. Mhaskar, N., Alabbad, M., & Khedri, R. (2021). A formal approach to network segmentation. *Computers & Security, 103*, 102162.

29. El Kafhali, S., El Mir, I., & Hanini, M. (2022). Security threats, defense mechanisms, challenges, and future directions in cloud computing. *Archives of Computational Methods in Engineering, 29*(1), 223–246.
30. Walkowski, M., Oko, J., & Sujecki, S. (2021). Vulnerability management models using a common vulnerability scoring system. *Applied Sciences, 11*(18), 8735.
31. Alyas, T., Ali, S., Khan, H. U., Samad, A., Alissa, K., & Saleem, M. A. (2022). Container performance and vulnerability management for container security using Docker engine. *Security and Communication Networks, 2022*(1), 6819002.
32. Vielberth, M. (2021). Security information and event management (SIEM). In *Encyclopedia of Cryptography, Security and Privacy* (pp. 1–3). Berlin, Heidelberg: Springer Berlin Heidelberg.
33. Sawhney, G., Kaur, G., & Deorari, R. (2022, October). CSPM: a secure cloud computing performance management model. In *2022 International Conference on Cyber Resilience (ICCR)* (pp. 1–5). Dubai, United Arab Emirates:IEEE.
34. Talluri, S., & Makani, S. T. (2023). Managing Identity and Access Management (IAM) in Amazon Web Services (AWS). Journal of Artificial Intelligence & Cloud Computing. 2(1), 2–5. *SRC/JAICC-159. doi.* 10.47363/JAICC/2023(2)147.
35. Pandit, P. (2021). Case Study on AWS Identity and User Management.
36. Zaal, S. (2022). *Azure Active Directory for Secure Application Development: Use Modern Authentication Techniques to Secure Applications in Azure.* Packt Publishing Ltd.
37. Subbarao, D., Raju, B., Anjum, F., Rao, C. V., & Reddy, B. M. (2023). Microsoft Azure active directory for next level authentication to provide a seamless single sign-on experience. *Applied Nanoscience, 13*(2), 1655–1664.
38. Talekar, R. (2023). Insider Threat For Service Account in Google Cloud Platform.
39. Aditya, D. J., Laxmanraj, S., & Krishna, R. S. B. (2023, June). Private document vault with server-side encryption in Cloud AWS S3 Bucket. In *2023 8th International Conference on Communication and Electronics Systems (ICCES)* (pp. 1774–1778). Coimbatore, Tamil Nadu: IEEE.
40. Moore, T. L., Conlon, S. S., Hewarathna, A. U., & Mailewa, A. B. (2022). Encryption methods and key management services for secure cloud computing: a review. Computer Systems Science & Engineering, 46(3), 967–984.
41. Antonopoulos, P., Arasu, A., Singh, K. D., Eguro, K., Gupta, N., Jain, R., ... & Zwilling, M. (2020, June). Azure SQL database always encrypted. In *Proceedings of the 2020 ACM SIGMOD International Conference on Management of Data* (pp. 1511–1525), Portland, Oregon, USA.
42. Bhardwaj, N., Banerjee, A., & Roy, A. (2021). Case study of Azure and Azure security practices. *Machine Learning Techniques and Analytics for Cloud Security,* 339–355.
43. Guptha, A., Murali, H., & Subbulakshmi, T. (2021, May). A comparative analysis of security services in major cloud service providers. In *2021 5th International Conference on Intelligent Computing and Control Systems (ICICCS)* (pp. 129–136). Madurai, India: IEEE.
44. Routavaara, I. (2020). Security monitoring in AWS public cloud.
45. Diogenes, Y., & Janetscheck, T. (2021). *Microsoft Azure Security Center.* Microsoft Press.
46. Roy, A., Banerjee, A., & Bhardwaj, N. (2021). A study on Google Cloud Platform (GCP) and its security. *Machine Learning Techniques and Analytics for Cloud Security,* 313–338.
47. DiCola, N., & Roman, A. (2021). *Microsoft Azure Network Security.* Microsoft Press.

48. Martseniuk, Y., Partyka, A., Harasymchuk, O., & Korshun, N. (2024). Automated conformity verification concept for cloud security. *Cybersecurity Providing in Information and Telecommunication Systems 2024, 3654,* 25–37.
49. Ali, M. I., Jha, P., Sharma, A., & Kaur, S. (2022). Security analysis and fraud prevention in cloud computing. In *The Role of IoT and Blockchain* (pp. 401–418). Apple Academic Press.
50. Hille, M., Klemm, D., & Lemmermann, L. (2018). Cloud computing vendor & service provider comparison. *Crisp Vendor Universe.*
51. Kortelainen, O. (2023). *Infrastructure Management in Multicloud Environments* (Master's thesis). https://trepo.tuni.fi/handle/10024/147777

Chapter 15

Mobile data anonymization in cyber security

Techniques, applications, and future directions

Anjali Yadav and Ajay Kumar

15.1 INTRODUCTION

15.1.1 Concept of mobile data anonymization and its importance in cyber security

PII (personally identifiable information) is hidden or deleted from mobile data as part of the anonymization process to safeguard individuals' privacy and prevent unauthorized access or usage [1]. Because more people are using mobile devices and more mobile data is being collected and stored, mobile data anonymization has become more crucial to cyber security.

By using mobile data anonymization techniques, businesses can comply with data privacy ordinances such as the General Data Protection Regulation (GDPR) and the California Consumer Privacy Act (CCPA) [2, 3]; which require businesses to protect personal information and give people the choice of how their data is used.

Mobile data anonymization [4] can also help avoid cyberattacks, such as identity theft and financial fraud because it is harder for attackers to get personal information now. Organizations can reduce the risk of data breaches and the impact of cyberattacks by anonymizing mobile data.

Reputation and confidence are largely dependent on anonymizing mobile data since people are more willing to share their information with businesses that prioritize security and privacy. Therefore, it is imperative that mobile data anonymization be properly considered and implemented as a crucial part of cyber security.

15.1.2 Need for data anonymization

Companies are gathering and keeping an increasing amount of data, which has increased the necessity for data anonymization. The risk of data breaches and unauthorized access to personal information has increased due to the exponential growth of data. By obscuring or deleting sensitive data, data anonymization increases the difficulty of identifying the individuals connected to PII. This aids in the protection of PII for enterprises [1].

DOI: 10.1201/9781032657264-15

To protect privacy and adhere to data protection laws like the CCPA and the GDPR, data anonymization is required. Organizations must abide by these rules to protect individuals' personal information and provide them control over how it is used [2, 3].

Also, mobile data anonymization is becoming a crucial component of cyber security due to the rising usage of mobile devices and the expansion of mobile data. Through increased difficulty in obtaining personal information by hackers, mobile data anonymization helps enterprises safeguard personal data and thwart cyberattacks, including identity theft and financial crime.

As a result, data anonymization has become more and more necessary as the amount of data that businesses collect and retain grows. Complying with data protection laws, protecting privacy, and thwarting cyberattacks all depend on data anonymization. Significant fines may be incurred for breaking these rules; under the GDPR, fines may total up to $20 million or 4% of worldwide revenue, whichever is larger.

Furthermore, mobile data anonymization is becoming a crucial component of cyber security due to the growing amount of mobile data and the growing use of mobile devices. By making it more difficult for attackers to get personal information, mobile data anonymization helps enterprises safeguard personal data and avoid cyberattacks, such as identity theft and financial fraud.

Data anonymization is becoming more and more necessary as the amount of data that businesses gather and retain grows. Ensuring privacy, adhering to data protection laws, and averting cyberattacks all depend on data anonymization.

15.1.3 Objectives of the chapter

This chapter aims to address the following aspects of the mobile data anonymization in cyber security:

1. Defining mobile data anonymization and its importance in protecting individual privacy.
2. Introducing and presenting a brief analysis of different techniques for mobile data anonymization, including data masking, pseudonymization, data aggregation, random data generation, data generalization, and data swapping.
3. Challenges and future scope of mobile data anonymization and related techniques.
4. Why Cyber Security is important in Mobile Data Anonymization?

15.2 UNDERSTANDING MOBILE DATA ANONYMIZATION

15.2.1 Mobile data anonymization and its goals

The process of converting or eliminating personally identifying information (PII) from mobile data in order to uphold data protection laws and safeguard individuals' privacy is known as mobile data anonymization. The following are the main objectives of anonymizing mobile data [5–7]:

15.2.1.1 Safeguard personal information and stop illegal access or use

Organizations can lessen the chance of data breaches and cyberattacks, including identity theft and financial fraud, by deleting or hiding personal information.

15.2.1.2 Respect privacy legislation regarding data

Data anonymization can assist organizations in adhering to data protection laws, such as the CCPA and the GDPR, which mandate that businesses protect personal information and provide individuals the ability to decide how their data is used.

15.2.1.3 Retain reputation and trust

People are more willing to share their data with companies that value security and privacy, therefore anonymizing mobile data is crucial to preserving reputation and trust.

15.2.1.4 Facilitate data sharing and utilization

Data anonymization safeguards individual privacy while enabling organizations to share and utilize mobile data. Without jeopardizing user privacy, it helps businesses to get value and insight from their data for purposes like marketing campaigns and user experience personalization.

The necessity for mobile data anonymization is rising as a result of the volume of data that businesses are gathering and storing. The risk of data breaches and unauthorized access to sensitive information has increased along with the exponential growth of data. Through mobile data anonymization, businesses can safeguard PII, thwart cyberattacks, guarantee privacy, adhere to data protection laws, and uphold customer confidence.

15.2.2 Data anonymization techniques

This section briefly introduces the various data anonymization techniques. A later section has a detailed discussion including use cases of each technique along with pros and cons. Below are the different data anonymization techniques [8, 9]:

15.2.2.1 Data masking

The process of hiding or changing values in a data set so that the data is still available but the original values cannot be re-engineered is known as data masking. By using false data that is nonetheless very persuasive but has no relation to the actual values, masking replaces original information. Static and dynamic data masking are the two main categories.

15.2.2.2 Data pseudonymization

In general, data pseudonymization is the act of changing "pseudonyms," or fake identities, to disguise direct identifiers in a data set. Pseudonymization is intended to be reversible and does not take indirect identifiers into account.

15.2.2.3 Data generalization

Data generalization "zooms out" on a data collection, creating a more comprehensive perspective of its contents and making it harder to identify certain traits and characteristics. Data sets large enough to guarantee that the data is sufficiently ambiguous without losing its utility are the ideal candidates for data generalization.

15.2.2.4 Data perturbation

Data perturbation, which adds ambiguity to data collection in a predictable and restorable manner without compromising analytical accuracy, is the purposeful randomization of data pieces.

15.2.2.5 Data swapping

Data swapping is the process of rearranging data in a data set so that attribute values no longer match the original data. An attribute from one row in the same column may be substituted with an attribute from another row using this data anonymization technique.

15.2.2.6 Synthetic

Although synthetic data is produced by machines, it closely resembles real, sensitive data. This generated data, which is widely utilized for model training and validation in machine learning (ML) and artificial intelligence (AI), is typically created by algorithms.

15.3 CHALLENGES OF MOBILE DATA ANONYMIZATION

Mobile data anonymization is a crucial process for protecting user privacy, but it comes with several challenges. These challenges include re-identification attacks, reduced data utility, data quality, compliance with regulations, technical challenges, and user trust and acceptance. Organizations must carefully evaluate these factors when implementing mobile data anonymization techniques and must be transparent and communicative with users about their data privacy practices [11, 12]. Some of the challenges are briefly explained in subsequent sub-sections.

15.3.1 Risk of re-identification and re-identification attacks

Even though data anonymization techniques can make it difficult to identify specific individuals, they do not eliminate the risk of re-identification completely. Attackers with access to additional information can still potentially re-identify individuals. An attacker can re-identify the people in the data using a variety of methods such as through linkage or inference attacks; even after the anonymization procedure has been completed. This can be achieved by comparing the data with other publicly accessible sources cross-referencing it or by analyzing data trends with ML methods.

15.3.2 Reduced data utility

When sensitive or distinctive data points are eliminated or obscured, anonymization processes may lead to a reduction in the usefulness of the data. As a result, the data may be less useful for analysis and understanding.

15.3.3 Data quality

The anonymization process can also impact the quality of the data, leading to inaccuracies, inconsistencies, or missing values. This can affect the reliability and validity of the data, making it difficult to draw meaningful conclusions.

15.3.4 Regulatory compliance

Organizations must ensure that they are complying with data privacy regulations while anonymizing mobile data. This can be challenging, as regulations vary across jurisdictions, countries, states, regions, and industries.

It can be difficult to figure out how to handle and abide by these regulations, particularly if a business operates in several jurisdictions or across different countries and they may have different requirements for anonymization and data retention.

15.3.5 Technical obstacles

The process of anonymizing mobile data can be difficult and complex. It calls for specialized knowledge, equipment, and infrastructure—all of which can be costly and resource-intensive.

15.3.6 User acceptance and trust

If users have privacy and security concerns, they can be reluctant to provide their data. It is imperative for organizations to maintain transparency regarding their data collecting and anonymization procedures, and to effectively communicate with users regarding the usage and protection of their data.

15.3.7 Integration with ML and AI models

Since anonymized data is not as rich as raw data, it is not as suitable for training ML and AI algorithms as original raw data, which need precise and comprehensive data to learn and then forecast.

15.3.8 Traditional or fragmented approach

To guarantee relational consistency and accuracy across many datasets and data repositories, a traditional or fragmented approach to anonymizing data may be problematic.

It's crucial to remember that despite these drawbacks, mobile data anonymization is still a useful technique for consumer privacy protection. Rather, they emphasize how crucial it is to properly weigh the trade-offs between data utility and privacy when employing anonymization approaches. To make sure that data is managed responsibly and securely, it's also critical to stay current on the newest laws and best practices for data anonymization.

15.4 DATA ANONYMIZATION TECHNIQUES

The data requirements and use cases of any organization vary, and in order to comply with regulatory requirements, it can be necessary to apply multiple data anonymization techniques. As a result, it is crucial to assess which strategy works best for the particular data stack. We have briefly introduced the most often used data anonymization techniques [13, 14]. This section presents a discussion of each of them.

15.4.1 Data masking

Data masking [14] is a data anonymization approach that entails removing private, confidential, or sensitive information from a dataset and substituting it with random characters, pseudo-information, or other fictitious data. Alternatively, data masking can be defined as the process of hiding or changing values in a data set so that the data may still be accessed but the original values can no longer be re-engineered. Data masking is carried out by using false data that is nonetheless very persuasive but has no relation to the actual values that it replaces. The data masking process is shown in Figure 15.1.

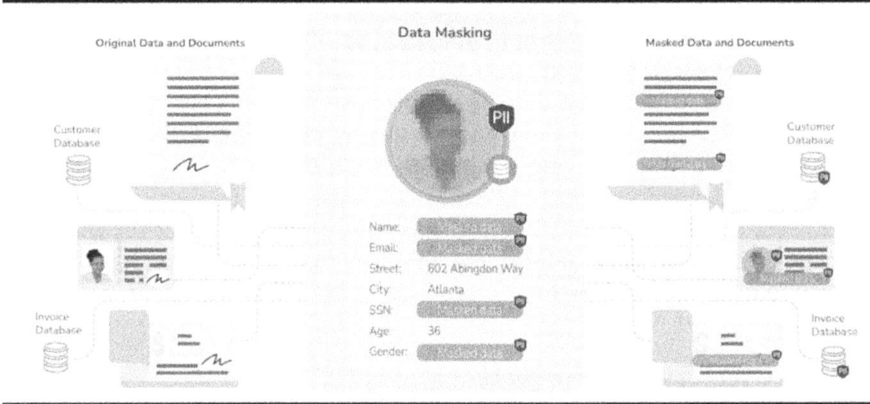

Figure 15.1 Data masking.

The fundamental principle behind data masking is to take the original data and turn it into a version that is still formatted and structurally sound but does not include any sensitive information that may be used to identify a specific person.

Data masking is replacing the original data with a modified version that seems realistic but doesn't include any crucial information [15]. For instance, a randomly generated number with the exact same format as a legitimate Social Security Number (SSN) could be used in place of an SSN. Similarly, a credit card number might be replaced with a randomly generated number that follows the same format as a real credit card number, including the same number of digits and the same checksum algorithm.

Data masking's main objective is to produce an untrue version of the data while maintaining the dataset's structural integrity. The way data masking technologies operate is by employing various data manipulation or shuffling techniques to replace sensitive values in a dataset with randomized values. This procedure safeguards the sensitive data while enabling its use for activities such as software testing and user training. Since data masking can be reversed, unlike data anonymization, it is still regarded as PII for several laws, including the GDPR.

Data masking can be broadly classified into static and dynamic data masking. Static data masking is a technique used in mobile data anonymization to protect sensitive information by replacing or masking static identifiers with fictional or anonymous values. An example of the static data masking is shown below:

Original data	Masked data
Jordan Philip	Anonymous User
1234567890	XXX-XXX-XXXX

Dynamic data masking is a technique used in mobile data anonymization to protect sensitive information by masking or altering dynamic data in real-time, as it is generated or transmitted.

Both static and dynamic data masking ensures that personal data is not exposed, even in cases where data is constantly changing or being updated.

Data masking is not a one-size-fits-all solution, and the best method to employ will depend on the particular use case and privacy requirements. Tokenization, character shuffles, and substitution are a few of the most used data masking strategies. Selecting a method that strikes the right balance between data utility and privacy is crucial.

15.4.1.1 Use-cases of data masking

There are numerous use cases for data masking, including:

1. **Achieving and Maintaining Compliance:** Data masking for PII and other sensitive data, such as financial or medical records, is required by data protection laws including GDPR, HIPAA, GLBA, PCI DSS, and others.
2. **Controlling Internal Access:** To ensure that employees who don't need access to sensitive information can't get it, organizations utilize data masking internally.
3. **Development Acceleration:** DevOps requires functioning datasets for its ongoing testing, however, version releases are slowed down when PII is manually removed, as it can be laborious and time-consuming. Development can proceed more quickly and effectively thanks to data masking.

15.4.1.2 Advantages of data masking

1. Data masking facilitates the usage of the original data for testing, training, and development by maintaining its format and structure [10].
2. Many different types of data, such as unstructured, semi-structured, and structured data, can be protected with data masking.
3. Because data masking can be automated, it is simple to apply to big datasets and to keep applying it whenever new data is added to a system.
4. Data masking can assist businesses in adhering to laws governing data privacy, such as GDPR and HIPAA [16].

15.4.1.3 Disadvantages of data masking

1. Data masking can be labor-intensive, including a large amount of manual labor in addition to computer resources
2. If the attacker has access to other information that might be used to re-identify the data, data masking might not be able to stop re-identification attempts [16].

3. If the masking procedure is not carried out appropriately or if the masked data is not adequately secured, data masking may be susceptible to re-identification attacks.
4. Data masking may lead to a loss of usefulness for the data, which makes it more challenging to analyze the data and draw conclusions from it [16].

To sum up, data masking is an effective data anonymization method that enables businesses to safeguard sensitive information as well as to safeguard sensitive data while maintaining data usefulness, data privacy, and legal compliance. Organizations can accomplish their goals while protecting the personal information of their clients by employing data masking.

However, it is important to carefully consider the trade-offs between data privacy and data utility when implementing data masking techniques and to ensure that the masking process is performed correctly and that the masked data is properly secured.

15.4.2 Data pseudonymization

The technique of hiding direct identities in a data set by substituting "pseudonyms" or "fictional" identifiers is known as data pseudonymization [5, 8]. Figure 15.2 shows an example of pseudonymization. Pseudonymization is intended to be reversible and does not take indirect identifiers into account [17].

Data pseudonymization is a data anonymization technique that replaces direct identifiers in a dataset with artificial identifiers, or pseudonyms, while keeping the indirect identifiers intact [17, 18]. This technique allows for the data to be re-identified if the additional information, or the "key" to the pseudonyms, is available.

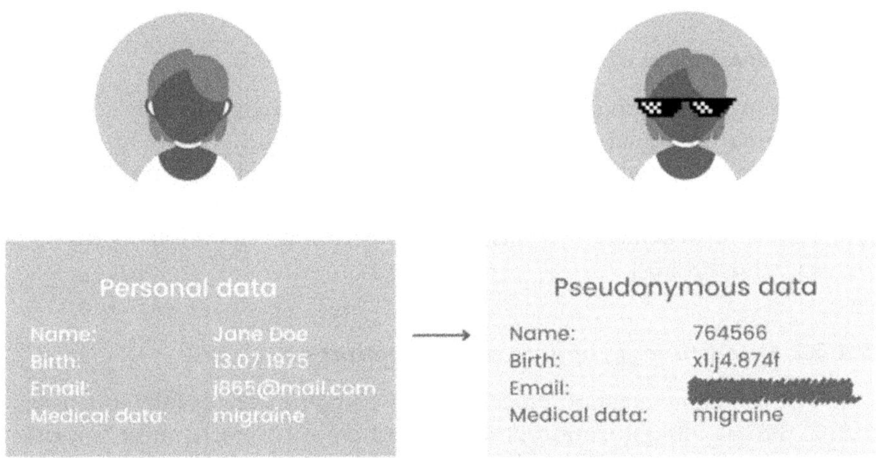

Figure 15.2 Example of data pseudonymization.

Pseudonymization is considered less restrictive than anonymization, as the GDPR recognizes that pseudonymous data can still be personal data if the key to re-identify the data is held by the data controller. However, pseudonymization can still provide a level of protection for personal data and help organizations comply with data protection regulations.

When processing data is required for particular goals but the organization prefers to restrict access or stop unwanted re-identification, pseudonymization may be helpful. Pseudonymization, for instance, can be utilized in the medical field to safeguard patient information while still enabling essential medical services. In order to target particular audiences without disclosing personal information, it can also be utilized in marketing.

While there are many ways to pseudonymize data, they all generally entail mapping the original data to the pseudonyms. The mapping key is then stored apart from the pseudonymized data, usually behind locked doors or with restricted access.

It is important to remember that, even though pseudonymization can shield sensitive information to some extent, anonymization cannot be guaranteed. Re-identifying the data would still be feasible in some circumstances, particularly if the pseudonyms' secret key is exposed or the indirect identifiers are distinct [18]. Thus, while choosing to employ pseudonymization as a data anonymization approach, it's crucial to take the particular use case and degree of risk into account.

It is worth understanding that data anonymization, or the process of eliminating all identifying information from a dataset, differs from data pseudonymization. While anonymization prevents data from being tracked back to the original identifying information, pseudonymization permits this. Furthermore, depending on the particular use case, extra privacy protection methods can be required in addition to data pseudonymization in order to safeguard the privacy of individuals inside a dataset.

15.4.2.1 Advantages of data pseudonymization

1. Improved privacy protection compared to leaving directly identifying information intact [19].
2. Allows for data to be linked back to the original identifying information if necessary, allowing for more flexible use of the data.
3. Can help organizations meet data protection regulations, such as the EU's GDPR [20].

15.4.2.2 Disadvantages of data pseudonymization

1. The pseudonymization process can be complex and resource-intensive.
2. If the pseudonymization process is not done correctly, there is a risk of re-identification of the individuals in the dataset [21].

3. The privacy of the individuals in the dataset may be jeopardized if the extra data required to connect the pseudonymized data to the original identifying information is compromised [21].

15.4.3 Data generalization

Data generalization is a kind of data anonymization technique that involves compressing data while maintaining its key characteristics and patterns. It is the process of taking data and using it to create a more generic or abstract representation. Different techniques like dimensionality reduction, sampling, and clustering can be used to accomplish this [22].

A data generalization may involve combining or aggregating data points to reduce the amount of detail in a dataset. Data is grouped or clustered using data generalization, which is based on the similarities between data points [23]. This substitutes a more generalized representation for particular values, hence reducing the complexity of the data. For instance, data generalization might aggregate ages into ranges like "20–30," "31–40," and so on rather than utilizing exact ages.

In essence, data generalization "zooms out" on a data collection, providing a more comprehensive perspective of its contents and lessening the capacity to identify specific traits and qualities. Large datasets to guarantee that the data is sufficiently ambiguous without losing its usefulness are ideal for data generalization.

In a dataset, data generalization can be used for both direct and indirect identifiers [23]. For instance, a data generalization strategy can employ a wider age group or a range of birth years in place of a given date of birth. Similarly, a generalization strategy can utilize a more general geographical identifier, like a city or zip code, in place of a specific address.

The amount of detail that can be eliminated from a dataset by different generalization approaches varies. Data may be grouped into broader categories, for instance, by a 1-level generalization, and even bigger categories by a 2-level generalization. The use case and the degree of risk involved in re-identification determine the proper degree of generalization.

When data needs to be analyzed statistically but the organization wants to minimize the possibility of re-identifying individuals in the dataset, data generalization might be helpful. For instance, generalization can be used to analyze consumer behavior in market research while maintaining privacy for everyone.

Data generalization, however, can also result in a dataset having less information, which may affect the data's correctness and utility. Therefore, while choosing to employ data generalization as a data anonymization technique, it's crucial to consider the particular use case and the trade-off between privacy and data utility.

The k-means algorithm can be used to implement data generalization. K-means implementation is available in many open-source libraries such as "scikit-learn" which provides an implementation of k-means in Python.

Data generalization is a strategy for data anonymization that entails combining or aggregating data points together to reduce the level of detail in a dataset. This method can not only affect the quality and usefulness of the data but also help prevent individual re-identification while preserving the overall statistical characteristics of the data intact. When choosing to employ data generalization as a data anonymization approach, it's critical to take the particular use case and degree of risk into account [23].

15.4.3.1 Advantages of data generalization

1. One of the main advantages of data generalization is that it can help protect individual privacy. By replacing specific values with more generalized representations, it becomes more difficult to identify individual data points. This is particularly useful when sharing data with third parties who may not have the same privacy concerns as the original data holder.
2. Another advantage of data generalization is that it can make data analysis easier and more efficient. By reducing the complexity of the data, it becomes easier to identify patterns and trends. This can be especially useful in large datasets where specific data points may be difficult to distinguish.

15.4.3.2 Disadvantages of data generalization

1. One of the main disadvantages is that it can lead to a loss of information. By replacing specific values with more generalized representations, some information is inevitably lost. This can result in less accurate or precise analysis.
2. Another limitation of data generalization is that it may not always be effective in protecting individual privacy. In some cases, it may still be possible to identify individual data points even after data generalization has been applied.

15.4.4 Data perturbation

Data perturbation is the kind of data anonymization that purposefully randomizes data pieces to introduce ambiguity in a data collection in a predictable and restorable manner without compromising analytical accuracy.

Data perturbation adds a certain amount of random noise or distortion to the data in order to mask PII while maintaining the data's usefulness for analytics and other uses [24]. With this method, it is possible to analyze trends and patterns in a dataset without identifying specific individuals. However, it is crucial to carefully assess the amount of noise introduced to the data and to combine this technique with other data anonymization methods.

It is possible to use data perturbation on both categorical and numerical data. When dealing with numerical data, for instance, it is possible to introduce a small amount of random noise into the data, which will make it challenging to identify certain persons but still permit insightful analysis. Data perturbation can be used to alter categorical data in a way that conceals individual identities while maintaining the data's statistical features.

Whenever sharing data between organizations or for research purposes is necessary but the revelation of personal information is undesirable, data perturbation is a helpful technique [25]. To add an extra degree of security for sensitive data, this method can potentially be combined with other data anonymization strategies like data masking and data generalization.

15.4.4.1 Disadvantages of data perturbation

Data perturbation does have certain restrictions including:

1. Carefully calibrating the level of noise supplied to the data is necessary to maintain its analytical value while maintaining an adequate level of identity protection.
2. If additional information about the individuals in the dataset is available or if the quantity of noise added is too little, it might be possible in some circumstances to reverse-engineer the original data from the perturbed data.

15.4.5 Data swapping

As a method of data anonymization, data swapping involves modifying the values of two or more entries in a dataset in order to safeguard private information while maintaining the overall usefulness of the data. To make sure that individual records cannot be linked to their original sources, this strategy entails choosing two or more records and switching one or more of their properties. Using this data anonymization method, an attribute from one row in the same column may be substituted with an attribute from a different row [26].

Hence, data swapping is the process of rearranging data in a dataset so that attribute values no longer match the original data. Swapping is a useful method for maintaining the dataset's statistical characteristics while protecting individual privacy. It is especially helpful in situations when the data must be kept valuable overall yet contains sensitive information that needs to be secured. For instance, swapping the income levels of two people from different regions can guarantee that their income information stays private while yet enabling analysis of income levels by location [27].

Both numerical and categorical data types can be switched [28]. However, it's crucial to make sure the data is not distorted or become biased because of the switching procedure. To maintain the overall data distribution and statistical features, swapping needs to be done methodically and properly.

To add extra levels of privacy, swapping can be used in conjunction with other data anonymization strategies like data masking or data generalization. Organizations can guarantee that their data is sufficiently protected while retaining its usefulness for analysis and other purposes by combining several data anonymization strategies.

15.4.6 Synthetic data

The process of creating generated data that closely matches the real data without containing any sensitive or private information is known as synthetic data anonymization. Although synthetic data is artificially created, it closely resembles real sensitive data. This method preserves the privacy of the individuals in the original dataset while producing new datasets that can be shared with other parties or used for testing and ML model training.

The synthetic data is widely utilized for model training and validation in ML and is typically created by algorithms [29]. Algorithms that build new data points based on statistical patterns identified in the original data are used to create synthetic data. Text, category, and numerical data are just a few of the many sorts of data that these algorithms may produce. The resulting synthetic data usually has the same distribution and structure as the original data, but any identifiable or sensitive information has been altered or eliminated.

15.4.6.1 Use cases of synthetic data

1. **Healthcare Industry:** In industries like healthcare, where access to actual patient data may be restricted or limited, this might be extremely helpful.
2. **Data Sharing with Partners:** Organizations can exchange information with partners or vendors without disclosing any sensitive or identifying information by creating synthetic datasets that closely mimic the real data.

15.4.6.2 Advantages of synthetic data

1. The ability to generate realistic datasets for training and testing without requiring access to or sharing actual user data is one of the main advantages of synthetic data. For instance, in the healthcare industry, researchers and developers can still test software applications and train ML models by using synthetic data without compromising the privacy of actual patients,
2. The ability to exchange data with other organizations or third parties without breaking privacy laws or jeopardizing personal information is another benefit of synthetic data.

15.4.6.3 Disadvantages of synthetic data

1. It is crucial to remember that synthetic data is not always an exact match for authentic user data.
2. Sometimes the original data complexity or nuance is not fully captured by the synthetic data, which can cause problems with the application (in fact, ML model) performance or accuracy.

3. Improper generation of the synthetic data could allow for the re-identification of persons or the deduction of sensitive information.

15.5 MOBILE DATA ANONYMIZATION

Let's say a mobile app company wants to collect location data from their users to provide location-based services and improve their app's user experience. However, the company is concerned about protecting the privacy of its users and wants to anonymize the location data before storing or sharing it.

To do this, the business could quietly modify the location data before storing or distributing it using data perturbation techniques. For instance, the business may add some random noise to the location data, which would make it less accurate but still helpful for location-based services, in place of maintaining the specific coordinates. By using this method, attackers may be prevented from using users' exact location information to identify specific individuals.

Furthermore, the organization may employ data-swapping strategies to enhance the anonymity of the location data. The business might, for instance, switch the location information of two users in a way that maintains the distribution of location information overall but makes it challenging to identify specific users. This technique can help prevent attackers from identifying individual users based on their location data.

However, both data perturbation and data swapping techniques have their own advantages and disadvantages. Data perturbation can make it difficult to identify individual users but can also make the data less precise and useful. Data swapping can help protect user privacy but can be computationally expensive and may require careful implementation to ensure that the data remains useful.

As a result, while utilizing data anonymization techniques, mobile app companies must carefully consider the trade-offs between privacy and utility. They should also consider additional methods to further safeguard user privacy while preserving the value of the data, such as data aggregation and differential privacy.

15.6 MOBILE DATA ANONYMIZATION IN CYBER SECURITY

15.6.1 Mobile data anonymization

An important part of cybersecurity in the age of mobile devices and apps is mobile data anonymization. Sensitive data is generated and communicated over mobile devices at an increasing rate due to the variety of personal and professional uses for these devices. Attackers may use this information to carry out many types of cyberattacks, including phishing, identity theft, and spying.

By altering or eliminating PII from the data, mobile data anonymization can help safeguard user privacy and stop illegal access to critical information [30]. This method can reduce the chance of privacy violations while ensuring that the data is still valuable for analytical purposes.

15.6.2 Why mobile data anonymization in cyber security?

Here are some reasons why mobile data anonymization is important in cybersecurity:

15.6.2.1 Protecting user privacy

Mobile devices often contain sensitive personal information, such as location data, financial information, and health records. Anonymization techniques can help protect user privacy by removing or modifying this information before storing or sharing it [31].

15.6.2.2 Preventing identity theft

By anonymizing mobile data, companies can prevent attackers from using the data to impersonate users and conduct various cyberattacks, such as phishing and identity theft [32].

15.6.2.3 Compliance with data protection regulations

Data protection laws are in place in many nations, and they oblige businesses to anonymize personal data and preserve customer privacy. Businesses can abide by these rules and stay out of trouble by using anonymization strategies [5, 7].

15.6.2.4 Improving data analytics

Mobile data anonymization can help improve data analytics by allowing companies to use large datasets without compromising user privacy. By removing PII, companies can use the data for various analytical purposes, such as market research, product development, and user behavior analysis [33].

15.6.2.5 Reducing cyberattack surface

By anonymizing mobile data, companies can reduce the cyberattack surface and make it more difficult for attackers to conduct targeted attacks. By removing or modifying sensitive information, companies can prevent attackers from using the data for reconnaissance and other malicious purposes [34].

Mobile data anonymization is a critical component of cybersecurity in the era of mobile devices and applications. By implementing anonymization techniques, companies can protect user privacy, prevent unauthorized access to sensitive data, comply with data protection regulations, improve data analytics, and reduce the cyberattack surface. Therefore, it's essential for companies to consider mobile data anonymization as part of their overall cybersecurity strategy [32, 34].

15.6.3 Enhancing personal privacy and regulatory compliance through mobile data anonymization

By altering or eliminating PII from mobile data before keeping or distributing it, a process known as "mobile data anonymization" helps protect individual privacy. By reducing the possibility of privacy violations, this method can assist in guaranteeing that the data is still valuable for analytical reasons [5, 7, 32].

The following are some ways that mobile data anonymization might support data privacy laws and safeguard individual privacy:

15.6.3.1 Removing or modifying PII

Mobile data anonymization techniques can help remove or modify PII, such as names, phone numbers, addresses, and other identifiers, from mobile data. This technique can help prevent attackers from using the data to identify individuals and conduct targeted attacks [5, 7].

15.6.3.2 Data aggregation

To make it harder to identify specific persons, data aggregation may be used as part of mobile data anonymization. For instance, businesses can aggregate the data to display population density or other statistical information rather than retaining individual location data points.

15.6.3.3 Differential privacy

By introducing noise into the data, this strategy safeguards the privacy of each individual. By introducing uncertainty into the data, this strategy can aid in preventing attackers from identifying specific individuals [35].

15.6.3.4 Complying with data protection regulations

Data protection laws are implemented in many nations, requiring businesses to anonymize personal data and preserve user privacy. By using anonymization strategies, businesses can comply with these rules and stay out of legal trouble [5, 7].

15.6.3.5 Building user trust

By implementing mobile data anonymization techniques, companies can build trust with users by demonstrating their commitment to protecting user privacy. This can help improve user engagement and loyalty.

One essential tactic for maintaining individual privacy and adhering to data privacy laws is mobile data anonymization. Companies can make sure that mobile data is used responsibly and ethically by eliminating or changing PII, aggregating data, adopting differential privacy, adhering to data protection laws, and cultivating user trust. As a result, businesses must incorporate mobile data anonymization into their overall data protection plan.

15.6.4 How different data anonymization techniques can be applied in cyber security contexts

The process of changing or eliminating PII from data in order to preserve individual privacy is known as data anonymization. Data anonymization techniques are employed in cybersecurity situations to safeguard sensitive information from theft, unauthorized access, and misuse. The following list of popular data anonymization methods includes examples of their use in cybersecurity:

15.6.4.1 Using a pseudonym—data pseudonymization

PII is substituted with aliases or pseudonyms during the pseudonymization process. Data warehouses and databases containing sensitive information can be secured with this method. Companies can reduce the risk of privacy breaches while still utilizing the data for analysis and other reasons by substituting pseudonyms for PII.

15.6.4.2 Data masking

Data masking is the process of substituting non-sensitive data for sensitive data. For instance, credit card numbers can be substituted with fictitious numbers as long as they have the same length and structure. Sensitive data can be safeguarded using this method while it is being tested or developed [14].

15.6.4.3 Data aggregation

The process of combining data sets makes it more challenging to identify specific persons. For instance, businesses can aggregate the data to display population density or other statistical information rather than retaining individual location data points. Sensitive data, including location data, can be safeguarded using this method.

15.6.4.4 Data perturbation

To preserve individual privacy, data perturbation includes introducing noise or randomness to the data. For instance, introducing random noise to the coordinates can skew location data. Sensitive data, including location data, can be safeguarded using this method.

15.6.4.5 Differential privacy

By introducing noise into the data, this strategy safeguards the privacy of each individual. Data warehouses and databases containing sensitive information can be secured with this method. Businesses can stop attackers from utilizing their data to identify specific persons by introducing noise into the data [35].

15.6.4.6 Cryptography

To prevent unwanted access to data, cryptography uses mathematical techniques. This technique can be used to protect sensitive data in transit or at rest. By encrypting data, companies can ensure that only authorized users can access the data.

Data anonymization techniques are critical for protecting sensitive data in cybersecurity contexts. By using techniques such as pseudonymization, data masking, data aggregation, data perturbation, differential privacy, and cryptography, companies can protect sensitive data from unauthorized access, theft, or misuse. Therefore, it's essential for companies to consider data anonymization as part of their overall cybersecurity strategy.

15.7 USE CASES OF MOBILE DATA ANONYMIZATION

An essential component of cybersecurity in the era of mobile devices and applications is mobile data anonymization. Businesses can guard sensitive data from theft, unauthorized access, and misuse by utilizing strategies including data aggregation, data masking, encryption, and data perturbation. As a result, businesses must incorporate mobile data anonymization into their overall cybersecurity plan. Below are some examples of real-world applications and use cases of mobile data anonymization in cybersecurity:

15.7.1 Healthcare mobile apps

Mobile healthcare apps such as telemedicine apps can collect and store sensitive health data from patients and can use anonymization techniques to protect patient privacy [36]. For instance, a health app could use data masking to replace patient names and other PII with pseudonyms [36]. Mobile data anonymization techniques, such as data aggregation and pseudonymization can be also used to protect this data from unauthorized access or theft.

15.7.2 Mobile payment apps

Mobile payment apps and platforms such as Google Pay, Apple Pay, etc. can use data anonymization to protect user financial information from theft or unwanted access. For example, a mobile payment app could use data generalization to replace specific user account numbers with more general information, such as the last four digits of the account number [37].

15.7.3 Mobile advertising

To target consumers with ads, mobile advertising organizations frequently gather data from mobile devices. To prevent theft or unwanted access, mobile advertising networks can use anonymization techniques like data aggregation and pseudonymization to protect user data while still delivering targeted ads [38]. For instance, an ad network could use data perturbation to add random noise to user location data, making it more difficult to pinpoint a user's exact location while still providing useful location-based targeting.

15.7.4 Mobile gaming

Mobile games can use data anonymization to protect user data while still delivering personalized gaming experiences. For example, a mobile game could use synthetic data generation to create fake user profiles for testing and training purposes, rather than using real user data.

15.7.5 Mobile browsing

Mobile browsers can use data anonymization to protect user browsing data. For instance, a mobile browser could use data pseudonymization to replace user IP addresses with pseudonyms, making it more difficult to trace user activity back to a specific individual [39].

15.7.6 Location-based services

Mobile devices are frequently used by location-based services, such as ridesharing and navigation apps, to gather location data. To prevent theft or unwanted access, mobile data anonymization techniques including data aggregation and perturbation can be applied [12].

15.7.7 Mobile device management

In business environments, mobile devices are managed and secured using Mobile Device Management (MDM) solutions. Sensitive data on mobile devices can be safeguarded with MDM solutions by utilizing mobile data anonymization techniques like data masking and encryption.

15.7.8 Mobile identity management

To authorize and authenticate users on mobile devices, mobile identity management systems are utilized. Sensitive user data can be safeguarded by mobile identity management systems using mobile data anonymization techniques like data masking and encryption [40].

15.8 FUTURE DIRECTIONS AND CHALLENGES

15.8.1 Emerging trends in mobile data anonymization

Here are some emerging trends in mobile data anonymization:

Machine Learning Integration: ML algorithms are being integrated into mobile data anonymization tools to automate the identification and anonymization of sensitive information [41].

Differential Privacy: Differential privacy is a method for gathering and evaluating information while maintaining privacy. To stop individual users from being identified, noise is added to the data [35, 42].

Homomorphic Encryption: The technology known as homomorphic encryption makes it possible to encrypt and analyze data without having to first decrypt it. This preserves sensitive information while enabling analysis to be done with it [43].

Federated Learning: This method eliminates the need for centralization by enabling decentralized data analysis across several servers or devices. This keeps data analysis possible while also helping to safeguard user privacy [44].

Real-Time Anonymization: This type of anonymization entails the generation or collection of data while it is still anonymous. In real-time applications like mobile gaming, live streaming, and Internet of Things (IoT) devices, this can assist in safeguarding user privacy [45].

Data Masking: Data masking is the process of substituting non-sensitive data with sensitive data while maintaining the structure of the original data. This preserves sensitive information while enabling analysis to be done with it [14].

Synthetic Data Generation: This process entails producing fictitious data that mimics the appearance and behavior of real data. This keeps data analysis possible while also helping to safeguard user privacy [46].

The overall goal of these new developments in mobile data anonymization is to increase the security and privacy of mobile data while maintaining the ability to analyze and get insights from it. As mobile data becomes increasingly valuable, it is important for organizations to prioritize data privacy and security to build trust with their customers and comply with regulations.

15.8.2 Future directions in mobile data anonymization research and development

Based on the current state of mobile data anonymization research and development, here are some potential future directions:

Improved User Control: As users become more aware of the importance of data privacy, there may be a growing demand for mobile data anonymization tools that give users more control over their data. This could include tools that allow users to selectively share data with third parties or that provide more transparency around how data is being used.

Integration with IoT: Mobile devices are frequently used to operate IoT devices, which are becoming more and more common [45]. As a result, the necessity for mobile data anonymization solutions that can safeguard the confidentiality of data produced by IoT devices is increasing.

Integration with AI: In order for AI algorithms to operate well, a lot of data is frequently needed. Mobile data anonymization technologies that can safeguard the privacy of data utilized in AI applications are therefore becoming more and more necessary [41, 46].

Real-time Anonymization: With the growing number of real-time applications for mobile devices, including live streaming, mobile gaming, and IoT devices, real-time anonymization is becoming more and more crucial. Future research in this area could focus on developing more efficient and effective real-time anonymization techniques [45].

Enhanced Privacy-Preserving Techniques: Although homomorphic encryption and differential privacy are two popular methods for anonymizing mobile data, more sophisticated methods are still required [11]. Subsequent investigations may concentrate on creating novel methods capable of offering even more robust privacy assurances [12].

Compliance with Regulations: With the increasing strictness of privacy regulations, there is a growing demand for mobile data anonymization solutions that can assist businesses in adhering to these laws. Subsequent investigations in this domain may concentrate on creating instruments that assist establishments in mechanizing the data anonymization procedure and guarantee adherence to statutes like the CCPA and GDPR [2, 3, 5, 7].

Overall, several factors, including technological advancements, user behavior shifts, and the changing constitutional surroundings, are likely to determine the future of mobile data anonymization. In order to safeguard user security and privacy, academics and practitioners must keep creating and improving methods for anonymizing mobile data as it becomes more and more useful.

15.8.3 Potential challenges of mobile data anonymization in cyber security

Here are some potential challenges of Mobile Data Anonymization in cyber security:

Trade-off between Privacy and Utility: Mobile data anonymization may make it less useful for analysis and insights. One of the main challenges in mobile data anonymization is striking the correct balance between privacy and usefulness.

Limited Computational Resources: Mobile devices have limited computational resources, which can make it challenging to implement complex anonymization techniques [12].

Heterogeneous Mobile Data: Mobile data is often heterogeneous and can come from different sources, making it challenging to apply consistent anonymization techniques.

Dynamic Mobile Data: It can be dynamic and change frequently, making it challenging to apply consistent anonymization techniques.

Lack of Standardization: There is a lack of standardization in mobile data anonymization techniques, making it difficult to compare and evaluate different approaches. This can also make it challenging to ensure that the anonymization techniques are effective and comply with regulations [6].

User Identity: It can be difficult to guarantee total privacy because anonymized data can occasionally be re-identified by merging it with other data sources.

Attacks on Anonymization Techniques: Anonymization techniques can be attacked by malicious actors, who can use various techniques to de-anonymize the data [12].

Data Quality: Anonymization methods may have an impact on the quality of the data, which makes it difficult to guarantee accurate and trustworthy analysis and insights.

Compliance with Regulations: Anonymization methods have to go by a number of laws, including GDPR and HIPAA, which can make the process more difficult and expensive [5, 7].

Ethical Considerations: Anonymization techniques must consider ethical considerations, such as user consent and transparency, to ensure that users' privacy is respected [5, 7].

Overall, these challenges highlight the need for ongoing research and development in mobile data anonymization techniques to ensure that they are effective, efficient, and ethical.

15.8.4 Identify limitations of mobile data anonymization in cyber security

Here are some limitations of mobile data anonymization in cyber security:

Risk of Re-identification: The possibility of the data being linked to other data sources and becoming re-identified exists even after anonymization. This phenomenon—referred to as the "mosaic effect"—occurs when several anonymous data points are pieced together to disclose a person's identity.

Loss of Data Utility: Using anonymized data for analytics or other purposes may become challenging if anonymization processes lead to a loss of data utility. For instance, data masking might eliminate crucial context, while data perturbation can lead to data inaccuracies.

Limited Scope: Mobile data anonymization techniques may not be effective in all contexts or for all types of data. For example, some data may be too sensitive to anonymize, while other data may not be suitable for certain anonymization techniques.

Complexity and Cost: Implementing mobile data anonymization techniques can be complex and costly, requiring significant resources and expertise. This can be a barrier for smaller organizations or for those with limited resources.

User Acceptance: Mobile data anonymization techniques may not be accepted by users, who may be concerned about the privacy implications of anonymization. This can make it challenging to implement anonymization techniques in a way that is both effective and user-friendly.

15.8.5 Recommendations for future research and development in mobile data anonymization techniques and applications

Here are some recommendations for future research and development in mobile data anonymization techniques and applications:

- Create more sophisticated anonymization methods that can strike a compromise between data utility and privacy. Present methods, such as pseudonymization and data masking, can occasionally lead to a loss of data utility. As a result, new methods that maintain data utility while maintaining privacy protection must be developed.
- Examine how AI and ML might be used to enhance anonymization methods. In addition to being trained to find patterns in data, ML algorithms may be utilized to create more advanced anonymization methods.
- Examine how blockchain technology can be used to anonymize data. Data anonymization with blockchain technology can be transparent and safe. By using blockchain, it is possible to create a decentralized

and distributed system for data anonymization that does not rely on a central authority.

- Examine how data anonymization affects ML and data analytics. Data anonymization can sometimes make ML and analytics more challenging. Consequently, it's critical to research how anonymization affects these domains and create strategies that can lessen its effects on ML and data analytics.
- Develop standardized anonymization techniques that can be widely adopted. Currently, there are no standardized anonymization techniques that are widely adopted. Developing standardized techniques can help ensure that data is anonymized consistently and can also make it easier to compare and evaluate different anonymization techniques.
- Investigate the potential of homomorphic encryption for data anonymization. Homomorphic encryption is a technique that allows data to be encrypted and analyzed without being decrypted. By using homomorphic encryption, it is possible to anonymize data while still allowing it to be analyzed.
- Explore the use of differential privacy to the anonymization of data. A technique called Differential Privacy involves adding noise to data in order to safeguard privacy. Differential privacy can be used to anonymize data while preserving its analytical capabilities.
- Explore how data anonymization affects cybersecurity and data breaches. In the case of a data breach, anonymization can aid in data protection. Studying how anonymization affects cybersecurity and data breaches is crucial as a result.
- Consider federated learning applications to data anonymization. A method called federated learning makes it possible to analyze data decentralized, spanning several servers or devices. Federated learning makes it feasible to anonymize data without compromising its analytical potential.
- Develop user-friendly anonymization tools that can be easily used by non-technical users. Currently, many anonymization tools require advanced technical skills to use. Developing user-friendly tools can help ensure that data is anonymized consistently and can also make it easier for non-technical users to protect their privacy.

REFERENCES

1. Klose, M., Desai, V., Song, Y., & Gehringer, E. (2020). EDM and Privacy: Ethics and Legalities of Data Collection, Usage, and Storage. International Educational Data Mining Society.
2. Naqvi, S. K. H., & Batool, K. (2023). A comparative analysis between general data protection regulations and California consumer privacy act. Journal of Computer Science, Information Technology and Telecommunication Engineering, 4(1), 326–332.

3. Blanke, J. M. (2020). Protection for 'Inferences drawn': a comparison between the general data protection regulation and the California consumer privacy act. Global Privacy Law Review, 1(2).

4. Malekzadeh, M., Clegg, R. G., Cavallaro, A., & Haddadi, H. (2019, April). Mobile sensor data anonymization. In Proceedings of the International Conference on Internet of Things Design and Implementation (pp. 49–58).

5. Štarchoň, P., & Pikulík, T. (2019). GDPR principles in data protection encourage pseudonymization through most popular and full-personalized devices-mobile phones. Procedia Computer Science, 151, 303–312.

6. Neves, F., Souza, R., Sousa, J., Bonfim, M., & Garcia, V. (2023). Data privacy in the Internet of Things based on anonymization: a review. Journal of Computer Security, 31(3), 261–291.

7. Murati, E., & Henkoja, M. (2019). Location data privacy on MaaS under GDPR. European Journal of Privacy Law & Technologies, 2(2), 115.

8. Marques, J. F., & Bernardino, J. (2020). Analysis of Data Anonymization Techniques. In KEOD (pp. 235–241).

9. Murthy, S., Bakar, A. A., Rahim, F. A., & Ramli, R. (2019, May). A comparative study of data anonymization techniques. In 2019 IEEE 5th Intl Conference on Big Data Security on Cloud (BigDataSecurity), IEEE Intl Conference on High Performance and Smart Computing,(HPSC) and IEEE Intl Conference on Intelligent Data and Security (IDS) (pp. 306–309). IEEE.

10. In, G. K. (2024, January 2). Data anonymization: Pros, cons & techniques in 2024. AIMultiple: High tech use cases &Amp; Tools to Grow Your Business. https://research.aimultiple.com/data-anonymization/

11. Rajan, A. A., & Rajan, A. A. (2020). Data anonymization techniques for preserving privacy in public release data model: a technical review. International Journal of Scientific Research in Computer Science and Engineering, 8(1).

12. Majeed, A., & Lee, S. (2020). Anonymization techniques for privacy preserving data publishing: a comprehensive survey. IEEE Access, 9, 8512–8545.

13. Puri, V., Kaur, P., & Sachdeva, S. (2021). ADT: anonymization of diverse transactional data. International Journal of Information Security and Privacy (IJISP), 15(3), 83–105.

14. Tachepun, C., & Thammaboosadee, S. (2020, July). A data masking guideline for optimizing insights and privacy under GDPR compliance. In Proceedings of the 11th International Conference on Advances in Information Technology (pp. 1–9).

15. Zhuo, W., Guowei, L. I. U., Yan, W., & Yuan, L. I. (2020). Research on the development and trend of data masking technology. Information and Communications Technology and Policy, 46(4), 18.

16. Baig, A. (2023, October 27). Data masking best practices and benefits - dataversity. Dataversity. https://www.dataversity.net/data-masking-best-practices-and-benefits/

17. Dimopoulou, S., Symvoulidis, C., Koutsoukos, K., Kiourtis, A., Mavrogiorgou, A., & Kyriazis, D. (2022, February). Mobile anonymization and pseudonymization of structured health data for research. In 2022 Seventh International Conference On Mobile and Secure Services (MobiSecServ) (pp. 1–6). IEEE.

18. Al-Zubaidie, M., Zhang, Z., & Zhang, J. (2019). PAX: using pseudonymization and anonymization to protect patients' identities and data in the healthcare system. International Journal of Environmental Research and Public Health, 16(9), 1490.

19. Rahim, Y. A., Sahib, S., & Abd Ghani, M. K. (2011, September). Pseudonmization techniques for clinical data: privacy study in Sultan Ismail Hospital Johor Bahru. In 7th International Conference on Networked Computing (pp. 74–77). IEEE.

20. Limniotis, K., & Hansen, M., & European Union Agency for Network and Information Security. (2019). Pseudonymisation: benefits and requirements

under GDPR. In A. Bourka & P. Drogkaris (Eds.), Recommendations on Shaping Technology According to GDPR Provisions. https://www.anonos.com/hubfs/ENISA_Pseudonymisation_Recomendations_GDPR_November_2018.pdf

21. What are the Differences Between Anonymisation and Pseudonymisation | Privacy Company Blog. (n.d.). https://www.privacycompany.eu/blog/what-are-the-differences-between-anonymisation-and-pseudonymisation

22. Yaseen, S., Abbas, S. M. A., Anjum, A., Saba, T., Khan, A., Malik, S. U. R., & Bashir, A. K. (2018). Improved generalization for secure data publishing. IEEE Access, 6, 27156–27165.

23. Devane, H. (2023b, April 5). What is data generalization? A complete overview. Immuta. https://www.immuta.com/blog/what-is-data-generalization/

24. Kargupta, H., Datta, S., Wang, Q., & Sivakumar, K. (2003, November). On the privacy preserving properties of random data perturbation techniques. In Proceedings of the Third IEEE International Conference on Data Mining (pp. 99–106). IEEE.

25. Kiran, A., & Shirisha, N. (2022). K-anonymization approach for privacy preservation using data perturbation techniques in data mining. Materials Today: Proceedings, 64, 578–584.

26. Eyupoglu, C., Aydin, M. A., Zaim, A. H., & Sertbas, A. (2018). An efficient big data anonymization algorithm based on chaos and perturbation techniques. Entropy, 20(5), 373.

27. Gunawan, D., & Mambo, M. (2019). Data anonymization for hiding personal tendency in set-valued database publication. Future Internet, 11(6), 138.

28. Satori Cyber. (2022, November 22). Data anonymization: use cases and 6 common techniques - Satori. Satori. https://satoricyber.com/data-masking/data-anonymization-use-cases-and-6-common-techniques/

29. Yoon, J., Drumright, L. N., & Van Der Schaar, M. (2020). Anonymization through data synthesis using generative adversarial networks (ADS-GAN). IEEE Journal of Biomedical and Health Informatics, 24(8), 2378–2388.

30 Yang, X., & Ardakanian, O. (2023, November). Privacy through diffusion: a white-listing approach to sensor data anonymization. In Proceedings of the 5th Workshop on CPS&IoT Security and Privacy (pp. 101–107).

31. Abd Razak, S., Nazari, N. H. M., & Al-Dhaqm, A. (2020). Data anonymization using pseudonym system to preserve data privacy. IEEE Access, 8, 43256–43264.

32. Zou, Y., Roundy, K., Tamersoy, A., Shintre, S., Roturier, J., & Schaub, F. (2020, April). Examining the adoption and abandonment of security, privacy, and identity theft protection practices. In Proceedings of the 2020 CHI Conference on Human Factors in Computing Systems (pp. 1–15).

33. Sharma, D. K., Sharma, R., Jeon, G., & Kumar, R. (2024). Data Analytics for Smart Grids Applications—A Key to Smart City Development. Springer International Publishing AG.

34. Yee, G. O. (2019, June). Attack surface identification and reduction model applied in scrum. In 2019 International Conference on Cyber Security and Protection of Digital Services (Cyber Security) (pp. 1–8). IEEE.

35. Aleroud, A., Yang, F., Pallaprolu, S. C., Chen, Z., & Karabatis, G. (2021). Anonymization of network traces data through condensation-based differential privacy. Digital Threats: Research and Practice (DTRAP, 2(4), 1–23.

36. Benjumea, J., Ropero, J., Rivera-Romero, O., Dorronzoro-Zubiete, E., & Carrasco, A. (2020). Privacy assessment in mobile health apps: scoping review. JMIR mHealth and uHealth, 8(7), e18868.

37. Tandon, R., Charnsethikul, P., Arora, I., Murthy, D., & Mirkovic, J. (2022). I know what you did on Venmo: discovering privacy leaks in mobile social payments. Proceedings on Privacy Enhancing Technologies.

38. Ullah, I., Boreli, R., & Kanhere, S. S. (2023). Privacy in targeted advertising on mobile devices: a survey. International Journal of Information Security, 22(3), 647–678.

39. Deußer, C., Passmann, S., & Strufe, T. (2020, May). Browsing unicity: on the limits of anonymizing web tracking data. In 2020 IEEE Symposium on Security and Privacy (SP) (pp. 777–790). IEEE.

40. Bernabe, J. B., David, M., Moreno, R. T., Cordero, J. P., Bahloul, S., & Skarmeta, A. (2020). ARIES: evaluation of a reliable and privacy-preserving European identity management framework. Future Generation Computer Systems, 102, 409–425.

41. Zuo, Z., Watson, M., Budgen, D., Hall, R., Kennelly, C., & Al Moubayed, N. (2021). Data anonymization for pervasive health care: systematic literature mapping study. JMIR Medical Informatics, 9(10), e29871.

42. Prokhorenkov, D. (2022). Anonymization level and compliance for differential privacy: a systematic literature review. 2022 International Wireless Communications and Mobile Computing (IWCMC), 1119–1124.

43. Iezzi, M. (2020, December). Practical privacy-preserving data science with homomorphic encryption: an overview. In 2020 IEEE International Conference on Big Data (Big Data) (pp. 3979–3988). IEEE.

44. Choudhury, O., Gkoulalas-Divanis, A., Salonidis, T., Sylla, I., Park, Y., Hsu, G., & Das, A. (2020). Anonymizing data for privacy-preserving federated learning. *arXiv preprint arXiv*:2002.09096.

45. Majeed, A., Khan, S., & Hwang, S. O. (2022). Toward privacy preservation using clustering based anonymization: recent advances and future research outlook. IEEE Access, 10, 53066–53097.

46. Jordon, J., Szpruch, L., Houssiau, F., Bottarelli, M., Cherubin, G., Maple, C., & Weller, A. (2022). Synthetic Data–what, why and how?. *arXiv preprint arXiv*:2205.03257.

For Product Safety Concerns and Information please contact our EU
representative GPSR@taylorandfrancis.com
Taylor & Francis Verlag GmbH, Kaufingerstraße 24, 80331 München, Germany